IGCSE Mathematics

A Complete Revision Guide

(Conceptual and Topical Approach)

2010 and 2011

Syllabus included

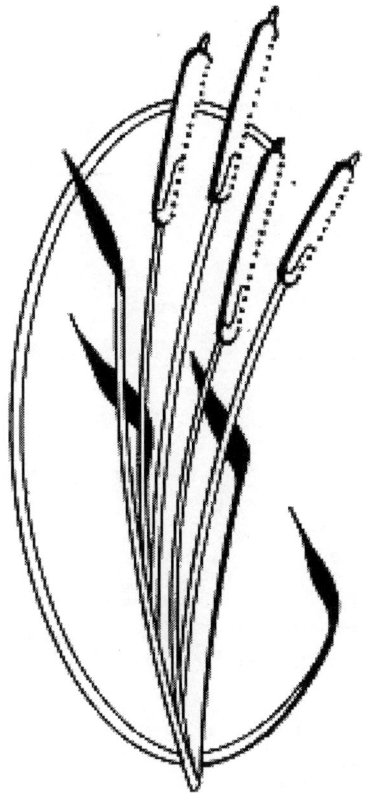

IGCSE Mathematics

A Complete Revision Guide

(Conceptual and Topical Approach)

by D. Sarat Chandra

2010 and 2011

Syllabus included

Editor
Johnathon D. Patterson

Senior Publisher
Steven Lawrence Hill Sr.

Awarded Publishing House
ASA Publishing Company

A Publisher Trademark Cover page

ASA Publishing Company
Awarded Best Publisher for Quality Books

105 E. Front St., Suite 205, Monroe, Michigan 48161
www.asapublishingcompany.com

Typeset: D. Sarat Chandra

Book: IGCSE Mathematics *A Complete Revision Guide*
Date Published: 02.17.10 /Edition 1 *Trade Paperback*
Book ASAPCID: 2380504

Expanded Global Distribution
ISBN: 978-0-9826490-2-2
ISBN: 978-1-886528-15-4
Library of Congress Cataloging-in-Publication Data

This book was published in the United States of America.
State of Michigan

A Publisher Trademark Title page

Contents

IGCSE Mathematics

A Complete Revision Guide

(Conceptual and Topical Approach)

2010 and 2011
syllabus included

Author - D. Sarat Chandra

This book **IGCSE MATHEMATICS A Complete Revision Guide (Conceptual and Topical Approach)** is a must aid for the students who are preparing for IGCSE Mathematics Examinations.

- ✓ Contains **more than 2000 problems** for practice upto A^*
- ✓ Covers the '**complete syllabus' contents**
- ✓ **Core curriculum** and **extended curriculum** are covered separately for easy to use by core and extended students
- ✓ Contains clear and concise work with illustrations and diagrams
- ✓ **'Working rules'** show how to use the best methods and techniques for solving problems
- ✓ Things to '**remember**' provide revision tips and examination techniques
- ✓ **'Don't do this'** explains the tricks to avoid common exam errors
- ✓ **'Solved examples'** show the method of working
- ✓ Big bank of '**Examination pattern questions'** are given in the '**Practice Exercises**' separately for core and extended students for the best preparation of IGCSE Examinations
- ✓ All the numerical '**answers**' are given at the end of the book

Cambridge International Examinations

CHAPTER 1 Number

1.01 Number facts

Core Curriculum

Remember

1. **Natural numbers:** The numbers used for counting. E.g. 1, 2, 3, 4,……
2. **Whole numbers:** The natural numbers together with zero. E.g. 0, 1, 2, 3, 4,……
3. **Integers:** The negatives of natural numbers together with the whole numbers.
   E.g. ……, -4, -3, -2, -1, 0, 1, 2, 3, 4,……
4. **Rational numbers:** All the integers and all the fractions with both numerators and denominators($\neq 0$) as integers are rational numbers. E.g. -2, $\frac{1}{3}$, $-\frac{4}{5}$, $\sqrt{9}$ etc.
5. **Irrational numbers:** The numbers whose root over is not exact are rational numbers.
   E.g. $\sqrt{2}, \sqrt[3]{5}, \sqrt[9]{10}, e, \pi$ etc.
6. **Real numbers:** The rational and the irrational numbers together form real numbers.
   E.g. 5, 0, -9, 3.7, $\frac{1}{2}$, $\sqrt{3}$ etc.
7. **Prime numbers:** The natural numbers(except 1) having exactly two factors.
   E.g. 2,3,5,7,11,13,17,19,23,29,31,37,41, 43 etc.
8. The terminating decimals are rationals. E.g. 1.2, 5.017, -40.0001, -0.713451 etc.
9. The non-terminating and repeating decimals are also rationals. E.g. 0.333…
10. The numbers of the form $\sqrt[n]{a^n}$ are rational numbers.
    E.g.(a) The square root of a perfect square is rational i.e. $\sqrt{1}$, $\sqrt{4}$, $\sqrt{9}$, $\sqrt{25}$ etc.
    (b) The cube root of a perfect cube is also rational i.e. $\sqrt[3]{1}$, $\sqrt[3]{8}$, $\sqrt[3]{27}$, $\sqrt[3]{125}$ etc.
    (c) $\sqrt{3^2}$, $\sqrt[3]{5^3}$, $\sqrt[7]{4^7}$, $\sqrt[10]{7^{10}}$ etc. are rational numbers.
11. The non-terminating and non-repeating decimals are also irrational numbers.
    E.g. 1.4101001000100001……, 0.16116111611116…… etc.
12. **Factors:** A number that exactly divides another number is called a factor of the given number. E.g. The factors of 6 are 1, 2, 3 and 6.
13. **Multiples:** A number that is the product of some other numbers is called a multiple.
    E.g. (a) 15 is a multiple of 5. (b) 319 is a multiple of 29.

► **Square numbers(Perfect squares):** The natural numbers those can be expressed as the product of two identical numbers (except 1) are square numbers.
E.g. 1,4,9,16,25,36,49,64,81,100,121,144 etc.

Remember

To find the square of a number just multiply the given number with itself.
E.g. (a) The square of 3 is $3^2 = 9$. (b) The square of 1.9 is $1.9^2 = 3.61$.

► **Square root:** When a square number is written as the product of two identical numbers, each identical number is called the square root of the given number.
E.g. The square root of 4 is 2 or -2.

$4 = 2 \times 2$ and $4 = -2 \times -2$

► **Cube numbers(Perfect cubes):** The natural numbers those can be expressed as the product of three identical factors (except 1) are cube numbers. E.g. 1, 8, 27, 64 etc.

Remember

To find the cube of a number just multiply the given number with itself twice.
E.g. (a) The cube of 3 is $3^3 = 27$. (b) The cube of 1.9 is $1.9^3 = 6.859$.

- **Cube root:** When a cube number is written as the product of three identical numbers, each identical number is called a cube root of the given number.

  E.g. The cube root of 8 is 2. $8 = 2 \times 2 \times 2$

Practise exercise 1(A) **Core Curriculum**

1 4 $\sqrt{4}$ π 2 -4 $\frac{1}{4}$

From the list of above numbers, write down
(a) a natural number, (b) a rational number, (c) a whole number,
(d) an irrational number, (e) an integer, (f) a prime number.

2 From the set of numbers 25, 26, 27, 28, 29, 30, 31, 32 and 33, write down
(a) a prime number, (b) a multiple of 7, (c) a square number,
(d) a multiple of both 5 and 6, (e) a cube number, (f) a factor of 64.

3 4 8 e $\sqrt{49}$ 64 70 72

From this list of numbers, write down one that is
(a) not a rational number, (b) a square number, (c) both a square and cube,
(d) a cube root of 64, (e) the product of 8 and 9, (f) between two consecutive primes.

4 Write down
(a) an integer, (b) a rational number, (c) an irrational number,
(d) a prime number, (e) a square number, (f) a cube number.

5 (a) List all the factors of 36.
(b) Write down the number of factors of 36.
(c) Write down the common factors of 36 and 198.
(d) Write down the prime factors of 91.
(e) Write 198 as a product of its prime factors.
(f) Write the number of prime factors of 198.

6 (a) Write down the common factors of 45 and 105.
(b)Write down the highest common factors of 45 and 105.
(c) Write down the highest common factors of 450 and 1050.
(d) Write down the lowest common multiples 45 and 105.
(e) Write down the lowest common multiples of 45000 and105000.

7 (a) Write down 60 as a product of prime numbers.
(b) What is the product of the two prime numbers just preceding 60 and following it?

8 (a) Write down a number, other than 1, which is a factor of both 21 and 56.
(b) Write down the least number, which is a multiple of both 21 and 56.

9 An integer n is such that $90 \leq n \leq 100$. Write down a value of n which is
(a) a prime number, (b) a multiple of 13, (c) a factor of 279.

Practise exercise 1(B) **EXTENDED CURRICULUM**

1 From the numbers between 40 and 60, write down
(a) the total number of prime numbers,
(b) the single number between two consecutive prime numbers,

(c) the prime factors to answer(b),
(d) a number which is the square of a prime number,
(e) a number which is a multiple of both 5 and 9.

2 3.14 $\frac{22}{7}$ π $\sqrt{441}$ $\sqrt[3]{8}$ 0.020020002...... 0.02020202......

From the set of numbers above, write down

(i) two integers, (ii) two irrational numbers,
(iii) five rational numbers (iv) the total number of real numbers,
(v) a prime number, (vi) a common multiple of 3 and 7.

3 A student is asked to list the natural numbers between 0 and 20, which are not divisible by 2.

(a) Write down the numbers to be listed by the student correctly.
(b) What is the special name of these numbers?
(c) From the answer to part(a),
  (i) find the difference between the first two numbers and the difference between the last two numbers and write down the special name of the difference,
  (ii) find the sum of the first four numbers,
  (iii) write the special name of the answer to part c(ii),
  (iv) find the sum of the last four numbers,
  (v) write the special name of the answer to part c(iv).

4 Twin primes are the prime numbers separated by 2.
For example, (3, 5) and (5, 7) are twin primes.

(a) Write down
  (i) a pair of twin primes between 40 and 50,
  (ii) a pair of twin primes whose product is 143,
  (iii) the prime number just followed by 143.
(b) Is (91, 93) a pair of twin primes? Give reason to your answer.

5 A prime number p such that $101 < p \leq 111$.

(a) Write down the values of p.
(b) When you find the difference between the two values of p, the result is a prime number. What is the name of this prime number?

1.02 Directed numbers

Core Curriculum

- Integers, decimals and fractions are directed numbers.

Operations on directed numbers

Working rule

1) Addition of directed numbers

(i) ***For positive directed numbers,*** simply add and the answer is positive.
(ii) ***For negative directed numbers,*** simply add and the answer is negative.
(iii) ***For a positive and a negative directed numbers,*** without considering the signs, simply subtract the numbers and put the sign of the number with greater absolute value.

2) ***To subtract a directed number,*** change its sign and add with the given number.

3) ***To multiply or divide two directed numbers,*** simply multiply or divide and
(i) same sign numbers give positive answer,
(ii) opposite sign numbers give negative answer.

Example

(1) (a) $^{+}2 + {}^{+}4 = {}^{+}6$ (b) $^{+}2.7 + {}^{+}1.5 = {}^{+}4.2$

(2) (a) $^{-}1 + {}^{-}5 = {}^{-}6$ (b) $^{-}1.2 + {}^{-}3.7 = {}^{-}4.9$

(3) (a) $^{+}7 + {}^{-}4 = {}^{-}3$ (b) $^{-}9.5 + {}^{+}15.6 = {}^{+}6.1$

(4) (a) $^{-}6 - {}^{+}2 = {}^{-}4$ (b) $^{-}4.5 - {}^{-}3.6 = {}^{-}0.9$

(5) (a) $^{+}9 \times {}^{-}8 = {}^{-}72$ (b) $^{-}4.1 \times {}^{-}1.5 = {}^{+}6.15$

(6) (a) $^{-}5 \div {}^{-}10 = {}^{+}0.5$ (b) $18 \div {}^{-}3 = {}^{-}6$

Practical use of directed numbers

Working rule

1) For rise/higher/more than/increased/above in water level, temperature etc, move towards positive numbers.
2) For fall/lower/less than/decreased/below in water level, temperature etc, move towards negative numbers.

Example

The table shows the maximum daily temperatures during one week.

Monday	Tuesday	Wednesday	Thursday	Friday	Saturday	Sunday
–3 °C	1.2°C	–1.7 °C	3 °C	1.2 °C	1 °C	–2.1 °C

(a) Which day has greatest temperature?
(b) What is the difference between the greatest and the least of these temperatures?
(c) What is the change in maximum temperatures between Tuesday and Sunday?
(d) Which days have same maximum temperatures?
(e) Which day has the maximum temperature 2.7 °C higher than that of Wednesday?

Working: (a) Thursday (b) 6 °C (c) 3.3 °C (d) Tuesday and Friday (e) Saturday

Practise exercise 2(A) **Core Curriculum**

1 Work out the following.

(a) $25 + 17$ (b) $6.7 + 9.213$ (c) $-14 + -21$

(d) $-5.21 - -2.94$ (e) $42 - -49$ (f) $-7.5 + 7.5$

(g) $\frac{1}{2} - \frac{2}{3}$ (h) $\frac{6}{11} + -\frac{3}{22}$ (i) $-\frac{5}{8} + -\frac{4}{5}$

(j) $-2 + 9 - 13$ (k) $2\frac{1}{5} - -13.25$ (l) $1 - \frac{1}{4} + 4.4$

2 Work out the following.

(i) $8 \times +2$ (ii) 7×-3 (iii) -13×5

(iv) -6×-9 (v) -45×0 (vi) 4.23×-5

(vii) $20 \div 4$ (viii) $-30 \div 6$ (ix) $-72 \div -9$

(x) $0 \div 9.9$ (xi) $-\frac{1}{2} \times \frac{3}{4}$ (xii) $-6\frac{1}{2} \times -2\frac{1}{6}$

(xiii) $\frac{12}{25} \div \frac{2}{5}$ (xiv) $-2\frac{1}{5} \div 5$ (xv) $-4\frac{2}{3} \div -\frac{7}{12}$

3 The temperature of a glass of normal water is 30°C. The temperature of a cup of tea is 13.5 °C higher. Find the temperature of the cup of tea.

4 The depth of water in a swimming pool was 10.4 feet. The depth of water is reduced by 3.2 feet. Find the new depth of water.

5 Mount Everest is 29028 ft above sea level and Dead Sea is 1312 ft below sea level. How much higher is Mount Everest than Dead Sea?

6 A measuring test tube contains 4 millilitres of chemicals.
(a) James used 2.4 ml of chemicals from the test tube for an experiment. Write down the new level of chemicals.
(b) Then he added 4.2 ml of chemicals in the test tube. Write down the new level of chemicals in the test tube.

7 On a particular day the maximum temperature in a City was –2°C. The minimum temperature was 7.4 °C fewer than this. Write down the minimum temperature.

8 At morning one day the temperature is –6.5°C. By evening the temperature has risen by 2.3°C. What is the temperature at evening?

Practise exercise 2(B) **EXTENDED CURRICULUM**

1 In the year 2007, one day the greatest temperature change occurred in Kashmir, when the temperature fell from 25°C to –12°C. How much did the temperature drop?

2 (a) On a particular day, the maximum temperature in town A was 27°C and in town B was 21.3°C lower than twice the temperature in town A.
Calculate the maximum temperature in town B.
(b) On the same day, the maximum temperature in town C was 2.7°C more than town B.
Calculate (i) the maximum temperature in town C,
(ii) the average of maximum temperatures in towns A, B and C.

3 The diagram shows the height of a building above sea level and the depth of a bore well below sea level.
(a) Calculate the difference in heights between the building and the bore well.
(b) A trench is dug 25 m below sea level.
(i) Calculate the difference in heights between the building and the trench.
(ii) How much lower is the bore well than the trench?

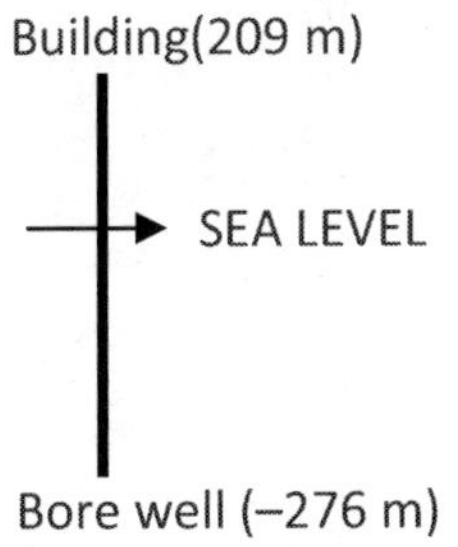

4 The table shows the average daily temperatures during one week in a city.

Monday	Tuesday	Wednesday	Thursday	Friday	Saturday	Sunday
4°C	–3°C	2°C	–2.9°C	2.4°C	2°C	2.2°C

(a) By how many degrees did the average temperature change between Tuesday and Thursday?
(b) What is the difference between the highest and the least of these temperatures?
(c) Which two days have same average temperatures?
(d) What is the average of these average temperatures?

Core Curriculum

1.03 Sequences

- **Recognising the pattern and continue a sequence**

Working rule

1) For an increasing sequence, add or subtract or perform both operations.
2) For a decreasing sequence, subtract or divide or perform both operations.

Example

1. Write down the next three terms in each sequence.

   (a) 2, 6, 18, 54, 162, ……

   The next three terms are

   162 x 3 = 486, 486 x 3 =1458, 1458 x 3 = 4374.

   > Starting at 2,
   > next term = Last term x 3

   (b) 1, 3, 7, 15, 31, ……

   The next three terms are

   2 x 31 +1 =63, 2 x 63 + 1 =127, 2 x 127 + 1 = 255.

   > Starting at 1,
   > next term = 2 x Last term + 1

2. Write the next three terms of each sequence.

   (b) 16, 8, 4, 2, 1, ……

   The next three terms are

   1÷2 = 0.5, 0.5 ÷ 2 = 0.25, 0.25 ÷ 2 = 0.125.

   > Starting at 16, each term is divided by 2.
   > Next term = Last term ÷ 2

   (c) 63, 31, 15, 7, 3, ……

   The next three terms are

   (3 – 1) ÷ 2 =1, (1 – 1) ÷ 2 = 0, (0 – 1) ÷ 2 = – 0.5.

   > Starting at 63,
   > next term = (Last term – 1) ÷ 2

Finding the nth term of a sequence

Working rule

1) If each term in a sequence is separated by a common difference(d), then
   n^{th} term = 1st term + (n – 1)d.
2) If each term in a sequence is separated by a common ratio(r), then
   n^{th} term = 1st term x r^{n-1}.
3) If a sequence does not have a common difference or a common ratio in the first attempt, then you need to find the repeated differences of consecutive terms until you get a common difference.

Example

Write down an expression in terms of n, for the nth term in each sequence.

(a) 7, 10, 13, 16, 19, ……

Working: 1st term = 7, d = 10 – 7 = 3

n^{th} term = 1st term + (n – 1)d

= 7 + (n – 1)3 = 3n + 4

> d = 2nd term – 1st term
> Use the formula for n^{th} term of a sequence.

(b) 2, 4, 8, 16, 32, ……

Working: 1st term = 2, r = 4 ÷ 2 = 2

n^{th} term = 1st term x r^{n-1}

= 2 x 2^{n-1} = 2^n

> r = 2nd term ÷ 1st term
> Use the formula for n^{th} term of a sequence.

(c) 2, 5, 10, 17, 26, ……

Working:

Terms	2		5		10		17		26 ……
1st difference		+3		+5		+7		+9 ……	
2nd difference			+2		+2		+2 ……		

n^{th} term = n^2 + 2

> Since the difference is repeated for twice, the nth term has n^2 and the common difference +2 is added with it.

Practise exercise 3(A) **Core Curriculum**

1 Write down the next two terms in each of the following sequences.

(i) 1, 3, 5, 7, ……
(ii) 0, –12, –24, –36, ……
(ii) 2, 4, 6, 8, ……
(iv) 128, 64, 32, 16, ……
(v) –7, –1, 5, 11,……
(vi) 9, 18, 27, 36, ……
(vii) 10, 100, 1000, 10000, ……
(viii) 4, 8, 14, 22, ……
(ix) –20, –10, 2, 16, ……
(x) 4, 7, 10, 13, ……
(xi) 10, 8, 6, 4, ……
(xii) 1, 1, 2, 3, 5, 8, ……
(xiii) 0, –4, –8, –12, ……
xiv) $10^9, 10^7, 10^5, 10^3$, ……
(xv) 3125, 625, 125, 25, ……
(xvi) 1, 4, 9, 16, ……
(xvii) 1, 8, 27, 64, ……
(xviii) 2, 10, 28, 65, ……
(xix) 0, 3, 8, 15, ……
(xx) $1, \frac{1}{4}, \frac{1}{9}, \frac{1}{16}$, ……
(xxi) $10x^9, 9x^8, 8x^7, 7x^6$, ……
(xxi) $4a^2, 6a^4, 8a^6, 10a^8$, ……

2 Find the first five terms in each sequence whose n[th] term is given below.

(a) $2n + 1$
(b) n^2
(c) n^3
(d) $n^2 + 1$
(e) $n^3 - 2$
(f) $\frac{1}{2}n(n+1)$
(g) $(3n + 1)^2$
(h) $4^n - 2$
(i) $40 - 2n$
(j) 2^{n-1}
(k) 5^{2n-2}
(l) $(-1)^n 3^{n-1}$
(m) $\left(\frac{n(n+2)}{2}\right)^2$
(n) $(-2)^n$
(o) $\frac{(n^2+n)(2n+1)}{6}$

3 6, 14, 22, 30, 38, …… For this sequence of numbers

(a) write down the next term,
(b) find the 12th term,
(c) find the nth term,
(d) which term has the value of 134?

4 Look at the patterns and fill in the empty boxes in the table below.

1st	2nd	3rd	4th	5th	10th	25th	100th	Nth
2	5	8	11		74			$3n - 1$
1	4	9		25	100			
1	8		64				1000000	
7	14	21			70			
				11				$2n + 1$
		6		10		50		
–1	2		–16					$(-1)^n 2^{n-1}$
2	5	10	17					

5 Afreen can spell 15 words in every two minutes and continued spelling words for 16 minutes at the same rate. Complete the table below.

Time in minutes	2	4	6	8	10	12	14	16
Number of words	15	30	45	60				

(a) Find the number of words she could spell in
(i) 40 minutes, (ii) 2 hours, (iii) n minutes.

(b) If she takes 16 minutes break during spelling words, how many words she could spell in 1 hour?

6 The diagram below shows a sequence of patterns made from dots.

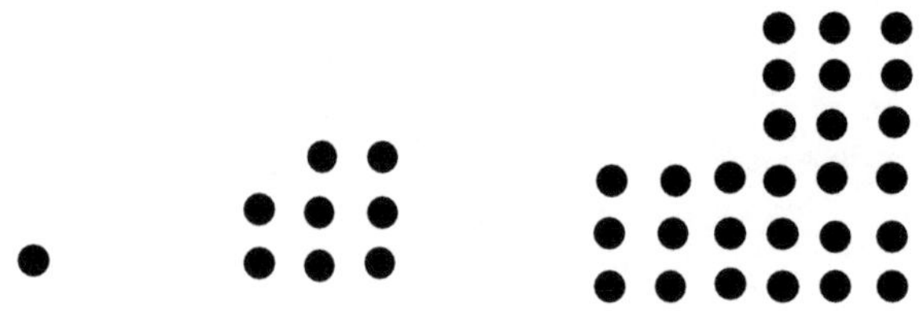

Pattern 1 Pattern 2 Pattern 3 Pattern 4

(a) Draw the fourth pattern in the space above.

(b) Complete the table below.

Pattern number	1	2	3	4	5	6	7
Number of dots	1	8	27				

(c) Write down the number of dots in the
(i) 11th pattern, (ii) 100th pattern (iii) n^{th} pattern.

(d) Write down the pattern number, that has 8000 dots.

Practise exercise 3(B) EXTENDED CURRICULUM

1 The nth term of the triangular number sequence is given by $\frac{n(n+1)}{2}$.

(a) Complete the table given below.

Term	1	2	3	4	5	6	7
Triangular number	1	3	6				

(b) Another sequence is formed by adding every two consecutive triangular numbers.
(i) Write down the first three terms of the new sequence.
(ii) What is the nth term of the new sequence?

2 (a) Write down the next four terms in the sequence 4.8, 4.5, 3.9, 3, ...

(b) Write down the next two terms and find an expression for the nth term for each.
(i) 15, 21, 27, 33, ... (ii) 29, 22, 15, 8, ...
(iii) $\frac{4}{9}, \frac{5}{11}, \frac{6}{13}, \frac{7}{15}, \ldots\ldots$ (iv) $\frac{2}{7}, \frac{3}{10}, \frac{4}{13}, \frac{5}{16}, \ldots\ldots$

3

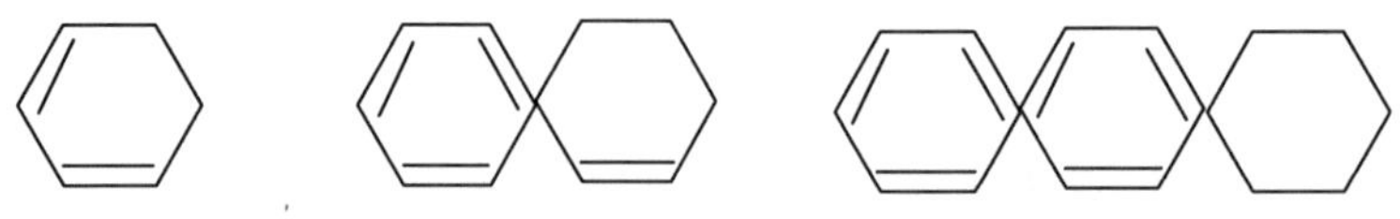

Diagram 1 Diagram 2 Diagram 3 Diagram 4

The first three diagrams in a sequence are shown. The diagrams are made up of lines of equal length.

(a) Complete the table.

Diagram	1	2	3	4	5	6	7
Number of lines	8	16	24				

(b) Find a formula for the number of lines in the nth diagram.

(c) Find a formula for the number of lines in the $(n+1)^{th}$ diagram.

(d) What is the difference between the number of lines in the nth diagram and $(n+1)^{th}$ diagram?

(e) Find the number of lines in the 2000th diagram.

(f) Which diagram has 1064 lines?

4 For each of the following sequences, write down the next term.
(a) 5, 8, 13, 21, 34, …… (b) $x^9, 2x^{11}, 4x^{13}, 8x^{15}, \ldots$ (c) 2, 8, 32, 128, 512, ……

5 For the sequence $18\frac{1}{2}$, 15, $11\frac{1}{2}$, 8, ……
(a) find an expression for n^{th} term, (b) work out the 100th term.

6 The table shows some terms of several sequences.

Term	1	2	3	4		n
Sequence A	–6	–11	–16	–21		A
Sequence B	3	10	29	66		B
Sequence C	$\frac{1}{5}$	$\frac{2}{6}$	$\frac{3}{7}$	$\frac{4}{8}$		C
Sequence D	8	27	64	125		D
Sequence E	–1	2	7	14		E

(a) Find the nth term of sequence (i) A, (ii) B, (iii) C, (iv) D, (v) E.
(b) Find the $(n-1)^{th}$ term of the sequence D.
(c) Which term in sequence A is equal to –3886?

1.04 Standard form

Core Curriculum

- The standard form or scientific notation of an ordinary number must be in the form $A \times 10^n$. Where n is an integer and $1 \le A < 10$.

Working rule

For writing the equivalent standard form of an ordinary number
1) ***between 0 and 1,*** move the decimal point towards right and the power of 10 is –ve,
2) ***between 1(inclusive) and less than 10,*** just put 10^0 after the number,
3) ***more than 10,*** move the decimal point towards left and the power 10 is +ve.

Example

Ordinary form	**Standard form**
0.000 000 00903	9.03×10^{-9}
4.672	4.672×10^0
7402 000 000 00	7.402×10^{11}

Remember

1. Use **EXP** or **EE** button to entre numbers in standard form in your calculator.
2. **EXP** or **EE** stands for **x10^.**
3. The standard form of a number more than 10 has positive power.
4. The standard form of a decimal number has negative power.

Example
The distance from Earth to Sun is 1.5×10^8 kilometres.
The distance from Uranus to Sun is 2.87×10^8 kilometres. Calculate in standard form,
(a) the total of both the distances, (b) the difference between both the distances,
(c) the product of both the distances.

Working: (a) $(1.5 \times 10^8) + (2.87 \times 10^8)$
$= 4.37 \times 10^8$
(b) $(2.87 \times 10^8) - (1.5 \times 10^8) = 1.37 \times 10^8$
(c) $(1.5 \times 10^8) \times (2.87 \times 10^8) = 4.31 \times 10^{16}$

> Use **EXP** or **EE** button for x10^.
>
> Write your answer in standard form, but not in ordinary form.

Practise exercise 4(A) **Core Curriculum**

1 Write down the following numbers in standard form.
(a) 4700 (b) 9 000 000 (c) 1060
(d)737 (e) 567 000 00 (f) 2 billion
(g) 0.003 trillion (h) 6 million (i) 0.000001
(j) 0.000000061 (k) 43×10^7 (l) 271×10^{19}
(m) 45.2×10^3 (n) 0.0093×10^{-8} (o) 4 quintillion

2 Calculate, giving your answer in standard form.
(i) $(1.2 \times 10^2) + (2.1 \times 10^3)$ (ii) $(3.4 \times 10^3) + (9 \times 10^{10})$
(iii) $(2.5 \times 10^4) - (1.3 \times 10^3)$ (iv) $(8 \times 10^{21}) - (6.72 \times 10^{-5})$
(v) $(4.4 \times 10^7) \times (2.35 \times 10^2)$ (vi) $(2 \times 10^{-6}) \times (2.573 \times 10^{-4})$
(vii) $(6.9 \times 10^7) \div (2.3 \times 10^7)$ (viii) $(8 \times 10^{-14}) \div (4.88 \times 10^{-5})$

3 Evaluate and give your answer in standard form.
(a) 5.25 x 24.78 (b) 0.00004 x 1000 000 00
(c) 0.75 x 0.0075 (d) 64000 00 ÷ 800
(e) (4 + 5.2) ÷0.002 (f) 14 h ÷ 7 min
(g) 0.6 min x 56 sec (h) 10 kg ÷ 0.5 g

4 The population of a city is approximately 147000 00. Write the number in standard form.

5 3500 million years ago, life began in water. Write this number in standard form.

6 The percentage of the metal gold in the earth's crust is 0.000 0002. Write this umber in standard form.

7 The mass of an atom is 1.42×10^{-20} grams. Find the mass of 80000 such atoms and give your answer in standard form.

8 The diameter of a micro organism is 0.0000381 mm. Calculate and express in standard form (a) the radius of the micro organism,
(b) the circumference of the micro organism.

9 The surface area of a country is approximately 730 000 000 square metres. Write this number in standard form.

Practise exercise 4(B) **EXTENDED CURRICULUM**

1 If $x = 1.42 \times 10^4$ and $y = 2.3 \times 10^{-2}$, evaluate and give your answer in standard form.
(a) $2x^2$ (b) $y^2 - y$ (c) $2x^2(y^2 - y)$

2 The speed of light is 300 000 000 metres per second.
(a) Write down this speed in kilometers per second and give your answer in standard form.
(b) Calculate and express in standard form the number of kilometers light can travel in
(i) 0.5 seconds, (ii) a day, (iii) 100 days.

3 There is 1.3 billion cubic kilometers of water on Earth and 97% of it is in Oceans.
(a) Calculate the volume water not in Oceans.
(b) Write your answer to part(a) in standard form.

4 An experiment shows that 254 grams of iodine contains 6.02×10^{23} iodine molecules.
(a) How many iodine molecules are there in 762 grams of iodine?
Express your answer in standard form, correct to three significant figure.
(b) How many grams of iodine contain 1.505×10^{25} iodine molecules?

5 In 2001 the population of a country was 1.7×10^8. It is estimated that in 2072, the population will be doubled. Calculate the estimated population in 2072, giving your answer in standard form.

6 Use the formula $S = UT + 0.5\ AT^2$ to calculate the value of S when $U = 4 \times 10^6$, $T = 5.2 \times 10^2$ and $A = 6.1 \times 10^7$.

7 The mass of a cake is 9 kilograms and is distributed equally among 400 pupils.
(a) Calculate, in kilograms, the mass of cake received by two pupils.
(b) Write your answer to part(a) in standard form.

8 Evaluate and write in standard form. (a) $\frac{5}{27}(4.26 \times 10^5)$ (b) $(9.86 \times 10^{-10}) \div 2\frac{1}{75}$

Core Curriculum

1.05 Fractions

- **Vulgar fraction:** It is a fraction with denominator other than a power of 10.

  E.g. $\frac{1}{2}, \frac{3}{5}, 2\frac{1}{9}$ etc.

- **Decimal fraction:** It is a fraction with denominator as a power of 10.

  E.g. $\frac{7}{100}, \frac{5}{10}, \frac{19}{10000}$ etc.

Remember

1. To change a fraction into a decimal, divide its numerator by the denominator.
2. Any vulgar fraction whose denominator is a multiple of 2 and/or 5 together with all decimal fractions can be changed into a terminating decimal.

   E.g. (i) $\frac{1}{2} = 0.5$ (ii) $\frac{2}{5} = 0.4$ (iii) $\frac{7}{20} = 0.35$ (iv) $\frac{19}{1000} = 0.019$
3. Any vulgar fraction whose denominator is not the only multiple of 2 and/or 5 can be changed into a non-terminating decimal.

   E.g. (i) $\frac{1}{3} = 0.333...$ (ii) $\frac{7}{15} = 0.4666666...$

- **Equivalent fractions:** These are group of fractions equal to the same amount.

  E.g. $\frac{1}{4}, \frac{2}{8}, \frac{3}{12}, \frac{4}{16}$ are equivalent fractions.

Remember

To find equivalent fraction, just multiply or divide both the numerator and the denominator of a fraction by same the number.

Fundamental operations on fractions:

Working rule

1) ***To add or subtract fractions,*** first change them equivalent and then add/subtract the numerators only with same denominator.
2) ***To multiply fractions,*** use the rule $\frac{\text{multiply top numbers}}{\text{multiply bottom numbers}}$
3) ***To divide fractions,*** use the rule **first fraction** x **reciprocal of other fraction**.

Example

Work out the following.

1. $2\frac{1}{3} + \frac{2}{5} = \frac{7}{3} + \frac{2}{5}$

$= \frac{35}{15} + \frac{6}{15}$

$= 2\frac{11}{15}$

Changing $2\frac{1}{3}$ into improper fraction as $\frac{7}{3}$.

Changing into equivalent fractions.

Adding numerators with same denominator.

2. $2\frac{1}{3} - \frac{2}{5} = \frac{7}{3} - \frac{2}{5}$

$= \frac{35}{15} - \frac{6}{15}$

$= 1\frac{14}{15}$

Changing $2\frac{1}{3}$ into improper fraction as $\frac{7}{3}$.

Changing into equivalent fractions.

Subtracting numerators with same denominator.

3. $2\frac{1}{3} \times \frac{2}{5} = \frac{7}{3} \times \frac{2}{5}$

$= \frac{14}{15}$

Changing $2\frac{1}{3}$ into improper fraction as $\frac{7}{3}$.

Using the rule $\frac{\text{multiply top numbers}}{\text{multiply bottom numbers}}$

4. $2\frac{1}{3} \div \frac{2}{5} = \frac{7}{3} \div \frac{2}{5}$

$= \frac{7}{3} \times \frac{5}{2}$

$= \frac{35}{6}$

$= 5\frac{5}{6}$

Changing $2\frac{1}{3}$ into improper fraction as $\frac{7}{3}$.

Using the rule

(first fraction) x **(reciprocal of other fraction)**.

Practise exercise 5(A) **Core Curriculum**

1 Convert each of the following into a decimal.

(a) $\frac{3}{10}$ (b) $\frac{47}{1000}$ (c) $\frac{7}{25}$ (d) $\frac{126}{125}$

(e) $1\frac{5}{8}$ (f) $5\frac{1}{50}$ (g) $\frac{2}{3}$ (h) $\frac{5}{7}$

2 Convert each of the following into a fraction.

(a) 0.4 (b) 0.09 (c) 0.00017 (d) 1.5

(e) 6.04 (f) 18.1234 (g) 12.0304 (h) –7.44

3 Fill in the blanks in each set of equivalent fractions.

(a) $\frac{1}{4} = \frac{4}{\ } = \frac{\ }{28} = \frac{12}{\ }$. (b) $\frac{6}{7} = \frac{\ }{21} = \frac{132}{\ } = \frac{\ }{350}$.

(c) $\frac{64}{128} = \frac{\ }{32} = \frac{8}{\ } = \frac{\ }{2}$. (d) $\frac{\ }{10} = \frac{50}{\ } = \frac{450}{180} = \frac{\ }{900}$.

4 Without using a calculator work out the following. Show all your working.

(a) $\frac{4}{11} + \frac{5}{11}$ (b) $6\frac{1}{12} + 2\frac{5}{12}$ (c) $\frac{2}{3} + \frac{4}{5}$ (d) $\frac{7}{9} + 1\frac{1}{2}$

(e) $2\frac{1}{4} + 1\frac{2}{3}$ (f) $\frac{3}{10} + 0.5$ (g) $5 + 3\frac{1}{2}$ (h) $\frac{6}{17} - \frac{4}{17}$

(i) $4 - \frac{1}{2}$ (j) $\frac{2}{7} - \frac{3}{4}$ (k) $9\frac{1}{2} - 3\frac{3}{4}$ (l) $\frac{2}{5} \times \frac{3}{7}$

(m) $\frac{4}{9} \times 6$ (n) $4\frac{2}{3} \times 18 \times \frac{1}{28}$ (o) $\frac{4}{5} \div \frac{2}{5}$ (p) $6 \div \frac{12}{13}$

(q) $3\frac{1}{4} \div 13$ (r) $4\frac{3}{7} \div \frac{9}{14}$ (s) $2\frac{1}{2} \times 3\frac{1}{3} \times 1\frac{1}{5}$ (t) $\frac{4}{5} + 3 - 1\frac{4}{10}$

5 Prove the following. Show all your working without using a calculator.

(a) $\frac{6}{7} + \frac{2}{21} = \frac{20}{21}$ (b) $4\frac{2}{5} + \frac{1}{2} = 4\frac{9}{10}$ (c) $6 - \frac{2}{3} = 5\frac{1}{3}$

(d) $3\frac{3}{4} - 1\frac{1}{2} = 2\frac{1}{4}$ (e) $\frac{25}{36} \times 3\frac{3}{5} = 2\frac{1}{2}$ (f) $\frac{1}{4} \times \frac{6}{7} \times 4\frac{1}{5} = 1\frac{4}{5}$

(g) $\frac{3}{10} \div \frac{2}{5} = \frac{3}{4}$ (h) $6\frac{2}{7} \div 7\frac{1}{3} \div 12 = \frac{1}{14}$ (i) $\frac{3}{8} \div \frac{4}{5} \times \frac{6}{7} = \frac{45}{112}$

6 A shopkeeper has a bag of $25\frac{1}{2}$ kg of sugar. He wants to make packets of $1\frac{7}{10}$ kg each.
How many packets does he need?

7 Edward paid $4096.64 for his furniture in 16 equal installments.
Calculate the amount money he paid in 2 installments.

8 A piece of rope is 85 metres. It has to cut into equal lengths of $4\frac{1}{4}$ metres.
How many of these lengths can be cut from this piece of rope?

Practise exercise 5(B) EXTENDED CURRICULUM

1 Micheal's school bag contains 6 text books and 9 note books.

Each text book weighs $\frac{3}{10}$ kg and each note book weighs $\frac{1}{5}$ kg.

(a) Calculate the total weight of all the books in his bag.

(b) Micheal takes away 4 text books from his bag. Calculate the weight of the remaining books in his bag.

(c) How could you compare the weight of the text books and the weight of note books in his bag?

2 Mona spent $\frac{4}{5}$ of her savings and had €260.25 left. Calculate the total amount she

(a) saved, (b) spent.

3 (a) Find the perimeter of a square with its side as $\frac{2}{5}$ metres long.

(b) Find the area of a square with its side as $12\frac{1}{3}$ centimeters long. Write your answer as a fraction.

4 A rectangular field is $104\frac{1}{4}$ metres long and $72\frac{3}{5}$ metres wide. Aminaa ran for four times around the field. Calculate the distance covered by Aminaa.

5 A stick of length $15\frac{3}{4}$ centimeters long is divided into two parts such that one part is $3\frac{1}{2}$ centimeters less than the other part. Calculate the length of each part.

6 Calculate $\frac{3^4}{4^3}$

(a) giving your answer as a fraction, (b) giving your answer as a decimal.

1.06 Percentages

- A percentage is the value obtained by taking the certain percent of a quantity. Percent means **out of 100** or **per hundred** or **hundredth** and is denoted by the symbol %.

Core Curriculum

Working rule

1) ***To convert an ordinary number into percentage,*** just multiply by 100 and put % on the right of the product.
2) ***To convert a percentage into an ordinary number,*** just divide by 100 and omit %.
3) ***To calculate a percentage of a quantity,*** just multiply both.
4) ***To express one quantity as a percentage of another,*** just multiply the ratio of first and the second quantity by 100.
5) ***To calculate the percentage increase or decrease,*** use a simple formula

$$\textbf{Change\%} = \frac{\text{change value}}{\text{original value}} \times 100.$$

[**Change** means increase/growth/profit/more/decrease/deprecition/loss/less/discount.]

Example 1

Convert into percentages. (a) $\frac{1}{4}$ (b) 0.245 (c) 5:8 (d) 13 out of 20

Working: (a) $\frac{1}{4} = \frac{1}{4}$ x 100% = 25%

Multiply by 100 and put % sign.

(b) 0.245 = 0.245 x 100% = 24.5%

(c) 5:8 = $\frac{5}{8}$ x 100% = 62.5% (d) 13 out of 20 = $\frac{13}{20}$ x 100% = 65%

Example 2

Convert into fractions. (a) 29% (b) 2.5% (c) $\frac{1}{2}$%

Working: (a) 29% = $\frac{29}{100}$

Divide by 100 and omit % sign.

(b) 2.5% = $\frac{2.5}{100} = \frac{25}{1000}$ (c) $\frac{1}{2}\% = \frac{1}{200}$

Example 3

Calculate 15% of $6000.

Working: 15% of 6000 = 15 x 600% = $900

Just multiply both.

Example 4

There are 45 teachers and 450 students in a school. Calculate the number of teachers as a percentage of the number of students in the school.

Working: $\frac{45}{450}$ x 100% = 10%

$\frac{\text{First}}{\text{Second}}$ x 100

Example 5

The cost of a ticket increases from $110 to $121. Calculate the percentage increase.

Working: Increase% = $\frac{\text{Increase value}}{\text{Original Value}}$ x 100

= $\frac{121-110}{110}$ x 100 = $10

Use **Change%** = $\frac{\text{change value}}{\text{original value}}$ x 100.

Example 6

Rani bought a box for $20 and later sold it for $13. Calculate the percentage loss.

Working: Loss% = $\frac{\text{Loss value}}{\text{Original value}}$ x 100

= $\frac{20-13}{20}$ x 100 = $35

Use **Change%** = $\frac{\text{change value}}{\text{original value}}$ x 100.

Simple Interest and Compound Interest

- Interest is the extra amount to be paid for the borrowed amount.

Working rule

1) ***To calculate the simple interest***, use the formula **I = PTR%** .
(Where I = Interest, P = Principal, T = Time period and R = Rate.)

2) **Amount = P + I.**

3) ***To calculate the compound interest***, use the same simple interest formula and the amount of first sub period becomes the principal for second sub period.

Remember

1. In simple interest, the Principal is same for the whole time period.
2. In compound interest, the Principal is different in each sub period.

Example 1

Poul invests \$2000 for 18 months at the rate of 7% simple interest per year.
Calculate (a) the interest, (b) the amount he receives.

Working: P = \$2000, T = 18 months = 1.5 years, R = 7%.

(a) I = PTR% = 2000 x 1.5 x 7% = \$210 — Use the formula for simple interest.

(b) A = P + I = 2000 + 210 = \$2210 — Use the formula for amount.

Example 2

James borrows £15000 from a bank at the rate of 12% compound interest per year.
How much interest will he get after 2 years?

Working: P = £15000, T = 2 years, R = 12%. — Use the formula for simple

For first year, I = PTR% = 15000 x 1 x 12% = £1800. — Use T = 1 for first year.

For second year, P = 15000 + 1800 = £16800 — Principal for second year = Amount of first year

For second year, I = PTR% = 16800 x 1 x 12% — Use the formula for simple interest.
= £2016 — Use T = 1 for second year.

Compound interest = 1800 + 2016 =£3816 — Compound interest = Sum of all interest.

Practise exercise 6

1 Convert the following into percentages.

(a) $\frac{3}{4}$ (b) $8\frac{1}{5}$ (c) 12.5 (d) 113.04

(e) 1 : 4 (f) 3 : 8 (g) 17 out of 25 (h) 21 out of 60

2 Convert the following into fractions.

(a) 24% (b) 145% (c) 34.5% (d) 114.25%

(e) $\frac{1}{5}$% (f) $75\frac{3}{8}$% (g) 200% (h) 0.0004%

3 Calculate the following.

(a) 10% of 25 (b) 27% of 810 (c) 15.2% of 304

(d) 90.50% of 200 (e) $\frac{4}{5}$% 0f 1 (f) $32\frac{3}{4}$% of 84

4 There are 240 girls and 360 boys in a school.
(a) Calculate (i) the total pupils in the school,
(ii) the fraction of girl pupils, (iii) the percentage of boys in the school.
(b) Change the answer to part(a)(ii) as a percentage.
(c) If 10% of the boys left school, find the number of boys remain in the school.

5 The cost price of a mobile phone is \$625. The shopkeeper gives a discount of 6% on it.
Calculate the selling price of the mobile.

6 Mica sold his television for Rs.7500. She got 4% profit on this transaction.
Calculate the cost price of the television.

7 Roma paid \$25 for a calculator. The shopkeeper purchased it for \$21.
Calculate the percentage profit.

8 In 2002 the number of teak trees in a forest land was 2004. In 2008 the number of teak

trees in the forest was 1503. Calculate the percentage reduce of trees.

9 Zara deposits $1250.50 for five years at the rate of 6.5% simple interest per year. Calculate the interest and the amount at the end of five years.

10 Shobha invests € 650.25 for 4 years. She earns an interest of €312.12. Calculate the rate of investment.

11 Ali borrows $9500 for two years at 8.5% compound interest to start a business. Calculate (a) the interest, (b) the amount at the end of two years.

12 A Mathematics book has 450 pages. A Physics book has 475 pages. Calculate the pages in the Mathematics book as a percentage of the pages in the Physics book.

13 The selling price of a car is fallen by 17%. The cost price of the car is £6420. What is the selling price of the car?

14 The cost price of a land is £4275. The land owner wants to sell it at 5% profit. Calculate the selling price of the land.

EXTENDED CURRICULUM

Reverse Percentage

- It is used to find the original value of an item when its present value is given.

Working rule

1) For an increased percentage, $\text{Original value} = \dfrac{\text{Present value}}{(100 + \text{Increased})\%}$

2) For a decreased percentage, $\text{Original value} = \dfrac{\text{Present value}}{(100 - \text{Decreased})\%}$

Example 1

There are 660 students in a school this year. This is 10% more than the number of students last year. How many students were there last year?

Working: No. of students last year $= \dfrac{660}{(100+10)\%}$

$= 600$

$\text{Original value} = \dfrac{\text{Present value}}{(100 + \text{Increased})\%}$

Example 2

The selling price of a computer is $900. This is 15% less than its cost price. Calculate the cost price of the computer.

Working: Cost price $= \dfrac{935}{(100-15)\%} = \$1100$

$\text{Original value} = \dfrac{\text{Present value}}{(100 - \text{Decreased})\%}$

Practise exercise 7

1 In 2008 the population of a City was 226856. This is 12% more than the population in 2007. Work out the population of the City in 2007.

2 Fara purchased a washing machine for $379.62 after a reduction of 7.5% on its original price. Calculate the original price of the washing machine.

3 In a sale, original price of each item is decreased by 9%. Ali bought a camera for £63.70.

(a) Calculate the original price of the camera.

(b) Later he sold it at a profit of 5% to Ahmed.
How much did Ahmed pay for the camera?

4 The selling price of a mobile phone depreciates every year by 4%.
At the end of the third year its selling price is \$483.84.
(a) Calculate the selling price of the mobile phone
(i) at the end of the second year, (ii) at the end of the first year.
(b) Calculate the percentage depreciate from the first year to the third year.

5 The total number of vehicles in a town grows every year by 6%.
In 2007 the total number of vehicles was106742.
(a) Calculate the total number of vehicles
(i) in 2006, (ii) in 2005.
(b) Calculate the percentage growth of vehicles from 2005 to 2007.

6 (a) In 2007 the cost of a laptop was \$1200.
In 2008 the cost was 10% less than 2007. Calculate the cost in 2008.
(b) The cost of \$ 1200 in 2007 was 10% more than the cost in 2006. Calculate the cost in 2006.

Core Curriculum

1.07 The four rules

▶ The four rules for calculations can be used to simplify expressions containing four fundamental operations along with brackets.

Working rule

To work out a numerical expression
1) ***without brackets***, use the order of operation **'DMAS'**
(i.e. Perform Division and Multiplication first, then Addition and Subtraction),
2) ***with brackets***, use the order of operation **'BEDMAS'**
(i.e. perform brackets and exponents first, then multiplication and division followed by addition and subtraction.)

Example 1
Without using a calculator work out the following.

(a) $7 + 12 \times 4 - 4$ (b) $1 - 5 \times 10 \div 5 + 2.1$

Working: $7 + 48 - 4$ $= 1 - 10 + 2.1$
$= 51$ $= -6.9$

Use **DMAS.** That is perform multiplication and division first, then addition and subtraction.

(c) $45.2 - (3 - 1.5) \times 3 + 5^2$

Working: $45.2 - 1.5 \times 3 + 25$
$= 45.2 - 4.5 + 25$
$= 65.7$

Use **BEDMAS.** That is perform brackets and exponents first, then multiplication and division followed by addition and subtraction.

Example 2
Insert brackets to make each statement true. (a) $4 \times 5 - 6 + 7 = 3$ (b) $6 - 4.5 \times 2.6 + 1.2 = 5.7$

Working: (a) $4 \times (5 - 6) + 7 = 3$
(a) $(6 - 4.5) \times (2.6 + 1.2) = 5.7$

Start from left side to get right side, using trial and error method by your calculator.

Practise exercise 8(A) **Core Curriculum**

1 Without using a calculator, work out the following.

(a) $9 - 2 \times 4$ (b) $6 + 10 \div 2$ (c) $2\frac{1}{2} - 1\frac{1}{4} \times \frac{1}{5}$

(d) $5\frac{1}{6} + \frac{1}{12} \div 1\frac{1}{2}$ (e) $1 - \frac{1}{4} \times \frac{1}{5}$ (f) $4 - 3 \times 2 + 2$

(g) $10 + 2 \times 4 - 1.3$ (h) $6 \times 4 - 10 \div 5$ (i) $2.4 - 4.2 \times 6 - 7.1$

(j) $29 + 20 \div 4$ (k) $25 \div 5 + 10 \div 2$ (l) $7 \times 5 - 6 \times 4^2$

(m) $6^2 - 5(4 - 3)$ (n) $2 + 7(2 + 1) - 24$ (o) $10 - (9 - (8 - 7)) + 3^3$

2 Insert brackets to make each of the following statements correct.

(a) $4 \times 5 - 6 + 7 = 3$ (b) $7 - 8 \times 9 + 4 = -5$ (c) $7 - 10 \div 2 + 3 = 5$

(d) $11 + 7 \times 4 - 5 = 4$ (e) $12 - 8 \times 5 + 4 = 36$ (f) $15 + 3 \div 3 \times 2 - 1 = 6$

Practise exercise 8(B) **EXTENDED CURRICULUM**

1 The unit's digit of a number is one less than the ten's digit. If the unit's digit is 4, find the number.

2 Half of the sum of two numbers is 21. If one number is 20. Find the other number.

3 Five more than twice a number is 57. Find the number.

4 A number is increased by 2 and then multiplied by 4 to get the result 28.

5 When a number is divided by 5 and then decreased by 1, the result is 3. Find the original number.

6 Work out the following. Showing your working.

(a) $2^3 + 2^2 - 12$ (b) $6^2 + 4 \times 3 - (10.4 + 2.1 \times 5)$ (c) $2^2(12 - (3 - 5^2))$

(d) $\frac{9+4}{9^2 - 7^2}$ (e) $\frac{\frac{2}{5} - \frac{3}{10} + \frac{1}{15}}{\frac{1}{5} \times \frac{2}{6}}$ (f) $\frac{2+12}{4+3 \times 8}$

1.08 Ordering

Core Curriculum

- The symbols used for ordering quantities by magnitude or size are
  - (a) $>$ is greater than
  - (b) $\geq$ is greater than or equal to
  - (c) $<$ is less than
  - (d) $\leq$ is less than or equal to
  - (e) $=$ is equal to
  - (f) $\neq$ is not equal to

Working rule

For ordering, first convert the fractions or percentages into decimals and then compare the digits starting from left.

Example 1

Use one of the symbols >, < or = to make each statement correct.

(a) 1.2 ......>.......... 12% (0.12)

(b) 0.75 ......=......... $\frac{3}{4}$ (0.75)

(c) 6.2×4 ...>...... 248×10^{-1}
(24.8) (2.48)

First change fractions and percentages into decimals and then compare.

Example 2

Write each of the statements in order of size with the smallest first.

0.4 $\frac{4}{1000}$ 4% 0.41 4.14 0.004%

(0.004) (0.04) (0.00004)

In order, 0.004% $\frac{4}{1000}$ 4% 0.4 0.41 4.14

First Change fractions and percentages into decimals and then compare.

Practise exercise 9(A) **Core Curriculum**

1 Use one of the symbols >, <, = to make each of the statements correct.

(a) 17% .... $\frac{5}{29}$
(b) 4 – 5 .... 5 – 4
(c) $\frac{1}{5}$ 0.5
(d) $\frac{1}{10}$ 9%
(e) $(0.7)^2$ 50%
(f) 7% 0.007
(g) $(6^2 - 2^2)$ 4x8
(h) 0.006 x $\frac{1}{5}$ .... $\frac{12}{1000}$
(i) (2 x 4 – 5.2) 28%
(j) $\frac{1}{4}$ .... $\frac{14}{56}$
(k) $\sqrt{0.0225}$......0.3 x 0.5
(l) 2.79^3 ...... 4.63^3

2 Write each of the statements in order of size with the smallest first.

(a) 0.23 0.231 0.157 0.203 2.00023
(b) 6.04 6.053 6.053 6.45 4.06
(c) $\frac{5}{4}$ $\frac{7}{6}$ $\frac{9}{8}$ $\frac{11}{10}$
(d) $\frac{11}{9}$ $1\frac{1}{3}$ 12.2% 1.02
(e) 21% $\frac{1}{2}$ 0.21% 0.021 2.1%
(f) 4 *l* 3999 ml 4002 cubic cm
(g) 24 tons 2400 kilograms 4200 grams
(h) $\frac{4}{5}$ 0.79 81%

Practise exercise 9(B) **EXTENDED CURRICULUM**

1 Write each of the statements in order of size with the smallest first.

(a) 1.5×10^4 5.1×10^3 1.2×10^5
(b) 6.27×10^{-2} 4.6×10^6 6.4×10^{-4}
(c) 17% 1.7×10^{-2} 17 x 17%
(d) 2.4% 2.4×10^{-4} 0.24 $\frac{24}{1000}$
(e) cos 75° tan 75° sin 75°

(f) 0.0008 8×10^{-5} 0.8% $\frac{1}{125000}$

2 $(0.4)^{\frac{1}{2}}$, 0.4, $\sqrt{0.4}$, $(0.4)^{-1}$, $(0.4)^2$.

From the numbers above, write down

(a) the smallest, (b) the largest, (c) the two numbers which are equal.

1.09 Estimation **Core Curriculum**

- Estimation is a way of approximating some numbers or calculations to the specified accuracy.

- **Estimating calculations**

Working rule

1) ***To estimate sums, differences and products,*** first round off each of the given numbers to its highest place value and then perform the required operation.
2) ***To estimate quotients,*** first round off the divisor to its highest place value and chose the closer value of the dividend that divides exactly, then perform division.

Remember

For many cases, you need to perform estimation to check whether your answer is sensible or not.

Example 1

Estimate the sum of 47 and 64.

Working: 47 is rounded to 50 and 64 is rounded to 60.
$50 + 60 = 110$

First round to the highest place value and then add.

Example 2

Estimate the area of a rectangle with 11.5 cm long and 4.2 cm wide.

Working: 11.4 cm is rounded to 10 cm and
4.2 cm is rounded to 5 cm.

Area = length x breadth = $10 \times 5 = 50$ cm^2

First round to the highest place value and then multiply.

Example 3

The cost of 49 books is \$ 427. Estimate the cost of each book.

Working: \$ 427 is rounded to \$ 400 and 49 is rounded to 50.

Estimated cost of each book = \$ 400 ÷ 50 = \$ 8

First round to the highest place value and then divide.

- **Rounding numbers**

We use rounding numbers to give approximations to the specified decimal places and to the nearest whole number, ten, hundred, thousand etc.

Working rule

To round a number, 1) locate the required place,
2) if the digit next to it is 5 or more, then **round up**(required place digit +1), otherwise **round down** (drop the digit after the required place).

Example

Write 4157.56149 correct to (i) 1 d.p. (ii) 2 d.p. (iii) 3 d.p.

(iv) nearest 10 (v) nearest 100 (vi) nearest whole number

Working: (i) 4157.565149 = 4157.6
(ii) 4157.565149 = 4157.57
(iii) 4157.565149 = 4157.565
(iv) 4157.565149 = 4160
(v) 4157.565149 = 4200
(vi) 4157.565149 = 400

> Use required place digit + 1 for > 5.
>
> Drop the digits after the required place for < 5.
>
> Use required place digit + 1 for = 5.

Significant figures

Working rule

To approximate a number correct to the specified significant figures,
1) locate the figure(digit) by starting from left and after the first non zero digit, count all the digits including zeros,
2) then round the digit next to it.

Example 1

Write 94.0714 correct to (i) 1 s.f. (ii) 2 s.f. (iii) 3 s.f. (iv) 4 s.f.

Working: (i) 94.0714 = 90
(ii) 94.0714 = 94
(iii) 94.0714 = 94.1 (iv) 94.0714 = 94.07

> Use required place digit + 1 for > 5.
>
> Drop the digits after the required place for < 5.

Example 2

Write 0.0004073695 correct to four significant figures.

Working: 0.0004073695 = 0.0004074

> Zeros before first non zero digit are not counted.
> Use required place digit + 1 for > 5.

Remember

If th e answer to any question is not exact and the degree of accuracy is not given, then give your answer to three significant figures.

Practise exercise 10(A) **Core Curriculum**

1 Estimate the cost of 122 pens at the rate of 49 cents each. Show your working.

2 Estimate the value of $1035 - 67 \times 11.1 + \sqrt{10}$.

3 The heights of four children are 142.5 cm, 144.4 cm, 148.4 cm and 147.6 cm.
(a) Estimate the total heights of all the four children.
(b) Use your answer to part(a) to estimate the average height.

4 Round each of the following correct to the number of decimal places(d.p) indicated.
(a) 4.92 (1 d.p.) (b) 6.952 (1 d.p.) (c) 0.0048 (2 d.p.)
(d) 9.999 (2 d.p.) (e) 25.940513 (3 d.p.) (f) 0.00009 (4 d.p.)

5 Round each of the following to the nearest degree of accuracy given in the brackets.
(a) 56 (nearest 10) (b) 495 (nearest 10) (c) 4075 (nearest 100)
(d) 12154 (nearest 100) (e) 5555 (nearest 1000) (f) 2997 (nearest 1000)
(g) 301296 (nearest 10 000) (h) 4951212 (nearest 100 000) (i) 44.73 (nearest ten)
(j) 0.0057 61.54 (nearest thousandth) (k) 61.54 (nearest whole number)

6 Write each of the following correct to the number of significant figures(s.f.) written in the brackets.
(a) 74.27 (1 s.f.) (b) 0.0079 (1 s.f.) (c) 49.624 (2 s.f.)
(d) 6.0543 (2 s.f.) (e) 13.6501 (3 s.f.) (f) 60.0909 (3 s.f.)

(g) 0.0040764 (3 s.f.) (h) 7.05059 (4 s.f.) (i) 0.01010574 (5 s.f.)

7 The height of a tower is 109.45 metres. Write this height correct to
(a) 1 decimal place, (b) 1 significant figure,
(c) two significant figures, (d) the nearest whole number.

8 The total expenditure to build a bridge is \$ 75094.60. Write down this amount correct to
(a) 3 significant number, (b) nearest whole number.

9 $$\frac{464 - 18 \text{ x } 23}{42 \div 12.4}$$

(a) Rewrite this expression with each number rounded to one significant figure.
(b) Use your answer to part(a) to estimate the value of the calculation.
(c) Write your estimation to part(b) correct to nearest ten.

10 $$\frac{7.92 \text{ x } 10.6}{39.41 \div 5.29 + \sqrt{10}}$$

(a) Write the above expression by rounding each number to 1 significant figure.
(b) Use your answer to part (a) to find the approximate value of the calculation.
(c) Showing your working, write your answer to part(b) correct to 2 decimal places.

Practise exercise 10(B) **EXTENDED CURRICULUM**

1 The height of a stack of 41 books of equal thickness is 89 centimetres.
Estimate the thickness of
(a) each book, giving your answer correct to 2 significant figures,
(b) 199 books, giving your answer correct to nearest 2 centimetres.

2 (a) The approximate area of a square is 26 square metres.
Estimate (i) the side of the square, (i) the diagonal of the square.

(b) $$\frac{\sqrt{50} - \sqrt[3]{9} \text{ x } 15^{\frac{1}{4}}}{\sqrt{99} \text{ x } 7.4}$$

(i) Estimate the value of the above calculation.
(ii) Write your estimation to part (a), correct to 5 d.p.

3 Each edge of a cube is 27.65 centimetres.
(a) Write down this number correct to
(i)one significant figure, (ii) the nearest whole number.
(b) Use your answer to part (a)(ii) to estimate the volume of the cube.
(c) Use your answer to part (a)(i) to estimate the diagonal, $d = \sqrt{3 \text{ x } (....)^2}$ of the cube.

4 The approximate diameter of a circular micro organism is $1.45 \text{ x } 10^{-7}$ millimetre.
Calculate the approximate circumference of the micro organism.

5

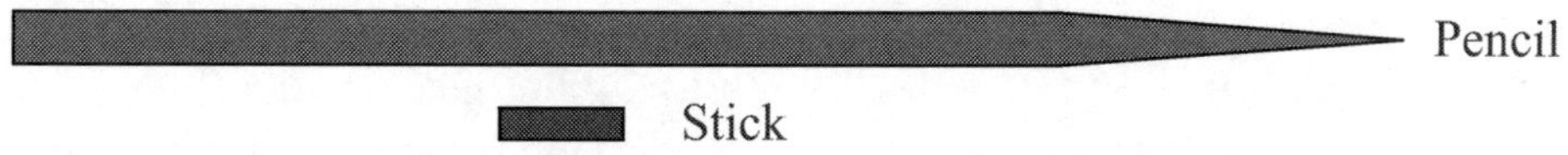

The diagram shows a pencil and a small stick. The length of the stick is 1.6 cm.
Using the length of the stick, estimate the length of the pencil and give your answer correct to 1 decimal place.

6 $$x = \frac{0.002751 \times 3400}{(9.8923 + 24.7777)^2}$$

(a) Rewrite each number in this calculation correct to 1 significant figure.
(b) Use your answer to part (a) to estimate the value of x.

1.10 Limits of accuracy

Core Curriculum

- Limits of accuracy are used to calculate the upper and lower bounds for data correct to a specified accuracy.

Working rule

The limits (or bounds) of a data, d, correct to the nearest accuracy, a, is given by the inequality

$$\left(\mathbf{d} - \frac{a}{2}\right) \leq \mathbf{d} < \left(\mathbf{d} + \frac{a}{2}\right).$$

↓ Lower limit or lower bound ↓ Upper limit or upper bound

Example 1
The length, l, of a pencil is 7 cm correct to the nearest centimeter.
Complete the statement

......cm $\leq l <$ ......cm.

Working: 6.5 cm $\leq l <$ 7.5 cm.

Lower bound $= 7 - \frac{1}{2} = 6.5$

Upper bound $= 7 + \frac{1}{2} = 7.5$

Example 2
The height of a mountain is 4.7 kilometres, to the nearest metre.
(a) Calculate the maximum and minimum possible height of the mountain.
(b) Use your answer to part(a), to write an inequality.

Working:
(a) Maximum possible height = 4.7005 km.
Minimum possible height = 4.6995 km.
(b) 4.6995 km $\leq h <$ 4.7005 km.

1 m = 0.001 km

Max. height $= 4.7 + \frac{0.001}{2} = 4.7005$

Min. height $= 4.7 - \frac{0.001}{2} = 4.7005$

Practise exercise 11(A) **Core Curriculum**

1 Write down the upper and lower bounds for each of the following data correct to the specified accuracy.
(a) 412 (nearest whole number)
(b) 50 (nearest 10)
(c) 1060 (nearest twenty)
(d) 900 (nearest 50)
(e) 7200 (nearest 100)
(f) 18500 (nearest 500)
(g) 69000 (nearest 1000)
(h) 24000 (nearest 10000)
(i) 45 mm (nearest millimetre)
(j) 7.5 cm (nearest centimeter)
(k) 64 meters (nearest centimeter)
(l) 27 g (nearest gram)
(m) 5.47 liters (nearest millimetre)
(n) 14.5 minutes (nearest second)
(o) 4.6 (correct to 1 decimal place)
(p) 9.74 (correct to 2 decimal place)
(q) 0.007 (correct to 1 significant figure)
(r) 6.3 (to 2 significant figures)

2 The total number of pupils in a school is 540, correct to the nearest 20.

Write down (a) the minimum possible number of pupils,
(b) the maximum possible number of pupils in the school.

3 The length, l, of a chalk piece is 8 centimeters, correct to the nearest centimetre.
(a) Write down (i) the smallest possible value of l,
(ii) the largest possible value of l.
(b) Use your answer to part(a) to complete the statement given below.
$\ldots\ldots\ldots\ldots \leq l < \ldots\ldots\ldots\ldots$

4 The height of a tower is 115 meters, correct to the nearest 5 meters.
(a) Calculate the upper and lower bounds to the height.
(b) Using, h, the height of the tower express the answer to part(a) as an inequality.

5 The population of a small country is 340000 correct to the nearest thousand.
Complete the statement ....≤ population < ....

6 The length of a river is 4200 kilometers, correct to the nearest 100 kilometers.
Complete the statement about the limits of the length. ....km ≤ length < ....km.

7 Rima weighs w, 47 kilograms, correct to the nearest kilogram.
Complete the statement ....kg ≤ w < ....

8 Daniel spent 7.4 hours to complete a project. The given time is correct to the nearest minute. Complete the statement about this time. ....hr ≤ Time spent < ....hr

9 A necklace weighs, w, 4.2 grams, correct to the nearest milligram. Complete the statement about the limits of w. ....g ≤ w < ....

10 The total cost of 17 kilograms of potatoes is \$5.3, correct to the nearest cents.
Complete the statement about the bounds of cost. \$.......≤ cost < \$ ....

11 The approximate value, v, of a sum is 16.24, correct to two decimal places.
Complete the statement about the range of v. ....≤ V < ....

12 In an examination, Karina's average mark is 70.8, correct to three significant figures.
Complete the statement about the limits of the average mark.
................≤ Average mark < ....

13

In the above diagram, the measurements are written correct to the nearest centimetre.
(a) Complete the following statements.
(i) The lowest possible measurement of length is .... cm.
(ii) The lowest possible measurement of breadth is .... cm.
(iii) The highest possible measurement of length is .... cm.
(iv) The highest possible measurement of breadth is .... cm.
(b) Calculate the
(i) lowest possible area, (ii) highest possible area,
(iii) minimum possible perimeter, (iv) maximum possible perimeter of the rectangle.

EXTENDED CURRICULUM

- Limits of accuracy can be used to obtain the upper and lower bounds to the solutions of problems in which the data are given to a specified accuracy.

Working rule

1) ***To obtain upper bound,***
(a) use both the maximum values (for addition or multiplication),
(b) use, in order, the maximum value and the minimum value (for minus or division).
2) ***To obtain lower bound,***
(a) use both the minimum values (for addition or multiplication),
(b) use, in order, the minimum value and the maximum value (for minus or division).

Example
Obtain the upper and lower bounds for each of the calculations, correct to the nearest places given in the brackets.
(a) 1200 + 700 (nearest 100) (b) 14.6 x 5.4 (one decimal place)
(c) 694 – 500 (nearest whole number) (d) 0.029 ÷ 4.1 (two significant figures)

Working: (a) Upper bound = 1250 + 750 = 2000
Lower bound = 1150 + 650 = 1800
(b) Upper bound = 14.65 x 5.45 = 79.8425
Lower bound = 14.55 x 5.35 = 77.8425

For addition and multiplication,
to find upper bounds, use both maximum values.
to find lower bounds, use both minimum values.

(c) Upper bound = 694.5 – 499.5 = 195
Lower bound = 693.5 – 500.5 = 193
(d) Upper bound = 0.0295 ÷ 4.05 = 0.0073
Lower bound = 0.0285 ÷ 4.15 = 0.0069

For minus and division,
to find upper bounds, use in order, max value and min value,
to find lower bounds, use in order, min value and max value.

Practise exercise 11(B) ⟵ **EXTENDED CURRICULUM** ⟶

1 Obtain the upper and lower bounds for each of the calculations, correct to the nearest places given in the brackets.
(a) 640 + 790 (nearest 10) (b) 17000 + 3000 (nearest 1000)
(c) 25.2 x (41.4 + 29.6) (1 d.p.) (d) 615 – 214 (nearest whole number)
(e) 14.45 ÷ 0.8 (nearest hundredth) (f) $\frac{4.9 \times 0.0027}{0.043}$ (2 s.f.)

2 The heights of two buildings are 90 m and 130 m, correct to the nearest 20 metres. For the total height of both the buildings, calculate
(a) lower bound, (b) upper bound.

3 Each side of a square shaped garden is 59.64 m, correct to two decimal places.
(a) Obtain the upper and lower bounds for
(i) the perimeter of the garden, (ii) the area of the garden.
(b) What is the difference between the maximum possible area and minimum possible area of the garden.

4 The area of a forest land is 84 square kilometers. The area of an agricultural land is 122 square kilometres. Both the measurements are correct to the nearest square kilometer.
(a) Calculate (i) the largest possible difference between both areas,
(ii) the smallest possible difference between both areas.
(b) Find the quotient, when the upper bound for the area of the agricultural land is divided by the upper bound for the area of the forest land.

5 Each edge of a cubical box is 60 centimetre correct to the nearest millimetre.

(a) Calculate the volume of the box.
(b) Obtain the upper and lower bounds for the answer to part(a).

6 The mass of eight books is 9.6 kilograms, correct to the nearest 0.02 kilogram.
(a) Calculate (i) the maximum possible mass of a book,
(ii) the least possible mass of a book.
(b) Obtain the appropriate upper and lower bounds for the mass of 25 such books.

7 The length and breadth of a rectangular play ground are 91.4 m and 54.5 m respectively. The above measurements are correct to three significant figures.
(a) Obtain (i) the minimum possible area of the play ground,
(ii) the maximum possible area of the play ground.
(b) Fathima runs twice around the play ground. Calculate the minimum possible distance covered by Fathima.
(c) The play ground management has planned to grow grass on the whole ground. The cost for growing grass in each square metre is \$2.45. Calculate the maximum possible cost for growing grass on the whole ground.

8 The diameter of a circle is 0.06 kilometre, correct to one significant figure.
Find the upper bound for
(a) the radius of the circle, (b) the circumference of the circle, (c) the area of the circle.

9 A plane flies an altitude of 9500 ft, correct to the nearest 100 ft.
A bird flies an altitude of 290 ft, correct to the nearest 10 ft.
(a) Obtain the maximum altitude difference between the plane and the bird.
(b) Obtain the least altitude difference between the plane and the bird.

Core Curriculum

1.11 Time

Twenty four – hour clock

- In 24 – hour clock, time is calculated after mid night as
  0100 0200 .... 1200 1300 ....2300 2400 hour.

a.m. **noon** **p.m.**

Remember

1. In twenty four – hour time, the first two digits stand for hours and the last two digits stand for minutes. E.g. 14 17 means 14 hours and 17 minutes.

2. 1 hour = 60 minutes and 1 minute = 60 seconds.

Example

A singing competition starts at 09 27 and ends at 13 12. How long does the competition take?

Working: Length of time = 13 12 – 09 27
= 3 hr 45 min

Use the time button on your calculator.
13.,,, 12.,,, – 09.,,,27.,,, = 3 hr 45 min

Example

Train	T_1	T_2	T_3
Departure time	02 25	07 45	10 30
Arrival time	16 47	18 14	23 15

The above table shows the departure and arrival times of three trains T_1, T_2 and T_3 on a particular day.
(a) Calculate the length of time for each train.
(b) Write the time difference between the T_1 and T_2.
(c) Which train takes longest time?
(d) Which train takes least time?

Working: (a) The length of time for
train T_1 = 16 47 – 02 25 = 14 hr 22 min
train T_2 = 18 14 – 07 45 = 10 hr 29 min
train T_3 = 23 15 – 10 30 = 12 hr 45 min
(b) Time difference between T_1 and T_2
= 14 22 – 12 45 = 1 hr 37 min
(c) Train T_1 takes longest time.
(d) Train T_2 takes least time.

> Use the time button on your calculator.
>
> 16.,,,47.,,, – 02.,,, 25.,,, = 14 hr 22 min
>
> 18.,,,14.,,, – 07.,,, 45.,,, = 10 hr 29 min
>
> 23.,,, 15.,,, – 10.,,, 30.,,, = 12 hr 45 min
>
> 14.,,, 12.,,, – 12.,,, 45.,,, = 1 hr 37 min

- **Twelve – hour clock**
  - In 12 – hour clock, we use a.m. for the time between after midnight and before midday and p.m. for the time between after midday and before midnight.

Remember

1. After midnight $<$ **a.m.** $<$ Before midday.
2. After midday $<$ **p.m.** $<$ Before midnight.
3. To perform calculations with 12 – hour clock time, first change the given times to 24 – hour clock time, then do the required calculation.
4. The a.m. time remain same when change to 24 – hour clock time.
5. The p.m. time added to 1200 when change to 24 – hour clock time.

Example
Rahul goes to school at 7.45 a.m. and comes back at 1.30 p.m.
How long does he stay in school?
Working: 7.45 a.m. = 07 45 and 1.30 p.m. = 13 30
The length of time = 13 30 – 0745 = 5 hr 45 min

> To change to 24 – hour clock time,
> a.m. remains same and p.m. adds to 1200.
> 13.,,,30.,,, – 07.,,,45.,,, = 5 hr 45 min

Practise exercise 12(A) **Core Curriculum**

1 Arya takes 4.5 hours to complete his Mathematics assignment.
Write this time in (a) minutes, (b) seconds, (c) hours and minutes.

2 Mouri won a running competition by finishing it in 79.4 seconds. Write this time in
(a) minutes, (b) hours, (c) minutes and seconds.

3 A Physics test starts at 11 30 and ends at 13 45. How long does the test take?

4 A bus departed at 05 57 and arrived its destination at 10 22. How many hours and minutes did the journey take?

5 A train has to travel for 23 hours and 10 minutes to reach the final destination. The train began its journey at 02 15. At what time did it complete the journey in the next day?

6 A cyclist reached its final destination at 14 35 after cycling for 5 hours and 25 minutes. At what time did he start the journey?

7 Change the following time to 24 – hour clock time.
(a) 4.20 a.m. (b) 0.48 p.m. (c) 2.27 p.m. (d) 4 hours and 15 min

8 Change the following 24 – hour clock time to 12 – hour clock time.
(a) 00 24 (b) 07 45 (c) 11 29 (d) 12 10 (e) 23 24

9 An English essay writing competition conducted for 1 hour 45 minutes. It ended at 1.10 p.m. At what time did it start?

10 Mr. Jack enters to his office at 8.50 a.m. and comes out at 16 25.
Write the length of time he spends in the office.

11 A plane started its journey at 10 22 and reached the final destination after 4 hours 15 minutes. Write in 12- hour clock time, the time at which it reached in the final destination.

Practise exercise 12(B) **EXTENDED CURRICULUM**

1 Reema, Seema and Nauma are asked to solve a mathematical puzzle.
Seema finished first in a time of 4 minutes 21 seconds. Nauma finished second, 12 seconds ahead of Reema. Reema finished last, 49 seconds behind Seema.
(a) Write down (i) Reema's time, (ii) Nauma's time.
(b) Reema finished at 10 05. Write down at what time
(i) Seema finished, (ii) Nauma finished.

2 Daniel wrote a test. She worked 15 questions in the first 45 minutes and the remaining 10 questions in the next 55 minutes.
(a) How long did she take to work out all the 25 questions? Give your answer in hour and minutes.
(b) The test started at 09 30. At what time did she work out
(i) the first 15 questions, (ii) all the 25 questions?
(c) Daniel finished work 25 minutes before the allotted time.
Write the allotted time for the test (i) in hours and minutes, (ii) in hours only.

3 A plane departs from A at 07 40 and arrives in B at 10 28. Than it departs from B at 10 50 and arrives in A at 15 45. Find the total time the plane takes for to and fro journey?

4 A train left Mumbai at 21 15 on Monday and arrived in Delhi at 09 40 on the next day. How long, in hours and minutes, was the journey?

1.12 Speed, distance and time

Core Curriculum

- The relationship between speed(S), distance(D) and time(T) can be represented by a single word **SeDoT** (means Speed equals Distance over Time).

Working rule

To change the units of speed, use the unit conversion formulae

1) x km/h = $\frac{5x}{18}$ m/s, 2) x m/s = $\frac{18x}{5}$ km/h.

Remember

1. 1 km = 1000 m (not 100 m)
2. 1 h = 60 min (not 100 min)
 = 3600 s (not 100 or 1000 s)
3. 1 min = 60 s (not 100 s)
4. To change minutes to hour divide by 60.
5. To change seconds to minutes divide by 60.

Example 1

A bus takes 15 minutes to cover a distance of 16 kilometres. Calculate the average speed of the bus in metres per second.

Working: T = 15 min = 15 x 60 s = 900 s

D = 16 km = 16 x 1000 m =16000 m

Now $S = \dfrac{1600\text{m}}{900\text{s}} = 17.8$ m/s

1 min = 60 s
1 km = 1000 m
Using **SeDoT**

Example 2

Change the speed of (a) 72.36 km/h into m/s, (ii) 15 m/s into km/h.

Working: (a) 72.36 km/h = $\dfrac{5 \times 72.36}{18}$ m/s = 20.1 m/s

(b) 15 m/s = $\dfrac{18 \times 15}{5}$ km/h = 54 km/h

x km/h = $\dfrac{5x}{18}$ m/s

x m/s = $\dfrac{18x}{5}$ km/h

Core Curriculum

Practise exercise 13(A)

1 Convert the units of the given average speeds as specified.

(a) 54 km/h into m/s (b) $7\frac{1}{5}$ km/h into m/s (c) 55 m/s into km/h

(d) 15 km/h into km/min (e) 4 m/s into m/min (f) 252.8 km/h into miles per hr

2 The highest speed of a missile is 880 kilometres per hour.
Write down this speed (a) in metres per second, (b) in kilometers per minute,
(c) in kilometres per second.

3 The average speed of an airbus is 75 metres per second.
Write down this average speed (a) in metres per minute,
(b) in kilometers per hour, (c) in kilometers per minute.

4 The average speed of a plane is 237 kilometres per hour.
Write down this speed in (a) miles per hour, (b) miles per minute.

5 A car takes 4 hours to cover a distance of 262 kilometres. Calculate the average speed of the car (a) in kilometers per hour, (b) in metres per second.

6 A bus took 2 hours 30 minutes to travel 140 kilometres.
(a) Write down, in hours, the total time taken for the journey.
(b) Calculate the average speed of the bus, in kilometers per hour.
(c) Change your answer to part(b) into metres per second.

7 Amrit cycles 16.45 km at an average speed of 9.4 kilometres per hour.
Calculate, how long does Amrit cycle (a) in hours (b) in hours and minutes.

8 Sukeerti drives her car at an average speed of 20 m/s for 2 hours and 15 minutes.
Calculate the distance covered by Sukeerti (a) in metres, (b) in kilometers.

9 A train departs at 10 20 and arrives its destination at 02 35 on the next day.
The train covers a distance of 1022 kilometres.
(a) Calculate, in hours, the length of time for the journey.
(b) Calculate, in kilomtres per hour, the average speed of the train.

10 A train took 13 hours to travel a distance of 710 kilometres and then another 11 hours for the remaining half distance to reach the final destination.
(a) Find out, for the whole journey
(i) the total distance travelled, (ii) the total time taken.
(b) Calculate in kilometers per hour, the average speed of the train.
(c) Convert your answer to part(b),
(i) into kilometers per minute, (ii) into metres per minute.

Practise exercise 13(B) EXTENDED CURRICULUM

1 The average speed of a train is 7.2×10^4 m/h. Write down this speed
(a) in metres per minute, (b) in metres per second, (c) in kilometers per hour.

2 A plane flies at the average speed of 2.34×10^5 m/h for 2 hours and 45 minutes to reach the destination. Calculate and write your answer, in standard form, the distance covers by the plane (a) in metres, (b) in kilometers.

3 A train of 250 metres long is running at an average speed of 54 km/h. How long does it take to cross an electric pole standing near the rail track?

4 A train of 240 metres long is travelling at a speed of 16 m/s. Another train of 260 metre long is travelling at a speed of 14 m/s. Both the trains are travelling in opposite direction.
(a) Find the length of time, in seconds, they take to cross a bridge of 1 kilometre long.
(b) Write your answer to part (a) in minutes.

5 A plane departs from City – A at 10 40 and after flying a distance of 342 kilometres reaches in City – B at 12 15. It takes half an hour halt in City – B. Then it starts flying to cover a distance of 626 kilomtres to reach in City – C at 15 10.
(a) Find, in kilomtres, the total distance flies by the plane.
(b) Find, in hours, the total time taken by the plane.
(c) Calculate in kilomtres per hour, the average speed of the plane for the whole journey.
(d) Write your answer to part(c) in miles per hour.

6 (a) A train completed a journey of 1050 kilomtres with an average speed of 64 kilometres per hour. Calculate, giving exact answers the time taken for this journey in
(i) hours, (ii) hours, minutes and second.
(b) Another train took 14 hours 24 minutes to complete the same 1050 km journey.
(i) It departed at 20 20. At what time on the next day, did this train complete the journey?
(ii) Calculate the average speed, in kilometers per hour, for the journey.

1.13 Variation

Core Curriculum

Rate

► Rate is the expression of one thing per the unit quantity of another thing.

Remember

1. To calculate rate, just divide.
2. Take care of the conversion units, especially capacity and time.

Example

The capacity of a water tank is 480 litres.
(a) Water flows through a pipe at the rate of 10 litres per minute. How long, in hour does it take to fill the whole tank.
(b) Another pipe can empty the full tank in 45 minutes. Calculate in litres per hour the rate at which water passes through this pipe.

Working: (a) Time $= \frac{480}{10} = 48$ minutes $= 0.8$ hours

(b) 45 min = 0.75 h

To calculate rate, just divide.

$48 \text{ min} = \frac{48}{60} \text{ h} = 0.8$

$45 \text{ min} = \frac{480}{0.75} \text{ h} = 0.75 \text{ h}$

$$\text{Rate} = \frac{480}{0.75} = 640 \text{ litres/h}$$

Practise exercise 14(A) **Core Curriculum**

1 An electric motor rotates 2590 rotations in 35 minutes. Calculate the rate of rotation
(i) per minute, (ii) per hour.

2 Neelam runs a distance of 2.4 kilometres at the rate of 6 kilometres per hour. How long, in minutes, does it take her to complete the total distance?

3 A small stream originates from a hill. Water comes out through the origin at the rate of 1380 litres per hour. How many litres of water comes out through the origin in 4 hours 36 minutes?

4 In a city, a drinking water tanker distributes 4950 litres of water in 2 hours 45 minutes. Calculate the rate, in litres per hour, the tanker distributes water.

5 In an oil filling station, petrol passes through the filling pipe at the rate of 2 litres per minute. Petrol runs through the filling pipe for 1 hour 12 minutes. How many litres of petrol comes out through the pipe?

6 Pinu and Nanu start to dig a trench. They both dig at the same rate.
(a) When they are half-way through the work, what fraction of the trench has Nanu dug?
(b) Tinu then arrives at help. All three dig at the same rate until the work is finished.
(i) What fraction of the trench did Pinu dig altogether?
(ii) What fraction of the trench did Tinu dig?

Practise exercise 14(B) **EXTENDED CURRICULUM**

1 The capacity of a cylindrical tank is 900 litres.
(a) Diesel flows through a pipe at the rate of 1500 cubic centimeters per minute. Calculate the length of time to fill the whole tank (i) in minutes, (ii) in hours.
(b) Another pipe can empty the full tank in 12 hours exactly. Calculate the rate at which diesel passes through this pipe
(i) in litres per hour, (ii) in litres per minute, (iii) in cubic centimeters per minute.

2 A book publisher uses 4 printers to print 46080 sheets of paper in 4 hours 30 minutes. All the four printers print at the same rate.
(a) Calculate the number of sheets of paper one printer prints per hour.
(b) All the four printers start working at the same time. One printer stops working just after first hour.
(i) How many sheets of paper do they print in the first hour?
(ii) How long will it take for the three printers to print the remaining sheets of paper?

3 A channel has a cylindrical cross-section of radius 1.4 m.
When the depth of water is 0.4 metres, the water flows along the channel at 3 cubic metres/minute.
Calculate the number of cubic metres which flow along the channel in
(i) one hour, (ii) two hours.

4 It takes 6 hours 45 minutes to fill a water tank using 3 pipes which supply water at the same rate. How long will it take to fill the tank if 2 pipes are used?

Ratio

- Ratio is a way to compare the sizes of like things.

Working rule

1) ***To simplify a ratio in its lowest terms,*** reduce all the terms by their highest common factors.
2) ***To write a ratio in the form 1: n,*** just divide both the terms by the first term.
3) ***To divide a quantity in a given ratio,***
   (a) first find the multiplier, $m = \dfrac{\text{Total quantity}}{\text{Sum of ratio parts}}$,
   (b) then multiply 'm' by each ratio part to get the respective share.

Remember

For ratios with mixed units, change all the terms to a common unit and then simplify.

Example 1

Write the following ratios in its lowest terms. (a) 45:240 (b) 65 tones : 520 kg

Working: (a) $45 : 240 = \dfrac{45}{15} : \dfrac{240}{15} = 3 : 16$

(b) 65 tones : 520 kg = 65000 kg : 520 kg

$= \dfrac{65000}{520} : \dfrac{520}{520} = 125 : 1$

> The highest common factor of 45 and 240 is 15.
> Divide both terms by 15.
> 1 tone = 1000 kg. So 65 tones = 65000kg.
> The highest common factor of 65000 and 520 is 520.
> Divide both terms by 15.

Example 2

Write 21 : 294 in the form 1: n.

Working: $21 : 294 = \dfrac{21}{21} : \dfrac{294}{21} = 1 : 14$

> Divide both the terms by the first term 21.

Example 3

Divide £ 2613 into three parts in the ratio 2 : 4 : 7.

Working: $m = \dfrac{2613}{(2+4+7)} = £\ 201$

First part = £ 201 x 2 = £402
Second part = £ 201 x 4 = £ 804
Third part = £ 201 x 7 = £ 1407

> Find the multiplier, $m = \dfrac{\text{Total quantity}}{\text{Sum of ratio parts}}$
> Then multiply each part by m.

Practise exercise 15(A) **Core Curriculum**

1 Write each of the following ratios in its lowest terms.
(a) 4 : 14 (b) 10 : 24 (c) 50 : 15
(d) 26000 : 65000 (e) 3.6 : 72 (f) 6 kg : 250 g
(g) 2.5 kg : 0.4 tone (h) 800 ml : 2 *l* (i) 9 *l* : 400 cm^3
(j) $ 5 : 75 cents (k) 10 : 12 : 22 (l) 2 tones : 40 kg : 800 g

2 A class has 45 girl students and 54 boy students.
Write down in its simplest form the ratio of
(a) girl to boy students, (b) boy to girl students, (c) girl to total students.

3 Write down each of the following ratios in the form 1: n.
(a) 7 : 28 (b) 175 : 1225 (c) 600 g : 24 kg (d) 750 cm^3 : 4.5 *l*

4 The number of participants of four different countries takes part in an Olympic game are shown in the table below.

China	India	Singapore	Japan
160	140	80	100

(a) Write down, in its simplest form, the ratio of the participants of
(i) China : Japan (ii) Japan : Singapore (iii) India : Japan : Singapore
(b) What is the ratio of the participants of
(i) India to China and Japan, (b) China and Singapore to India and Japan?

5 (a) Divide \$ 195 into two in the ratio 5 : 8.
(b) Divide € 328 into three parts in the ratio 2 : 5 : 9.

6 Sonali and Mithali Share £ 550 in the ratio Sonali : Mithali = 2 : 3.
(a) How many pounds does Mithali receive?
(b) How many pounds does Sonali receive?

7 The number of pupils, parents and teachers participated in a Mathematics seminar are in the ratio pupils : parents : teachers = 6 : 5 : 7.
The total participants in the seminar is 234.
(a) Find the number of pupils, (b) Find the number of teachers.

8 A number is divided in the ratio 3 : 4 : 5. The larger part is 85.
(a) Calculate the smaller part. (b) What is the original number?

9 The length of a line AC is 12.6 cm and point B is on the line AC.
If AB : AC = 5 : 4, Calculate (a) the length of AB, (b) the length of AC.

Practise exercise 15(B) **EXTENDED CURRICULUM**

1 In 2007, a school has 262 boys and 269 girls. In 2008, 12 boys were left and six new girls are joined in the school.
(a) Write in its simplest form, the ratio of girls to boys in 2008.
(b) What is the ratio between
(i) the total number of students in 2007 and the total number of students in 2008,
(ii) the total number of girls in 2008 and the total number of boys in 2007?

2 The sides of a triangle are in the ratio 3:2:4. The perimeter of the triangle is 40.5 cm.
Calculate the length of (i) the longer side, (b) the shorter side.

3 The interior angles of a quadrilateral are in the ratio 3:5:4:6.
Calculate each of the interior angles of the quadrilateral.

4 (a) Lal, Bal and Pal start a business. They invested £ 2250 in the ratio 7:10:8.
(i) How much does Pal invest? (ii) How much does Bal invest?
(b) The investment amount and the profit reach to £ 3825 after two months.
They share the profit same as the investment ratio.
(i) Show that Lal receives profit of £ 441. (ii) How much profit does Pal receive?
(iii) How much more profit does Bal receive than Pal?

5 The age of a mother is equal to the ages of her son and daughter.
The ratio between the ages of her son and daughter is 6:5.
The age of her daughter is 20 years.
(a) Find the age of her son. (b) Find the age of mother.

6 In an Examination, Selina scored twice as many as Amina's marks and Alina scored one-third of Selina's marks.
(a) Find the ratio of marks scored in the order Amina : Selina : Alina.

(b) Amina and Alina together scored 75 marks.
Calculate the marks scored by (i) Alina, (ii) Selina.

Proportion

- **Direct proportion:** In direct proportion, when one quantity increases, the other also increases and when one quantity decreases, the other also decreases.
- **Inverse proportion:** In inverse proportion, when one quantity increases, the other decreases and when one quantity decreases, the other quantity increases.

Working rule

1) ***For the questions on direct proportion,*** use cross multiplication.
2) ***For the questions on inverse proportion,*** use straight multiplication.

Example 1
The cost of 15 note books is \$ 26.25. Work out the total cost of 152 such note books.

Working: $15 : 26.25 = 152 : x$

$15x = 152 \times 26.25$

$x = \dfrac{3990}{15} = \$\ 266$

When the number of note books increases, then the total cost also increases.
So this is a direct proportion question.
For direct proportion, use cross multiplication.

Example 2

Four men can complete a work in 20 days. How many days will it take for 10 men to complete the same work at the same rate?
Working:

Men	Days
4	20
10	x

$10x = 4 \times 20$ or $x = \dfrac{80}{10} = 8$ days

When the number of men decreases, then the number of days increases. So this is a question on inverse proportion.

For inverse proportion, use straight multiplication.

Core Curriculum

Practise exercise 16(A)

1 Twenty four chocolates costs \$ 18.72. Find the total cost of 42 such chocolates.
2 A car uses 17 litres of petrol to travel 238 kilometres of distance.
(a) How many litres of petrol will it need to travel a distance of 434 kilometres?
(b) How many kilomatres will it travel using 6.5 litres of petrol?
3 An examiner can mark 12 answer scripts in 2 hours.
(a) How much time will he take to mark one answer script?
(b) How many answer scripts can he mark in 6 hours 40 minutes?
4 Six students can complete a decorating work in 4 hours.
(a) How long will it take for 8 students to complete the same work?
(b) How many students can complete the work in 1 hour?

5 A group of 15 people shares equally a box of sweets and each get 7 sweets. How many sweets will each people gets, if the same box of sweets shares equally among 21 people?

6 It takes 8 months for 42 men to build an apartment.
(a) How many men can build the apartment in (i) one month, (ii) three months?
(b) How many months will it take for 24 men to build the same apartment?

Practise exercise 16(B) ⟵ **EXTENDED CURRICULUM** ⟶

1 An exporter ordered 5000 television sets at the cost of $ 600 000. Due to high demand in the international market, he increased the number of television sets by 1400. Find the total cost of the television sets ordered by the exporter.

2 In a news paper publishing company, 6 technicians can type 48 pages in one hour. On a certain day, 2 of the technicians are absent. How many pages can the remaining technicians type in one hour?

3 In a hostel, 60 pupils can use the available food items in 50 days. Fifteen more pupils are joined in that hostel. How many days will it take for all the pupils to finish the available food items?

4 Ahmed plans to use 15 hours in a day to complete his assignment in 88 hours.
(a) How long will he take to complete the assignment, if reduces the planning time in a day by 3 hours?
(b) Write your answer to part(a) in days.

Map scales and proportions

Remember

1. The scale of a map 1: n means **1 cm map length = n cm actual length.**
2. Map scale is based on direct proportion.
3. 100 cm = 1 m or 1 cm = 0.01m 4. 100 000 cm = 1 km or 1 cm = 0.000 01 km

Example

The scale of a map is 1 : 4000 000.
(a) The length of a river on the map is 29.4 cm. Find the actual length of the river, in km.
(b) The actual distance between two cities is 720 kilometres. Find this distance on the map.

Working: (a) $1:40 = 29.4: x$
$1\,x = 40 \times 29.4$ or $x = 1176$ km
The actual length of river is 1176 km.
(b) $1:40 = x:720$

$40\,x = 1 \times 720$ or $x = \dfrac{720}{40} = 18$ cm

So the map distance is 18 cm.

First change 4000 000 cm into km.

4000 000 cm = 40 km.

Use cross multiplication as in direct proportion.

Practise exercise 17(A) **Core Curriculum**

1 Use the given map scale to complete each statement.
1 : 100 000 means
(i) 1 cm map length = ... cm actual length.
(ii) 1 cm map length = ... m actual length.

(iii) 1 cm map length = ... km actual length.
(iv) 1 mm map length = ... m actual length.

2 A map has a scale of 1:200 000. On the map the length of a bridge is 2.4 cm. Calculate the actual length of the bridge, in kilometres.

3 The scale of a map is 1:50 000. The real distance between two villages is 13.2 kilometres. Find this distance on the map.

4 The altitude of a tower is 260 metres. On a map this altitude shows 6.5 cm. Find the scale of the map, in the form 1: n.

5 On the open ocean, the actual distance between Australia and Africa is 10 000 kilometres. On a map this distance shows as 12.5 cm.
(a) Find the scale of the map, in the form 1: n.
(b) On the same map the length of a high way is 3.1 cm. Calculate the real length of the high way.

6 A model of a toy has a scale of 1:20. The model is 16.5 cm long. Calculate, in metres, the actual length of the toy.

7 The scale of a map is 1:45000 000. On the map the distance between two Cities is 17.25 cm. Calculate the actual distance between these two cities. Give your answer in kilometres.

EXTENDED CURRICULUM

Practise exercise 17(B)

1 The scale of a map is 1 : 20. Complete the following statements.
(a) 1 cm^2 map length = ... cm^2 actual length.
(b) 1 cm^2 on the map = ...m^2 in real.
(c) 1 cm^2 on the represents .... km^2 in actual.
(d) 1 cm^3 on the map represents .... m^3 on the actual model.

2 On a map of scale 1:250 000, the distance between two wireless towers is 14 cm.
(a) Calculate the actual distance between both the towers
(i) in centimeters, (ii) in metres, (iii) in kilometres.
(b) The real distance between another pair of wireless towers is 45 kilometres. Find this distance on the map.

3 The plan of a small play ground is drawn to a scale of 1:200. The actual area of the play ground is 600 square metres. What is the area in cm^2, of the play ground on the map?

4 The area of a house plot shown on a map of scale 1:1000 is 270 square centimetres. Calculate the real area of the house plot
(a) in square metres, (b) in square kilometres.

5 The model of an auditorium hall is drawn using a scale of 1:100. The actual volume of the auditorium is 800 m^3.Calculate the volume of the model in
(a) cubic centimeters, (b) cubic metres.

EXTENDED CURRICULUM

Proportion and algebra

- You need to know the statement forms and their respective equations for proportion and algebra.

Statement	Proportionality	Equation
y is directly proportional to x	$y \alpha x$	$y = kx$
y is proportional to the square of x	$y \alpha x^2$	$y = kx^2$
y is proportional to the square root of x	$y \alpha \sqrt{x}$	$y = k\sqrt{x}$
y is proportional to the cube of x	$y \alpha x^3$	$y = kx^3$
y is proportional to the cube root of x	$y \alpha \sqrt[3]{x}$	$y = k\sqrt[3]{x}$
y is inversely proportional to x	$y \alpha \frac{1}{x}$	$y = \frac{k}{x}$
y varies inversely as the square of x	$y \alpha \frac{1}{x^2}$	$y = \frac{k}{x^2}$
y varies inversely as the square root of x	$y \alpha \frac{1}{\sqrt{x}}$	$y = \frac{k}{\sqrt{x}}$
y varies inversely as the cube of x	$y \alpha \frac{1}{x^3}$	$y = \frac{k}{x^3}$
y varies inversely as the cube root of x	$y \alpha \frac{1}{\sqrt[3]{x}}$	$y = \frac{k}{\sqrt[3]{x}}$

Working rule

To solve the problems based on proportion and algebra
1) first form an equation by writing '= k' in place of 'α',
2) then solve the equation for x,
3) at last put the given values in the equation to get the required value.

- **Increasing or decreasing a quantity by a given ratio**

Remember

1. To increase or decrease a quantity by a given ratio, **use direct proportion**.

Example 1
The area A, of a rectangular land is proportional to the square of its one side, S. The area of the land is 112 m^2, when one of its sides is 4 m.
(a) Find an equation connecting A and S.
(b) When one side of the land is 6 m, calculate its area.
(c) Find the length of the proportional side, when its area is 567 m^2.

Working: (a) $A \alpha S^2$

$A = kS^2$

$112 = k \times 4^2$ or $k = 7$

The equation is $A = 7S^2$.

Write '= k' in place of 'α'.
Solve the equation for k.
Put the value of k in the equation.

(b) $A = 7S^2 = 7 \times 6^2 = 252\ m^2$

Area = 252 m^2

Write down the equation.
Put S = 7 in the equation.

(c) $A = 7S^2$

$567 = 7 \times S^2$ or $S^2 = 81$ or $S = 9$ m

Side = 9 m

Write down the equation.
Put A = 567 in the equation.

Example 2

The frequency(f) of a pendulum varies inversely as the square root of its length(l).
A pendulum of 64 cm long has a frequency of 0.4 Hz.
(a) Find an equation connecting f and l.
(b) Calculate (i) f when $l = 256$ cm, (ii) l when f = 0.5 Hz.

Working: (a) $f \alpha \frac{1}{\sqrt{l}}$

Write '= k' in place of 'α'.

$f = \frac{k}{\sqrt{l}}$

$0.4 = \frac{k}{\sqrt{64}}$ or $k = 3.2$

Solve the equation for k.
Put the value of k in the equation.

The equation is $f = \frac{3.2}{\sqrt{l}}$

Write down the equation.
Put $l = 256$ in the equation.

(b) (i) $f = \frac{3.2}{\sqrt{l}} = \frac{3.2}{\sqrt{256}} = 0.2$ Hz

Write down the equation.
Put f = 0.5 in the equation.

(ii) $0.5 = \frac{3.2}{\sqrt{l}}$ or $l = 40.96$ cm

Example 3

The radius of a circle of 15 cm is increased in the ratio 3:7. Calculate the new radius.

Working: $3:7 = 15: x$

Use direct proportion

$3 \times x = 7 \times 15$ or $x = 35$

The new radius is 35 cm.

Practise exercise 18

1 A quantity, q, is directly proportional to another quantity, r. When q = 6 r = 18.
(a) Find an equation connecting q and r.
(b) Calculate (i) r when q = 24, (ii) q when r = 111.

2 Given that x is proportional to the square of (y+1). When x = 144 y = 3.
(a) Find the value of x when y = 6.
(b) Find the value of y when x = 50.

3 Given m is inversely proportional to n. When m = 5 n= 20.
(a) Find a formula that connects m and n.
(b) Calculate (i) m when n = 25, (ii) n when m = 2.

4 The thickness(T) of a book and the total number of pages (P) in it are in direct proportion. The thickness of a book of 400 pages is 2.5 cm.
(a) Write down an equation for T, on terms of P.
(b) Calculate the thickness of a book of 192 pages.
(c) Find the number of pages in a book whose thickness is 3.7 cm.

5 A number, p, varies inversely as the cube root of another number q. If q=1, then p=5
(a) Write down a formula for p in terms of q.
(b) Find (i) p when q = 8, (ii) q when p = 1.

6 At constant temperature, the pressure, P Nm^{-2}, applied on a given mass of gas is inversely proportional to its volume, V m^3. P = 15 when V = 4.
(a) Calculate pressure when volume is 12.
(b) Calculate volume when pressure is 10.

7 The edge of a cubical box is 18.The edge of the box is increased in the ratio 6:10. Calculate the new edge of the box.

8 A rope of 120m long is to be reduced in the ratio 4:5. Find the new length of the rope.

9 The cost price of an article is $ 240.If the cost price of the article is decresed in the ratio 4 : 6, find the new cost price.

10 A rectangular garden is 45m long and 30m wide. The length is increased and the breadth is decreased in the ratio 4:3.
(a) What is the length of the garden after increase?
(b) What is the breadth of the garden after decrease?
(c) What can you say about the original area and the new area?

11 An author plans the size of a book. First he thought that the length of the book will be 28 cm and its breadth will be 18 cm. He then decided to reduce the length in the ratio 6:7 and increase the width in the ratio 9:10. Calculate the new dimension of the book?

1.14 Money

Core Curriculum

Remember

1. Any two units of money are changed in **direct proportion**.
2. Exchange rate is the worth of one unit of a currency in term of another currency.
3. For the questions on money, give your answer up to 2 decimal places.
4. For the answer correct to nearest currency, round to the nearest whole number.
5. For the answer correct to nearest sub unit(e.g. cent), round up to 2 decimal places.

Example

Ameeth changed 2008 Euros(€) for 2429.68 US Dollars($).
(a) Write down the exchange rate in dollars per euro. (b) How many euros worth $199?

Working: (a) $2008:2429.68 = 1:x$

$2008x = 1 \times 2429.68$

$x = 1.21$

So 1 € = $ 1.21

As in direct proportion, use cross multiplication.

(b) $1:1.21 = x:199$

$1.21x = 1 \times 199$

$x = 164.46$

Therefore $199 = € 164

For 'how many euros', give your answer correct to nearest whole number.

Practise exercise 19(A) **Core Curriculum**

1 Mayan changed 15550 Euros into Indian Rupees at the rate of 1 Euro = 55.965 Indian Rupees. How many Indian Rupees did he receive?

2 In September 2007, one US dollar equals 115.46 Japanese yen.
(a) Change 450 US dollars into Japanese Yen.
(b) Change 2 million Japanese Yen into dollars.

3 Given that $ 1 = 3.141 Argentine Pesos.
(a) Change $ 150.50 into Argentine Pesos, giving your answer up to 2 decimal places.
(b) What is worth of one Argentine Peso into dollars?

Give your answer correct to 4 significant figures.

4 In 2007, Alok went from USA to Russia. He changed \$2070 into Russian Rouble at an exchange rate of \$ 1 = 25.635 Russian Roubles.
(a) How much money did he receive exactly?
(b) Write down your answer to part(a), to the nearest Rouble.

5 Alice changed 400 Euros(€) into dollars(\$), when the rate was € 1 = \$ 1.409.
(a) How much money did he receive in dollars?
(b) Write your answer to part(a), to the nearest ten dollars?

6 In 2006, on a certain day,
US \$1 = 10.95 Mexican Pesos and US \$1 = 921.04939 Korean Won.
(a) Calculate the value of (i) one Mexican Peso in Korean Wons,
(ii) one Korean Won in Mexican Pesos, correct to 5 decimal places.
(b) Use your answer to part(a), to change
(i) 2200 Mexican Pesos into Korean Wons, (ii) 7000 Korean Wons into Mexican Pesos.

7 Edward visited USA with 24939 British Pounds(£). He exchanged his money at the rate of £ 1 = \$ 2.022. How much did he receive in dollars? Give your answer to the nearest cent.

Practise exercise 19(B) **EXTENDED CURRICULUM**

1 Chung sent \$ 12000 from USA to China through American Express to his parents. American express took \$ 45 as commission and delivered the remaining at the rate of \$ 1 = 7.499 Chinese Yuan.
(a) Calculate the amount of money received by his parents. Give your answer correct to the nearest Yuan.
(b) What is the total value of money that Chung sent in Chinese Yuan?

2 In an international 'Super market', a digital camera costs 600.18 Canadian dollars or 692.58 Australian dollars.
(a) What is the worth of one Canadian dollar in Australian dollars? Give your answer correct to 2 decimal places.
(b) In the same 'Super market' a laptop costs 1100 Canadian dollars. Calculate the cost of the laptop in Australian dollars.

3 William wants to convert £ 10000 into euros(€) in a bank. In the bank, he has two options.
First option: Convert total pounds at the rate of £ 1 = € 1.425.
Second option: Give commission of £ 50 and then convert the remaining at £ 1 = € 1.435.
(a) How much euros does he receive from
(i) the first option, (ii) the second option.
(b) Which option gives him less euros and by how much?
(c) Find the value of one euro comparing to pounds in the
(i) first option, (ii) second option.
Give your answer up to 5 significant figures.

4 Given that \$ 1 = € 0.714 and € 1 = £0.724.
(a) Find the exchange rate, correct to 5 significant figures
(i) per pound(£) in dollar(\$), (ii) per dollar(\$) in pound(£).
(b) Use your answer to part(a), to change £ 7001 into dollars correct to the nearest cent.

5 (a) In 2006 the cost of a car in euros was € 25 000. The cost of this car in pounds was £ 16020. Calculate the exact value of the exchange rate in 2006.

(b) In 2007 the car still cost € 25 000 and the exchange rate was € 1 = £ 0.7416.
Calculate the difference, in pounds, between the cost in 2006 and 2007.

1.15 Use of an electronic calculator

Core Curriculum

Remember

1. Calculator does multiplication and division before addition and subtraction.
2. While working with numbers in standard form, keep your calculator in SCI mode.
3. Your calculator must be in degree mode while working with trigonometric calculations.
4. Your calculator must be in gradient mode while working with normal calculations.

Practise exercise 20

Use a calculator to work out the questions(1 to 45) and give exact answers.

(1) $121 + -212$ (2) $-713 + 42.7$ (3) $9.8 - -64$

(4) $-1204 - 61.6$ (5) $-415 \text{ x } 69$ (6) $-125 \text{ x } -521$

(7) $74.25 \text{ x } -16.9$ (8) $625 \div (-12.5)$ (9) $-1363 \div -47$

(10) $45 + -54 - -74$ (11) $4 - 0.8 \text{ x } 7$ (12) $17 - 9 \text{ x } 2 + 4$

(13) $5502 \div 917 \text{ x } 6$ (14) $(14 + 4) \div 7$ (15) $(1111 - 101) \text{ x } 1010$

(16) $(1.9 + 19.3) \text{ x } (4.7 - 2.1)$ (17) $\dfrac{21 \text{ x } 45}{35}$ (18) $\dfrac{51 - 34}{17}$

(19) $\dfrac{45 - 3 \text{ x } 5}{5 \text{ x } 3}$ (20) $2\frac{1}{3} - \frac{3}{7}$ (21) $9.7 + \frac{7}{5} - 5$

(22) 7^2 (23) -4^2 (24) $(-12)^2$

(25) 5^3 (26) -9^3 (27) 11^4

(28) $(-2)^2$ (29) $(16 - 9)^3$ (30) $(3 - 5)^6$

(31) $3^2 - 2^3$ (32) $5^4 - 5^3 - 5^2$ (33) $64^{-1/3}$

(34) $343^{4/3}$ (35) $1024^{1/10}$ (36) $\sqrt{44.89}$

(37) $3\sqrt{1089}$ (38) $\sqrt[3]{64}$ (39) $2\sqrt[3]{2197}$

(40) $3\sqrt{1764} - \sqrt[3]{17576}$ (41) $9.4 \text{ x } 10^4 - 3.7 \text{ x } 10^3$ (42) $\cos 60^\circ + \sin 30^\circ$

(43) $\tan 45^\circ - \cos 60^\circ$ (44) $2\cos 60^\circ + 3(\tan 30^\circ)^2$ (45) $\sqrt{\cos 60^\circ + (\sin 45^\circ)^2}$

Use a calculator to work out the questions(46 to 54) and give your answer to the specified accuracy.

(46) $\dfrac{1}{6}$ (2 d.p.) (47) $\dfrac{2}{3} + \dfrac{4}{5}$ (3 d.p.) (48) $6.7 \div 1.4$ (1 s.f.)

(49) $\dfrac{1.67 - 0.97}{0.45}$ (3 s.f.) (50) $\dfrac{4 \text{ x } 9}{0.007}$ (1 d.p.) (51) $\dfrac{10 + 4}{13}$ (1 d.p.)

(52) $\dfrac{55}{11 - 6}$ (1 s.f.) (53) $\dfrac{3^3 - 2^3}{3^2 - 2^2}$ (1 s.f.) (54) $\dfrac{7^2 + 5^2}{7 + 5}$ (3 s.f)

(55) Work out the reciprocal of the following.

(a) 5 (b) 2^2 (c) 3^2 (d) 2^4 (e) 4^{-3}

(56) (a) Change the following into hr, min and sec.

(i) 1.2 hrs (ii) 6.5 hrs (iii) 0.432 hr (iv) 4.291 hrs

(b) Change the following into hours only.

(i) 3 hrs 25 min (b) 7 hrs 35 min 33 sec (c) 45 min 30 sec

CHAPTER 2 Algebra

2.01 Algebraic representation and formulae **Core Curriculum**

Term: It is constant alone or a variable alone or their mixed form without + or – sign.

E.g. x, 4x, pq^2, $-9pq^2$ etc.

- **Like terms:** These are the terms with same letter(s) and same power.
  E.g. x, 2x, –5x are like terms.

Arithmetic operations in algebra

Working rule

1) ***To add or subtract algebraic terms,***
   (i) first arrange the like terms together (ii) then add (or subtract).
2) ***To multiply algebraic terms,***
   (i) first add the powers of the same letter in the terms,
   (ii) then multiply the remaining factors.
3) ***To divide algebraic terms,***
   (i) first write the division as a fraction, (ii) then cancel out the common factors.

Example 1
Simplify $4x - 9 + 7x + 5$
Working: $4x + 7x - 9 + 5 = 11x - 4$

Arrange like terms together.
Add the like terms.

Example 2
Multiply out $(2x^2) \times (-3x^3y)$
Working: $2 \times -3 \times x^2 \times x^3 \times y = -6x^5y$

Arrange the same letter terms together.
Add the powers of same letter terms, i.e. $x^2 \times x^3 = x^5$
Then multiply the other factor terms.

Example 3
Simplify $-10x^2y \div 2xy^2$

First write as a fraction and then cancel out the common factors.

Working: $\dfrac{-10x^2y}{2xy^2}$

$= \dfrac{-5x}{y}$

Don't do this.
$\dfrac{-10x^2y}{2xy^2} = -5$

- **Formula:** A formula is an equation which is true for all the numbers that replace the letters. E.g. The formula for y is four more than twice of x is $y = 2x + 4$.

Substitution

Working rule

To substitute in an expression or a formula,
1) replace the letters by the given numbers,
2) then use BEDMAS to simplify the resulting expression or formula.

Example

Evaluate (a) $4ab - b^2$ when a = 2, b = –3. (b) $9x^2 + 10y^3$ when x = –5, y = –4.

Working: (a) $4ab - b^2$

$= 4 \times 2 \times -3 - (-3)^2 = -33$

(b) $9x^2 + 10y^3$

$= 9 \times (-5)^2 + 10 \times (-4)^3 = -415$

Replace the letters by numbers.
Use bracket for negative number to raise any power.

Example Find P when q = 64 for the formula $P = 1 + 3\sqrt{q}$.

Working: $P = 1 + 3\sqrt{q}$

$= 1 + 3\sqrt{64}$

$= 25$

Replace the letter by number.

Don't do this. $1 + 3\sqrt{64} = 1 + \sqrt[3]{64} = 5$

Practise exercise 1

1 Simplify each of the following algebraic expressions.

(i) $6a + 13a$ (ii) $2x^3 + 1 + 5x^3$ (iii) $4x - 3x$

(iv) $x^2 - 8x^2$ (v) $5t + 5 - 9t$ (vi) $4p - 10 + 6p - 4p$

(vii) $-14a - 10b^2 - a$ (viii) $2a \times 3b$ (ix) $-6x \times 5y$

(x) $-3xy \times (-x)$ (xi) $\left(\frac{1}{2}x^2y\right) \times \left(\frac{5}{4}xy^2\right)$ (xii) $(-7.9p^2) \times \left(\frac{3}{4}pq^4\right)$

(xiii) $8xy \div 2x$ (xiv) $-12x^2y^2 \div 4xy^2$ (xv) $20p^2 \div (-5p)$

(xvi) $(-0.6\,m^6) \div (-1.2m^5)$ (xvii) $\frac{1}{2}abc \div \left(-\frac{1}{4}bc\right)$ (xviii) $\frac{-100x^6y^7z^8}{20x^4y^5z^6}$

(xix) $0.5x + y - 5x - 1$ (xx) $\frac{a}{4} - \frac{b}{4} + 0.4b$ (xxi)

$\frac{x - y}{10} + 10x - 0.001y$

2 Using x = 2, y = –3 and z = 4, evaluate the following expressions.

(a) $3x$ (b) $-5y$ (c) $x + y$ (d) $x - y$

(e) $\frac{1}{2}z$ (f) $y - z$ (g) $x + y - z$ (h) $2x + y$

(i) $4y - 4$ (j) $\frac{3}{z} - y$ (k) $\frac{x + y}{2}$ (l) $\frac{y - z}{7}$

(m) x^2 (n) y^3 (o) $(x - z)^2$ (p) $3\sqrt{z}$

(q) $2x^2 + y$ (r) $\frac{3z + 2}{x}$ (s) $\sqrt[3]{2x} - y$ (t) $-x^2 - y^2 + z^2$

3 Find the value of 'y' in each of the following.

(a) $y = 2a + b^2$ when a = –4, b = 2 (b) $y = -6x^2 + (6x)^2$ when x = 5

(c) $y = m - 3\sqrt[3]{n}$ when m = 7.7, n = 8 (d) $y = -p^2 - (-p)^2$ when p = 14

(e) $y = 10x^2 - 10$ when x = 1.5 (f) $y = \frac{1}{2}\left(p - \sqrt[3]{p}\right)$ when $p = \frac{1}{8}$

4 Given a = 9 and b = –5.

(a) Find the value of (i) $(a + b)^2$ (ii) $a^2 + b^2 + 2ab$.

(b) What do you conclude from the answers to part(a)(i) and (a)(ii)?

5 $P = \frac{1}{4}t^2 - qr$. Calculate the value of

(a) p when $t = -4$, $q = 10$, $r = -7$,
(b) q when $p = \frac{1}{4}$, $t = -2$, $r = 6$,
(c) r when $p = 20$, $t = 0.2$, $q = -1$,
(d) t when $p = 9$, $q = -1$, $r = 0$.

6 Kinetic energy, E, is related to mass, m, and velocity, v, by the formula $E = \frac{1}{2}mv^2$.

(a) Calculate E when $m = 5.6$ and $v = 15$.
(b) Calculate v when $m = 10$ and $E = 1024$.

7 Given $a = \frac{b-c}{d}$, find the value of a when $b = \frac{7}{9}$, $c = \frac{3}{5}$ and $d = \frac{2}{7}$.
Show all your working clearly.

Transforming simple formulae

Working rule

To change the subject of a formula
1) remove the brackets(if any).
2) use the rule ***'side change, sign change"*** to bring the subject term(s) to one side of the formula and reduce its coefficient to 1.

Remember

The single letter(variable) in one side of a formula is called its subject.
E.g. In the formula $y = mx + c$, the subject is y.

Example

Make 'p' the subject for the formula $x = 4p - 4q$.

Working: $x = 4p - 4q$

Rearrange the formula using the rule 'side change, sign change' to make 'p' as its subject.

$4p = x + 4q$

$p = \frac{x + 4p}{4}$

Don't do this.
$x = \cancel{4}p - \cancel{4}q$
$\cancel{4}p = x + \cancel{4}q$
$p = \frac{x + \cancel{4}p}{\cancel{4}}$

Practise exercise 2

1 Make the 'letter' in the brackets as the subject of the formula.

(a) $x = a + b$ [b]
(b) $d = c - e$ [c]
(c) $x = pq$ [p]
(d) $y = \frac{m}{n}$ [m]
(e) $y = mx + c$ [x]
(f) $y = ax - 4$ [a]
(g) $ax + by = c$ [y]
(h) $c = a - 2b$ [b]

(i) $p = 9 - qd$	[d]	(j) $m = \frac{c-d}{2}$	[c]
(k) $f = g - \frac{1}{2}e$	[e]	(l) $10 + \frac{x}{10} = y$	[x]
(m) $p = 2l + 2b$	[b]	(n) $p = 4l$	[l]
(o) $C = 2\pi r$	[r]	(p) $S = \frac{a+b+c}{2}$	[b]
(q) $x = 18 - y - z$	[z]	(r) $C = \pi r l$	[l]

2 (a) Make 'p' the subject of the formula $m = n - \frac{p}{5}$.
(b) Make 'b' the subject of the formula $w = x(2a - b)$.
(c) Change 'x' as the subject of the formula $y = 2x^2$.
(d) Write 'y' as the subject of the formula $p = y^2 - 1$.

3 Make 'x' as the subject of each formula.
(i) $y = \sqrt{x}$ (ii) $y = k\sqrt{x}$ (iii) $y = \frac{k}{\sqrt{x}}$ (iv) $y = \frac{k}{x^2}$

4 (a) The area of a square with length, l, is $A = l^2$. Change l as the subject of this formula.
(b) The area of a triangle with base(b) and height(h) is $A = \frac{1}{2}bh$. Rearrange this formula to find 'h' in terms of A and b.
(c) The formula for area of a trapezium with parallel sides a and b, and height, h, is $A = 0.5h(a+b)$. Change the subject of the formula as 'a'.
(d) The formula for volume of a cube with each edge, l, is $V = l^3$. Make l the subject of this formula.

5 The area of a shape is given by the formula $A = \frac{11\pi r^2}{4}$.
Make r the subject of the formula.

EXTENDED CURRICULUM

Transforming complex formulae

Remember

To transform complex formulae, use the same working rule as transforming simple formulae with emphasis on
1. square and square roots are opposite,
2. cube and cube roots are opposite,
3. 4th power and 4th root are opposite to each other and so on.

Example For the formula $x = 2b^2 - 3\sqrt{c}$,
(a) make 'b' as the subject, (b) make 'c' as the subject.

Working: (a) $x = 2b^2 - 3\sqrt{c}$

$2b^2 = x + 3\sqrt{c}$ or $b^2 = \frac{x + 3\sqrt{c}}{2}$

Use the rule 'side change, sign change'.
Square changes to square root.

Don't do this.

$b^2 = x + 3\sqrt{c}/2$ and $b = \frac{\sqrt{x + 3\sqrt{c}}}{2}$

or $b = \sqrt{\dfrac{x + 3\sqrt{c}}{2}}$

(b) $x = 2b^2 - 3\sqrt{c}$ or $3\sqrt{c} = 2b^2 - x$

or $\sqrt{c} = \dfrac{2b^2 - x}{3}$ or $c = \left(\dfrac{2b^2 - x}{3}\right)^2$

Don't do this.

$3\sqrt{c} = x - 2b^2$

$\sqrt{c} = x - 2b^2/3$

$c = \dfrac{(2b^2 - x)^2}{3}$

Practise exercise 3

1 For the formula $a = b^2 + c^2$,
(a) make 'b' the subject of the formula, (b) write down the subject of the formula as c.

2 $x = \sqrt{y^2 - z^2}$
(a) Make 'y' the subject of the formula. (b) Change the subject of the formula to z.

3 Make r the subject of the formula $V = \dfrac{4}{3}\pi r^3$.

4 $c = ax^2 - bx$. Rearrange this formula for b in terms of x, c and a.

5 $3\sqrt{l} + 2\sqrt[3]{m} = p$
(a) Make l the subject of the formula. (b) Change m the subject of the formula.

6 $a = x + \dfrac{y^2}{6}$, make y as the subject of this formula.

7 $\dfrac{1}{x} + \dfrac{1}{y} = p$. Rearrange this formula to make (i) x as the subject, (ii) y as the subject.

8 $a = \dfrac{bc}{b + c}$
(i) Express 'b' in terms of 'a' and 'c'. (ii) Write down 'c' in terms of 'a' and 'b'.

9 $x = \dfrac{4}{5}y^3 - 5$. Transform the subject of the formula to y.

10 $m = pq - \dfrac{1}{6}rt^4$. Write down the subject of the formula (i) as p, (ii) as t.

11 Given $\dfrac{4a + b}{a - 4b} = c$.
(a) Make the subject of the formula as 'a'.
(b) Write down the subject of the formula as 'b'.

12 (a) $t = 3\sqrt[3]{s} - \dfrac{1}{3}$
Rearrange the formula to make 's' the subject.
(b) $u = \sqrt[3]{\sqrt{v} + \sqrt[3]{w}}$
(i) Change 'v' as the subject of the formula.

(ii) Make the subject of the formula as 'w'.

2.02 Algebraic manipulation

Core Curriculum

Expanding single bracket

Working rule

For single bracket, multiply the term just outside the bracket by all the terms inside the bracket.

Example 1

Expand the following.

(a) $2(3x - 7)$
$= 6x - 14$

(b) $-5(-4y + z)$
$= 20y - 5z$

(a) Multiply 2 by 3x to get 6x and multiply 2 by −7 to get −14.
(b) Multiply −5 by −4y to get 20y and multiply −5 by z to get −5z.

Don't do this. (a) $6x - 7$ (b) $20y + z$

Example 2

Expand the bracket and simplify.

$6a^2 - a(2a - 4b^2)$
$= 6a^2 - 2a^2 + 4ab^2$
$= 4a^2 + 4ab^2$

Multiply −a by 2a to get $-2a^2$ and multiply −a by $-4b^2$ to get $4ab^2$.
Simplify $6a^2$ and $2a^2$ to get $4a^2$.

Don't do this. $6a^2 - 2a^2 - 4ab^2$ and last step as $4a^2 + 4ab^2 = 8a^3b^2$

Simple factorisation

Working rule

To factorise a simple algebraic expression
1) take away the 'highest number' and 'highest letter(s)' that go into every number,
2) then write down the remaining terms inside the bracket.

Example

Factorise completely.

(a) $5mn + 15n$
$= 5n(m + 3)$

(b) $14x^2y^3 - 21x^3y^2$
$= 7x^2y^2(2y - 3x)$

(a) 5 goes in both 5 and 15 and n is present in both terms.
(b) 7 goes in both 14 and 21, x^2 in both x^2 and x^3. Also y^2 in both y^3 and y^2.

Don't do this. (a) Last step $n(5m + 15)$ or $5(mn + 3n)$
(b) Last step $7(2x^2y^3 - 3x^3y^2)$ or $7x^2(2y^3 - 3xy^2)$ or $7y^2(2x^2y - 3x^3)$

Practise exercise 4

1 Expand the following brackets.

(a) $2(x + 3)$ (b) $5(x - 7)$ (c) $8(4 - a)$ (d) $-(x - 4)$
(e) $6(2x + 1)$ (f) $10(3x - 2)$ (g) $-4(x - 3)$ (h) $-7(y + 5)$
(i) $-12(6 - y)$ (j) $x(x + 1)$ (k) $x(2x - 1)$ (l) $5x(4 - 3x)$
(m) $-2x(3x - 4y)$ (n) $7(3a - 2b + c)$ (o) $-6x^2(9x - y + 1)$ (p) $\frac{1}{2}(4x - 6y)$
(q) $-\frac{3}{4}(12x^2 - 4 + 4y)$ (r) $-6x(-\frac{x^2}{2} + \frac{x}{3} - 2)$

2 Expand the following brackets and simplify.

(a) $6 + 2(x - 3)$ (b) $19 - 9(a + 1)$ (c) $15 - (x + 2)$

(d) $24 - (y - 20)$ (e) $4x - (2x + 1)$ (f) $-6x - (5 - 7x)$
(g) $7x + x(x - 4)$ (h) $5b - b(12 + 10b)$ (i) $5x + 2y - (4x + 5)$
(j) $9p - q + 4(p - 2q)$ (k) $3x - 7y - 6x(y + 1)$ (l) $10p^3 - 3p^2 - (6p^3 + 5p^2)$
(m) $2(x + 1) + 3(x + 2)$ (n) $4(x - 2) + 7(x + 3)$
(o) $5(2x - 1) - 4(2x + 4)$ (p) $7(2 - 3a^2) - (10 - 9a^2)$
(q) $x(x - 5) + 2x(x - 2)$ (r) $-p^2(4 - p) + p(2p^2 + 3)$
(s) $4(t^2 - 2t^4) - 3(2t^2 - 4t^4)$ (t) $\frac{4}{5}x(10x^2 + 5) - \frac{5}{4}x^2(8x - 12)$

3 Factorise the following.
(a) $7a + 14$ (b) $10 - 5x$ (c) $x^2 + 2x$
(d) $x^2 - 7x$ (e) $-6a - 10$ (f) $3x + 6y - 3z$
(g) $4x^2 + 4y^2$ (h) $2a - 2b + 2c$ (i) $0.2p - 0.4q - 0.8r$

4 Factorise the following completely.
(a) $4a - 8a^2$ (b) $5x^2 - 15x$ (c) $64a^2b + 12ab$
(d) $1.5xy - 4.5yz$ (e) $2x^3 + x^2 - x$ (f) $10y - 5y^2 - 5y^3$
(g) $6x^2 - 12x + 18xy$ (h) $7x^2 - 49x^3$ (i) $9y^2 - 3xy$
(j) $5x^2y + 10xy^2$ (k) $4a^2b - 2a$ (l) $x^3y^2 + 2x^2y^3$
(m) $24pq^2 - 60p^2qr$ (n) $16x^4y^3z^2 + 12x^2y^3z^4$ (o) $\frac{1}{4}x^2 - \frac{1}{2}xy$
(p) $x^2y - xy^2 + 2x^2y^2$ (q) $2x(x + y) + (x + y)$ (r) $6p(q - 1) - q(q - 1)$

EXTENDED CURRICULUM

Expanding double brackets

Working rule

To expand double brackets, you must multiply every term in one bracket by every term in other bracket.

Example

Expand and simplify. (a) $(2a^2 + 5)(a^2 - 2)$ (b) $(x - 7)^2$

Working: (a) $2a^2(a^2 - 2) + 5(a^2 - 2)$
$= 2a^4 - 4a^2 + 5a^2 - 10$
$= 2a^4 + a^2 - 10$

Multiply out brackets. Then combine like terms.

Don't do this. The last answer as $2a^4 - 9a^2 - 10$

(b) $(x - 7)(x - 7)$
$= x(x - 7) - 7(x - 7)$
$= x^2 - 7x - 7x + 49$
$= x^2 - 14x + 49$

Use $(x - 7)^2 = (x - 7)(x - 7)$.
Multiply out brackets and then combine like terms.

Don't do this. $(x - 7)^2 = x^2 - 7^2$

Practise exercise 5

1 Expand the brackets and simplify.
(a) $(x + 1)(x + 2)$ (b) $(a + 4)(2 + a)$ (c) $(x - 1)(x + 3)$
(d) $(a + 2)(a - 3)$ (e) $(x - 4)(x - 2)$ (f) $(a - 5)(3 - a)$
(g) $(2x + 3)(4x + 5)$ (h) $(7a - 2)(a + 4)$ (i) $(9 - a)(3a - 6)$
(j) $(x^2 + 4)(x^2 - 4)$ (k) $(p + 10)(p - 10)$ (l) $(m^2 - 2)(m^2 - 2)$
(m) $(a - b)(a^2 + ab + b^2)$ (n) $(x^2 + 2x - 1)(x^2 - 3x + 2)$ (o) $(2p^2 - 3p + 4)(4p^2 + p - 6)$

2 Expand the following.

(a) $(x + y)^2$ (b) $(x + 1)^2$ (c) $(a + 10)^2$
(d) $(x - 2)^2$ (e) $(x - 8)^2$ (f) $(4 - p)^2$
(g) $\left(x + \frac{1}{2}\right)^2$ (h) $\left(a - \frac{3}{4}\right)^2$ (i) $\left(x + \frac{1}{x}\right)^2$
(j) $(6t - 5)^2$ (k) $(10 - 10q)^2$ (l) $(a^2 + b^2)^2$
(m) $(x^2 - 4)^2$ (n) $(16x^2 + 25)^2$ (o) $(2x + 3)^2$
(p) $(x - 3)^2$ (q) $(x^2 - 1)(x^2 + 1)$ (r) $(a - 0.5)^2$

Factorising expressions of the form ax + bx + kay + kby

Working rule

To factorise expressions of the form **ax + bx + kay + kby,** choose the common factors by grouping.

Example Factorise completely.

(a) $ax + bx + kay + kby$ (b) $mx + 6np - 2nx - 3mp$

Working: (a) $x(a + b) + ky(a + b)$
$= (a + b)(x + ky)$

(b) $mx - 2nx - 3mp + 6np$
$= x(m - 2n) - 3p(m - 2n)$
$= (m - 2n)(x - 3p)$

(a) Choose x from first two terms and ky from last two terms. Choose the common factor (a + b). Write the remaining terms x and +ky in another bracket.

(b) Choose x from first two terms and 3p from last two terms. Choose the common factor (m – 2n). Write the remaining terms x and –3p in another bracket.

Factorising expressions of the form $a^2x^2 - b^2y^2$

Working rule

To factorise expressions of the form $a^2x^2 - b^2y^2$**,** use the formula
$a^2x^2 - b^2y^2 = (ax + by)(ax - by)$.

Example

Factorise completely.

(a) $4x^2 - 9y^2$ (b) $8a^2 - 2$

Working: (a) $2^2x^2 - 3^2y^2$
$= (2x + 3y)(2x - 3y)$

Rewrite each term as a square.
Use the formula $\mathbf{a^2x^2 - b^2y^2 = (ax + by)(ax - by)}$.

Don't do this. The last answer as $(4x + 9y)(4x - 9y)$

(b) $2(4a^2 - 1)$
$= 2(2^2a^2 - 1^2)$
$= 2(2a + 1)(2a - 1)$

First choose 2 as a common factor.
Rewrite each remaining term as a square.
Use the formula $\mathbf{a^2x^2 - b^2y^2 = (ax + by)(ax - by)}$.

Don't do this. The last answer as $2(4a^2 - 1)$

Practise exercise 6

1 Factorise the following completely.

(a) $x(x - 2) - 5(x - 2)$ (b) $x(x^2 + 3) - y(x^2 + 3)$
(c) $2x + 2y + 5ax + 5ay$ (d) $3x - 6y + ax - 2ay$
(e) $ax - bx - ay + by$ (f) $ay + cy - 5a - 5c$
(g) $4a - 10 + 6pa - 15p$ (h) $12x + 18y - 6ax - 9ay$

(i) $5xy + 10y + 2x + 4$ (j) $6 - 5ab + 15b - 2a$
(k) $2p^2 + pq^2 - 4p - 2q^2$ (l) $m^2 - 2n - mn + 2m$
(m) $6x^2 - y + 2xy - 3x$ (n) $\frac{1}{2}x - \frac{1}{2}xy - \frac{1}{4}p + \frac{1}{4}py$

2 Factorise the following.
(a) $a^2 - b^2$ (b) $a^2 - 2^2$ (c) $m^2 - 9$
(d) $4 - x^2$ (e) $x^2 - 1$ (f) $7^2p^2 - 9^2q^2$
(g) $x^2 - \frac{1}{4}$ (h) $\frac{9}{25} - y^2$ (i) $(x + 2)^2 - y^2$
(j) $4x^2 - 16$ (k) $x^2 - 289$ (l) $3m^2 - 2523$

3 Factorise the following completely.
(a) $2x^2 - 2y^2$ (b) $600 - 6p^2$ (c) $9x^2 - 144y^2$
(d) $4a^2 - b^2c^2$ (e) $m^2 - 169n^2$ (f) $5b^2 - 1445$
(g) $x^3 - 49x$ (h) $x^4 - 1$ (i) $64 - x^4$
(j) $10x^4 - 10y^4$ (k) $-81 + t^2$ (l) $625n^2 - 529m^2$
(m) $2x^2 - 2(y - 1)^2$ (n) $\frac{x^2}{16} - \frac{y^2}{4}$ (o) $p^2 - \frac{q^2}{961}$
(p) $2x^4 - 512$ (q) $12x^2 - 3y^2$ (r) $5n^2 - 1280$

Factorising expressions of the from $a^2 + 2ab + b^2$

Working rule

To factorise expressions of the from $a^2 + 2ab + b^2$ use the formulae
$a^2 + 2ab + b^2 = (a + b)(a + b)$ and $a^2 - 2ab + b^2 = (a - b)(a - b)$.

Example
Factorise fully. (a) $9a^2 + 54ab + 49b^2$ (b) $4m^2 - 20mn + 25n^2$

Working: (a) $3^2a^2 + 2 \times 3a \times 7b + 7^2b^2$
$= (3a + 7b)(3a + 7b)$

Rewrite $9a^2$ as 3^2a^2 and $49b^2$ as 7^2b^2.
Use the formula $\mathbf{a^2 + 2ab + b^2 = (a + b)(a + b)}$.

(b) $2^2 m^2 - 2 \times 2m \times 5n + 5^2n^2$
$= (2m - 5n)(2m - 5n)$

Rewrite $4m^2$ as $2^2 m^2$ and $25n^2$ as 5^2n^2.
Use the formula $\mathbf{a^2 - 2ab + b^2 = (a - b)(a - b)}$.

Factorising expressions of the from $ax^2 + bx + c$

Working rule

To factorise expressions of the form $ax^2 + bx + c$,
1) first find a pair of factors of **ac** that add up to **b**,
2) then use the pair of factors to break the x term ,
3) at last choose the common factors.

Example

Factorise completely $3x^2 + 7x - 6$

Working: $3x^2 + 9x - 2x - 6$

$= 3x(x + 3) - 2(x + 3)$

$= (x + 3)(3x - 2)$

Comparing $3x^2 + 7x - 6$ with $\mathbf{ax^2 + bx + c}$, we have
$a = 3$, $b = 7$ and $c = -6$.
$ac = 3 \times -6 = -18$
$18 = 1 \times 18$
$= 2 \times 9$ $\{ +9 + {}^-2 = +7$ (b)
$= 3 \times 6$ $\{ +9 \times {}^-2 = -18$ (ac)

Practise exercise 7

1 Factorise the following fully.

(a) $x^2 + 2xy + y^2$ (b) $p^2 - 2pq + q^2$ (c) $x^2 + 4xy + 4y^2$
(d) $a^2 - 10ab + 25b^2$ (e) $4a^2 + 12ab + 9b^2$ (f) $9x^2 - 30xy + 25y^2$
(g) $m^2 + 34m + 289$ (h) $n^2 - 16n + 64$ (i) $100x^2 + 4y^2 + 40xy$
(j) $169 + 144f^2 - 312f$ (k) $-66at + 121a^2 + 9t^2$ (l) $50x^2 - 140x + 98$
(m) $a^2 + 2 + \frac{1}{a^2}$ (n) $x^2 + \frac{1}{x^2} - 2$ (o) $1 + \frac{1}{4x^2} + x^2$

2 Factorise the following completely.

(a) $x^2 + 9x + 20$ (b) $x^2 + 6x + 9$ (c) $a^2 + 40a + 375$ (d) $x^2 - 7x + 12$
(e) $y^2 - 20y + 99$ (f) $m^2 - 50m + 600$ (g) $x^2 + 3x - 4$ (h) $x^2 + 10x - 39$
(i) $b^2 - 11b - 80$ (j) $x^2 - 4x - 5$ (k) $p^2 - 21p - 72$ (l) $m^2 - 15m - 100$

3 Factorise the following completely.

(a) $3x^2 + 7x + 4$ (b) $5x^2 + 6x + 1$ (c) $2a^2 + 25a + 72$
(d) $2x^2 - 11x + 15$ (e) $4x^2 - 15x + 14$ (f) $9t^2 - 36t + 20$
(g) $3x^2 + 7x - 6$ (h) $5x^2 + 12x - 9$ (i) $7a^2 + 18a - 40$
(j) $6x^2 - 7x - 3$ (k) $2a^2 - a - 15$ (l) $8b^2 - 10b - 7$

Simplifying algebraic fractions

Working rule

To simplify algebraic fractions, use the same working rules stated for fundamental operations on fractions.

Example 1

Simplify and write as a single fraction. $\frac{2}{x-4} + \frac{x-4}{4}$

Working: $\frac{8}{4(x-4)} + \frac{(x-4)(x-4)}{4(x-4)}$

$= \frac{8 + (x-4)(x-4)}{4(x-4)}$

$= \frac{8 + x^2 - 4x - 4x + 16}{x(x-4)} = \frac{x^2 - 8x + 24}{4(x-4)}$

First make the denominators same. Then add the numerators and keeping a common denominator.

Expand the brackets and then simplify the numerators.

Don't do this. $\frac{2}{x - \cancel{4}} + \frac{x - \cancel{4}}{4}$

Example 2

Factorise and simplify $\frac{x^2 - 4x}{x^2 - 7x + 12}$.

Working: $\frac{x(x-4)}{(x-3)(x-4)}$

$= \frac{x}{(x-3)}$

First factorise the numerator and the denominator.Then cancel out the common factors.At last write down the remaining factors.

Don't do this. $\frac{\cancel{x}}{(\cancel{x} - 3)}$

Practise exercise 8

1 Simplify and write your answer as a single fraction.

(a) $\frac{6x}{7} \times \frac{14}{xy}$ (b) $\frac{21xy}{15} \times \frac{10y}{3x}$ (c) $\frac{2x}{x+5} \times \frac{x+5}{10y}$

(d) $\frac{p-1}{9} \times \frac{9p}{2p-1}$ (e) $\frac{8}{x-1} \times \frac{(x-1)(x-2)}{4(x+2)}$ (f) $\frac{x^2-4}{x+4} \times \frac{2a}{x+2}$

(g) $\frac{m^2}{m^2-9} \times \frac{m-3}{3m}$ (h) $\frac{x}{5} \div \frac{5}{x}$ (i) $\frac{7-x}{6} \div \frac{12(7-x)}{5x}$

(j) $\frac{a+2}{10} \div \frac{a}{2}$ (k) $1 \div \frac{x(x^2-1)}{x-1}$ (l) $\frac{x-y}{x^2-y^2} \div \frac{(x+y)^2}{x-y}$

2 Write as a single fraction in its simplest form.

(a) $\frac{2}{x} + \frac{x}{6}$ (b) $\frac{p}{2} + \frac{p}{6} + \frac{p}{12}$ (c) $\frac{2}{x+4} + \frac{x+4}{2}$

(d) $\frac{x-10}{14} + \frac{7}{x-10}$ (e) $(p-2) + \frac{p}{p-2}$ (f) $\frac{x}{x^2-1} + \frac{5}{x+1}$

(g) $\frac{1}{(x+1)(x-2)} + \frac{2}{x-2}$ (h) $\frac{(x-1)}{(x+3)(x+4)} + \frac{(x+1)}{(x+3)}$ (i) $\frac{2}{x+2} + \frac{4}{x+4} + \frac{3}{x+2}$

3 Write as a single fraction in its simplest form.

(a) $\frac{y}{7} - \frac{7}{y}$ (b) $m - \frac{m}{3} - \frac{m}{6}$ (c) $\frac{5}{x-5} - \frac{x-5}{10}$

(d) $\frac{p+7}{14} - \frac{21}{p+7}$ (e) $\frac{6}{x+6} - \frac{2}{x-6}$ (f) $\frac{x}{x-2} - \frac{1}{x-2} + 1$

(g) $\frac{3x^2-9}{3} - \frac{2x+1}{2}$ (h) $\frac{x-1}{(x+1)(x+2)} - \frac{x-2}{x+2}$ (i) $\frac{x}{x-1} + \frac{4}{x-1} - \frac{2}{x-2}$

(j) $\frac{5}{x} - \frac{4}{x+1}$ (k) $\frac{x+2}{x} - \frac{x}{x+2}$ (l) $\frac{x-3}{4} + \frac{4}{x-3}$

4 Simplify the following algebraic expressions.

(a) $\frac{x(x+1)}{(x-2)(x-1)(x+1)}$ (b) $\frac{(x+4)}{x^2+7x+12}$ (c) $\frac{(x+3)}{x^2+5x+6}$

(d) $\frac{x+1}{x^2-6x-7}$ (e) $\frac{x(x-5)}{x^2+10x-75}$ (f) $\frac{x^2-4x}{x^2-9x+20}$

(g) $\frac{2x-60}{x^2-50x+600}$ (h) $\frac{x^2-4}{x^2-4x-12}$ (i) $\frac{(x+5)(x+2)}{2x^2+9x+10}$

(j) $\frac{2x^3-6x^2}{2x^2-11x+15}$ (k) $\frac{9x^2-16}{3x^2+5x-12}$ (l) $\frac{15x+9}{5x^2-12x-9}$

2.03 Indices **Core Curriculum**

- Indices are the short way of writing repeated multiplication.

Remember

Rules of indices

Product rule: (1) $a^m \times a^n = a^{m+n}$ (2) $a^m \times b^m = (a \times b)^m$

Quotient rule: (1) $a^m \div a^n = a^{m-n}$ (2) $\frac{a^m}{b^m} = \left(\frac{a}{b}\right)^m$

Power rule: (1) $(a^m)^n = a^{m \times n}$ (2) $((a^m)^n)^p = a^{m \times n \times p}$

Reciprocal rule: (1) $a^{-m} = \frac{1}{a^m}$ (2) $\left(\frac{a}{b}\right)^{-m} = \left(\frac{b}{a}\right)^m$

Zero power rule: $a^0 = 1$

Exponential equation rule: (1) If $a^m = a^n$ then m = n. (2) If $a^m = b^m$ then a =b.

Example 1

Simplify (a) $p^6 \times p^5 = p^{6+5} = p^{11}$
(b) $3^x \times 2^x = (3 \times 2)^x = 6^x$
(c) $b^5 \div b^{-7} = b^{5- -7} = b^{12}$
(d) $\frac{5^p}{10^p} = \left(\frac{5}{10}\right)^p = \left(\frac{1}{2}\right)^p$

Use $a^m \times a^n = a^{m+n}$
Use $a^m \times b^m = (a \times b)^m$
Use $a^m \div a^n = a^{m-n}$
Use $\frac{a^m}{b^m} = \left(\frac{a}{b}\right)^m$

Example 2

Find the exact value of each.

(a) $(4^3)^2 = 4^{3 \times 2} = 4^6 = 4096$

(b) $((3^2)^2)^2 = 3^{2 \times 2 \times 2} = 3^8 = 729$

(c) $9^{-2} = \frac{1}{9^2} = \frac{1}{81}$

(d) $\left(\frac{3}{5}\right)^{-3} = \left(\frac{5}{3}\right)^3 = \frac{5^3}{3^3} = \frac{125}{27}$

(e) $99^0 = 1$

Use $(a^m)^n = a^{m \times n}$
Use $((a^m)^n)^p = a^{m \times n \times p}$
Use $a^{-m} = \frac{1}{a^m}$

Use $\left(\frac{a}{b}\right)^{-m} = \left(\frac{b}{a}\right)^m$
Use $a^0 = 1$

Example 3

Find the value of x in each case.

(a) $2^{x-1} = 1024$ (b) $3^x = \frac{1}{81}$

Working: (a) $2^{x-1} = 2^{10}$
$x - 1 = 10$
$x = 9$

$1024 = 2^{10}$
If $a^m = a^n$ then m = n.

(b) $3^x = \frac{1}{3^4}$
$3^x = 3^{-4}$
$x = -4$

$81 = 3^4$
$a^{-m} = \frac{1}{a^m}$
If $a^m = a^n$ then m = n.

Practise exercise 9

1 Simplify the following expressions.
(a) $a^3 \times a^2$ (b) $a^4 \times a^{-4}$ (c) $2^x \times 2^y$ (d) $5^{-x} \times 5^{x+1}$
(e) $p^a \times p^b \times p$ (f) $a^5 \div a^3$ (g) $a^2 \div a^{-6}$ (h) $3^x \div 3^y$

(i) $7^{p-1} \times 7^{1-p}$ (j) $\frac{x^7}{x^9}$ (k) $\frac{x^5}{y^5}$ (l) $\frac{4^3}{6^3}$

(m) $y^a \times z^a$ (n) $2^4 \times 3^4$ (o) $(x^{a+1})^{a-1}$ (p) $(4x^2)(5x^3)$

(q) $9x^5 \div 3x^3$ (r) $(x^3)^6$ (s) $((a^2)^4)^8$ (t) $\left(\left(\frac{1}{x}\right)^2\right)^{10}$

(u) d^0 (v) $(n^2)^4$ (w) $\left(\frac{8}{k}\right)^{-2}$ (x) $(((r)^2)^4)^{-3}$

2 Find the exact value of each of the following.

(a) 2^4 (b) 3^{-3} (c) 11^{-1} (d) 9^0

(e) $\left(\frac{3}{8}\right)^0$ (f) $(7^0 - 2 \times 3^0) - 8^0$ (g) $\left(\frac{2}{3}\right)^4$ (h) $\left(\frac{22}{7}\right)^2$

(i) $(2^3)^4$ (j) $(5^0)^5$ (k) $\left(2\frac{1}{3}\right)^{-3}$ (l) $5^3 \times 5^2$

(m) $3^{-4} \div 3^2$ (n) $\frac{7 \times 7^{-3}}{7^3}$ (o) $\frac{2^4 \times 3^3}{2^3 \times 3^2}$ (p) $((-4)^2)^3$

3 Write down the value of x in each of the following.

(a) $2^x = 2^3$ (b) $2^x = 16$ (c) $2^{x+1} = 32$ (d) $2^{x-2} = 4^3$

(e) $512 = 2^{2x-1}$ (f) $3^{3+x} = 243$ (g) $\frac{1}{4} = \left(\frac{1}{2}\right)^x$ (h) $3^x = \frac{1}{27}$

(i) $\left(\frac{2}{3}\right)^x = \frac{8}{27}$ (j) $5^x = 1$ (k) $7^{x-2} = 1$ (l) $2^x \times 3^x = 36$

(m) $6^x \times 6^2 = 216$ (n) $5^3 \div 5^x = 5^{-3}$ (o) $a^{2x} \times a = a^{-1}$ (p) $4^{x-1} = 8^{x-1}$

(q) $\frac{1}{6} = 6^x$ (r) $\left(\frac{1}{5}\right)^{-x} = \frac{1}{625}$ (s) $6^r = \frac{1}{36}$. (t) $3^x = \frac{1}{81}$.

EXTENDED CURRICULUM

Remember

You need to know and use the **fractional index rules** along with the **rules of indices** stated earlier in this section.

Fractional index rule: (1) $a^{\frac{1}{n}} = \sqrt[n]{a}$ (2) $(a^p)^{\frac{m}{n}} = a^{p \times \frac{m}{n}}$

Example

Simplify (a) $49^{\frac{1}{2}}$, (b) $\left(\frac{a^4}{16}\right)^{\frac{3}{4}}$

(a) Use $a^{\frac{1}{n}} = \sqrt[n]{a}$

(b) Use $\left(\frac{a}{b}\right)^m = \frac{a^m}{b^m}$

Use $(a^p)^{\frac{m}{n}} = a^{p \times \frac{m}{n}}$

Working: (a) $\sqrt{49} = 7$

(b) $\left(\dfrac{a^4}{2^4}\right)^{\frac{3}{4}} = \dfrac{a^{4 \times \frac{3}{4}}}{2^{4 \times \frac{3}{4}}}$

$= \dfrac{a^3}{8}$

Don't do this.

$\left(\dfrac{a^{\cancel{4}}}{16}\right)^{\frac{3}{\cancel{4}}}$ and $\left(\dfrac{a^{\cancel{4}}}{2^4}\right)^{\frac{3}{\cancel{4}}}$

Practise exercise 10

1 Find the exact values of the following.

(a) $\left(7^4\right)^{\frac{1}{2}}$ (b) $(16)^{\frac{1}{2}}$ (c) $(25)^{-\frac{1}{2}}$ (d) $8^{\frac{2}{3}}$

(e) $49^{-\frac{3}{2}}$ (f) $(81)^{2\frac{1}{2}}$ (g) $(243)^{-\frac{3}{5}}$ (h) $(256)^{-\frac{3}{4}}$

(i) $(512)^{-\frac{4}{3}}$ (j) $(1024)^{-\frac{2}{5}}$ (k) $\left(\dfrac{4}{9}\right)^{\frac{1}{2}}$ (l) $\left(\dfrac{27}{64}\right)^{\frac{1}{3}}$

(m) $\left(\dfrac{3}{2}\right)^{-3}$ (n) $\left(1\dfrac{3}{4}\right)^{-2}$ (o) $\left(\dfrac{625}{81}\right)^{\frac{3}{2}}$ (p) $\left(\dfrac{16}{729}\right)^{-\frac{2}{3}}$

(q) $\left(3\dfrac{3}{8}\right)^{-\frac{1}{3}}$ (r) $\left(1\dfrac{61}{64}\right)^{-2\frac{2}{3}}$ (s) $\dfrac{8^{\frac{1}{3}} \times \sqrt{27}}{\sqrt[3]{81}}$ (t) 7^{-3}

2 Simplify the following.

(a) $\left(\dfrac{a^2}{11^4}\right)^{\frac{1}{2}}$ (b) $\left(\dfrac{169}{a^2}\right)^{-\frac{1}{2}}$ (c) $\left(\dfrac{x^4}{361}\right)^{\frac{3}{2}}$ (d) $\left(\dfrac{64}{x^6}\right)^{-\frac{5}{6}}$

(e) $\left(\dfrac{x^8}{8}\right)^{-\frac{1}{2}}$ (f) $\left(\dfrac{27}{x^{27}}\right)^{\frac{4}{3}}$ (g) $\left(\dfrac{p^{-3}}{27}\right)^{\frac{1}{3}}$ (h) $\left(\dfrac{x^{10}}{y^{20}}\right)^{-\frac{3}{5}}$

(i) $\left(\sqrt{x}\right)^{\frac{1}{2}}$ (j) $\left(\sqrt{x}\,\sqrt[3]{x}\right)^3$ (k) $\sqrt{x^{-3}y^{-2}} \times \sqrt{\dfrac{y^2}{x^3}}$ (l) $\left(\sqrt[3]{x}\right)^{-\frac{3}{2}}$

(m) $\left(\dfrac{a^{\frac{2}{3}} \times a^{\frac{1}{6}}}{b^{\frac{2}{3}}}\right)^{-6}$ (n) $\left(\dfrac{x^{-\frac{4}{5}}}{y^{\frac{2}{3}} \times y^{\frac{3}{5}}}\right)$ (o) $\left(\dfrac{x^{27}}{27}\right)^{\frac{2}{3}}$ (p) $\left(\sqrt{x}+2\right)\left(\sqrt{x}-2\right)$

3 Solve the following exponential equations.

(a) $27^{3x-2} = 1$ (b) $64^{\frac{2}{3}x-3} = 16^{-3}$ (c) $25^{\frac{x}{2}+\frac{1}{3}} = 125^{\frac{x}{2}-2}$ (d) $\sqrt{x} = \sqrt{5}$

(e) $\sqrt[3]{a} = 7^{\frac{1}{3}}$ (f) $\sqrt{x}\,\sqrt[3]{x} = \sqrt[6]{32}$ (g) $5^x = \sqrt{125}$ (h) $3^p = 3\sqrt[3]{9}$

(i) $2^x = \dfrac{1}{\sqrt{32}}$ (j) $2^{\frac{x}{2}} = \sqrt[3]{16}$ (k) $7^x = \dfrac{1}{\sqrt[3]{49}}$ (l) $x^{-\frac{4}{3}} = \sqrt[3]{\dfrac{1}{16}}$

(m) $4^{\sqrt[5]{x}} = 1$ (n) $64^{a} = 4$ (o) $9 = 27^{-2x}$ (p) $\sqrt{32} = 2^{x}$

2.04 Equations

Core Curriculum

Simple equation

- Solving a simple equation means finding a number that replaces the letter to make the equation true.

Working rule

To solve a simple equation, 1) remove the brackets(if any),
2) Use the rule 'side change, sign change',
(a) to move the terms containing letters to one side (of the = sign),
(b) to make the coefficient of the letter as 1.

Remember

The rule of 'side change, sign change' means **LHS** = **RHS**

LHS		RHS
+ term	⟷	– term
x term	⟷	÷ term

Example

Solve (a) $3x - 5 = 15$ (b) $3(2x - 1) + 2(x + 7) = 10x - 3$

Working: (a) $3x = 15 + 5$

$$x = \frac{20}{3} = 6\frac{2}{3}$$

Use the rule 'side change, sign change'.
Give your answer as a fraction, since $\frac{20}{3}$ is non

Don't do this. $3x = 15 - 5$ and last answer as $x = 6.67$ or 6.66

(b) $6x - 3 + 2x + 14 = 10x - 3$

$6x + 2x - 10x = -3 + 3 - 14$

$-2x = -14$

$$x = \frac{-14}{-2} = 7$$

First remove the brackets.
Then use the rule 'side change, sign change'.

Don't do this.
Expanding brackets as $6x - 1 + 2x + 7 = 10x - 3$
Last answer as -7

Practise exercise 11

1 Solve the following simple equations.

(a) $x + 4 = 0$ (b) $x + \frac{1}{2} = \frac{1}{4}$ (c) $x + \frac{2}{3} = -6$ (d) $x - 7 = 42$

(e) $y - 5 = -10$ (f) $y - \frac{2}{3} = -\frac{3}{2}$ (g) $5x = 55$ (h) $1.2x = -72$

(i) $-9x = \frac{1}{3}$ (j) $\frac{x}{3} = 12$ (k) $16 = \frac{x}{-8}$ (l) $\frac{-p}{11} = \frac{11}{-12}$

2 Solve each of the equations.

(a) $2x + 3 = 9$ (b) $5x + 7 = -14$ (c) $-3x + \frac{1}{6} = -\frac{1}{6}$

(d) $9x - 4 = 5$ (e) $-4x - 4 = 0$ (f) $7x - \frac{5}{7} = -1$

(g) $\frac{x}{4} + 1.6 = 5$ (h) $\frac{-x}{7} + 7 = 21$ (i) $\frac{3x}{-5} + 1 = -11$

(j) $\frac{y}{2} - \frac{1}{2} = 1$ (k) $\frac{4y}{9} - \frac{2}{3} = \frac{-4}{9}$ (l) $\frac{-y}{6} - \frac{1}{6} = 10$

(m) $4x + 2x = 3$ (n) $2x - 3x = 6$ (o) $x + \frac{1}{4} = -3x$

(p) $3x + 2 = 5x - 2$ (q) $\frac{x}{2} + \frac{x}{5} = -10$ (r) $x = -4 - \frac{x}{2}$

(s) $\frac{2x}{3} + \frac{x}{6} = 1 - \frac{x}{2}$ (t) $4.5x - 0.5 = 4x - 4.5$ (u) $\frac{10x + 7}{5} = 1$

(v) $\frac{4x - 3}{2} - \frac{x}{3} = -6$ (w) $\frac{2x + 1}{2} + 5 = \frac{x}{4} - 5$ (x) $0 = 4 - \frac{4x + 2}{8}$

3 Solve the equations.

(a) $4(x + 3) = 6$ (b) $2(x - 3) = 7$ (c) $3(2 - x) = 5$

(d) $5(x - 4) = x - 2$ (e) $-7(3 - x) = 1$ (f) $3(2y - 1) = 24 - y$

(g) $9x - (7x - 4) = 6$ (h) $12 - (8 - 2t) = -1$ (i) $0.4(2x + 3) = 1 - x$

(j) $4(3x + 2) = 3x - 2$ (k) $2(3x - 1) + 3(x + 2) = 6$ (l) $6(x - 6) = 4(x + 2)$

(m) $7 - 10x = 5(2x - 1) + 2(x + 1)$ (n) $0.9(5x + 3) = 7.9 - 1.5x$

(o) $6.4 - (2.9t - 4) = 4.6(2t + 8)$ (p) $\frac{-3}{4}\left(10 - \frac{x}{3}\right) = \frac{1}{2} - x$

(q) $\frac{1}{5}(10x - 5) = \frac{1}{4}(8 - 2x)$ (r) $5\left(\frac{x}{5} - 3\right) + 2 = 10 - (5x - 4)$

4 The measures of the angles in a triangle are x, $2x - 15^\circ$ and $4x - 15^\circ$. Construct an equation in x and find the angles.

5 The cost of x biscuit boxes at \$2.25 each is \$ 10.50 more than the cost of x chocolate boxes at \$1.75 each.

(a) Find the number of biscuit boxes.

(b) What is the total number of biscuit and chocolate boxes?

6 The interior angles of a pentagon are $4x$, $4x - 10^\circ$, $6x - 5^\circ$, $7x - 10^\circ$ and $7x + 5^\circ$.

(a) Write down an equation in terms of x.

(b) Solve the equation.

7 Poul thought a number. He then added 4 to it. After doubling the sum, when he takes away 31from it the result is 8 less than the original number. Find the original number.

8 A shopkeeper purchases and then sells some boxes of biscuits.

(a) The cost price of x boxes of biscuits is \$ 30. Write down the cost of each box of biscuits, in terms of x.

(b) The selling price of each box of biscuits is \$ 1.50. What is the selling price of all boxes of biscuits, in terms of x?

(c) Write down in terms of x, the profit receives by the shopkeeper in selling all boxes of biscuits.

9 When two-fifth of a number, y, is trebled, the product is 8 more than the original number.

(a) In terms of y, write down an equation. (b) Solve the equation for y.

10 In an examination, Saleem scored x marks. Salman scored 4 marks less than Saleem. Salva scored 96 marks, which is same as the mean mark scored by Saleem and Salman.
(a) Write down an equation in x. (b) How many marks did Salman score?

11 In a cyclic quadrilateral, the opposite angles are $(2x + 6^\circ)$ and x.
Find each angle of the cyclic quadrilateral.

Simultaneous linear equations

- Solving simultaneous equations means finding a pair of numbers that replaces the letters to make the equations true.
- Simultaneous equations can be solved by substitution method or elimination method or graphical method.

Working rule

Substitution method
1) Express one of the letters(say y) in terms of the other letter(say x) from equ(1).
2) Replace y in terms of x in equ(2).
3) Solve the resulting simple equation for x.
4) Then substitute the x-value in any of the equations to get y-value.

Elimination method
1) If any of the letters have same coefficient(if not, make the coefficient of any letter same), then eliminate one letter by adding(for opposite signs) and by subtracting(for same signs).
2) Solve the resulting simple equation.
3) Then substitute the value of the letter from step-2 in any of the equations to get the value of the other letter.

Remember

It is easy to use the substitution method when both the coefficients of a letter is 1 or −1.

Example
Solve the simultaneous equations.

(a) $3x - y = 9$ ....(1)

$\frac{1}{2}x + 2y = 8$ ....(2)

Working: From equ(1), we have

$y = 3x - 9$

Expressing y in terms of x from equ(1).

Don't do this. $y = 9 - 3x$

$\frac{1}{\not 2}x + \not 2(3x - 9) = 8$

Replacing y in terms of x in equ(2), we have

$\frac{1}{2}x + 2(3x - 9) = 8$

$\frac{1}{2}x + 6x - 18 = 8$ or $\frac{13}{2}x = 26$ or $x = 4$

Solving the simple equation

Substituting $x = 4$ in equ(1), we have

$3 \times 4 - y = 9$ or $y = 3$

So $x = 4$, $y = 3$.

(b) $4x - 3y = 23$ ....(1)

$5x - 3y = 25$ .... (2)

Working: From equ(1) – equ(2), we have

$-x = -2$ or $x = 2$

Subtract equ(1) from equ(2), since y has same coefficients and same signs.

Substituting the $x = 2$ in equ(1), we have

$4 \times 2 - 3y = 23$ or $-3y = 23 - 8$ or $y = -5$

So $x = 2$, $y = -5$

Practise exercise 12(A) **Core Curriculum**

1 Solve the simultaneous equations.

(a) $2x + y = 7$, $y = 3x - 8$

(b) $5x - y = 6$, $y = 3x - 2$

(c) $3x + 2y = 4$, $y = 3 - 2x$

(d) $y = x - 3$, $4x - 3 = 7y$

(e) $x + 2y = 15$, $x = y + 3$

(f) $x = 10 - 9y$, $10x - 9y = 1$

(g) $x + y = 11$, $x - y = 3$

(h) $x + y = 14$, $x - 2y = -13$

(i) $3x - y = -5$, $x - y = -3$

(j) $4x - 7y = 5$, $4x - 3y = 25$

(k) $2x + 3y = 6$, $x - 3y = 0$

(l) $4x + y = 13$, $2x + 3y = 9$

(m) $7x - 4y = 29$, $3x - 2y = 13$

(n) $5x + 3y = -8$, $3x + 5y = 8$

(o) $2m - n = 15$, $m - 2n = -6$

(p) $9p + 5q = -9$, $5p - 2q = -44$

(q) $-x + 3y = 18$, $2x - y = -11$

(r) $2a - 4 = 7b$, $6a + 4 = 29b$

(s) $6x - 2y = 4$, $3x + y = 1$

(t) $1.5c - d = 2.55$, $c + 1.5d = 4.18$

(u) $3x = 4y$, $2x - y = 10$

3 The sum of the length(l) and width(w) of a rectangle is 14 cm. The difference between twice the width and the length of the rectangle is 1.

(a) Write down two equations in terms of l and w.

(b) Solve the equations to find the length and the width of the rectangle.

4 The cost of a pen is \$ x. The cost of a pencil is \$ y.
Five pens and four pencils together costs \$ 12.
Three pens and two pencils together costs \$ 7.

(a) Construct two equations in x and y. (b) Find the values of x and y.

5 Thrice a number, x, equals five times another number, y. When 3 is added to the number, x, it results twice of y.

(a) Write down two equations in x and y. (b) Solve the equations to find the numbers.

6 The ten's digit of a two-digit number is 3 less than the unit's digit. The unit's digit of the number and twice of its ten's digit together gives 15.
The unit's digit of the number is m. The ten's digit of the number is n.

(a) Write down two equations in terms of m and n.

(b) Find the ten's digit and the unit's digit of the number. (c) What is the given number?

7

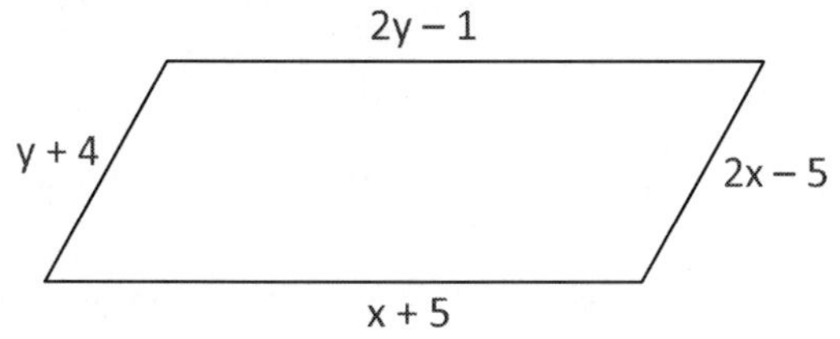

The above diagram shows a parallelogram.

(a) Using the information given on the diagram, write down two simultaneous equations.

(b) Solve the simultaneous equations. (c) Find the sides of the parallelogram.

Practise exercise 12(B) EXTENDED CURRICULUM

1 Solve the simultaneous equations.

(a) $\frac{1}{2}x + 2y = 5$
$\frac{1}{3}x - 2y = 0$

(b) $\frac{1}{4}x - \frac{1}{2}y = -2$
$x + \frac{2}{5}y = 4$

(c) $\frac{1}{6}m - \frac{2}{5}n = -4$
$\frac{1}{3}m + \frac{1}{5}n - 7 = 0$

(d) $\frac{2}{7}p - \frac{3}{7}q = \frac{1}{2}$
$\frac{1}{5}p - \frac{2}{7}q = \frac{2}{3}$

(e) $\frac{2x+y}{2} - 4 = 0$
$\frac{3x-y}{3} = 6$

(f) $\frac{5x-2y}{5} = 1$
$\frac{5y-3x}{5} = -2$

(g) $x + 6y = 1\frac{1}{3}$
$4x = \frac{3y-1}{3}$

(h) $\frac{x+y}{3} - \frac{x-y}{2} = 6$
$\frac{x-y}{4} - \frac{x+y}{3} = 2$

(i) $0.4x + 2y = 10$
$0.3x + 5y = 18$

(j) $\frac{1}{2}x + 2y = 16$
$2x + \frac{1}{2}y = 19$

(k) $\frac{1}{2}x + y = 5$
$x - 2y = 6$

(l) $4x + 5y = 0$
$8x - 15y = 5$

3 Half of a number, p and one-third of another number, q together give 15.
The difference between twice of the number, p and one-sixth of the number, q is 42.
(a) Write down two equations in terms of p and q.
(b) Find the two numbers that denote p and q.

4 The total number of boys(b) and girls(g) in a class is 60. Two-third of the boys and one-nineth of the girls add up to 25.
(a) Construct two equations in b and g to show this information.
(b) Solve the equations and write down the number of girls in the class.

5 The unit's digit of a number is 1 more than thrice of its ten's digit. When the digits are interchanged, the new number formed is 9 less than three times of the original number.
(a) Using this information, write down two equations. (b) Find the original number.

6

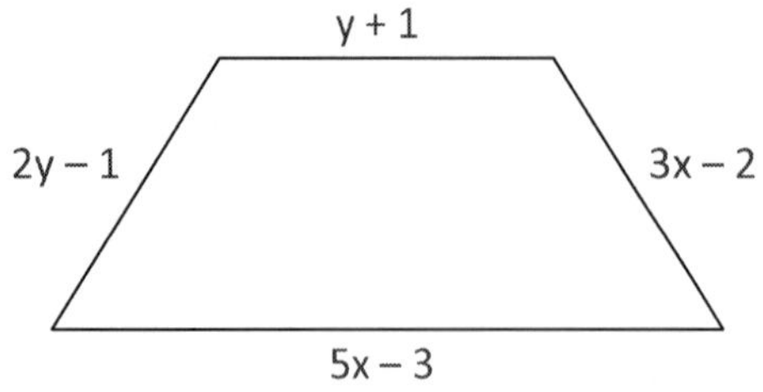

The diagram shows an isosceles trapezium. Among the parallel sides, the shorter side is half of the longer side.
(a) Write down two equations in x and y using the given information in the diagram.
(b) Solve the equations. (c) Find the equal sides of the trapezium.

Quadratic equations

EXTENDED CURRICULUM

- Solving a quadratic equation means finding the two numbers that replace the letter to make the equation true.

Remember

Before solving a quadratic equation, rearrange it in the form $ax^2 + bx + c = 0$.

Working rule

A quadratic equation can be solved by any of the three methods described below.

Factorisation method

1) Factorise the quadratic expression(i.e. $ax^2 + bx + c$).

2) Equate each factor to 0. 3) Solve for the letter(i.e. x).

Completing the square method

1) Divide each f the terms by a(if $a \neq 1$).

2) Rearrange the equation in the form $\left(x + \frac{b}{2}\right)^2 = \left(\frac{b}{2}\right)^2 - c$.

3) Solve the equation in step-2 by taking over the square root. (Don't forget the ± sign)

Direct formula method

Use the direct formula $x = \frac{1}{2a}(-b \pm \sqrt{(b^2 - 4ac)})$.

Example

Solve the equation $2x^2 - 7x + 6 = 0$.

Working: Factorisation method

$$2x^2 - 7x + 6 = 0$$
$$2x^2 - 4x - 3x + 6 = 0$$
$$2x(x - 2) - 3(x - 2) = 0$$
$$(x - 2)(2x - 3) = 0$$
$$(x - 2) = 0 \text{ or } (2x - 3) = 0$$
$$x = 2 \text{ or } x = 1.5$$

Choosing two numbers –4 and –3 which add up to –7 and multiply up to 12.
Take out the common factors.
Equate each factor to 0.
Solve the resulting equations.

Direct formula method

$$2x^2 - 7x + 6 = 0$$
$$x = \frac{1}{2a}(-b \pm \sqrt{(b^2 - 4ac)})$$
$$x = \frac{1}{2 \text{ x } 2}\left(--7 \pm \sqrt{((-7)^2 - 4 \text{ x } 2 \text{ x } 6)}\right)$$
$$x = \frac{1}{4}(7 \pm 1)$$
$$x = \frac{1}{4}(7 + 1) \text{ or } \frac{1}{4}(7 - 1) = 2 \text{ or } 1.5$$

Comparing $ax^2 + bx + c = 0$ with

$2x^2 - 7x + 6 = 0$, we get $a = 2, b = -7$ and $c = 6$.

Substituting the values of a, b and c in the formula.

Don't do this.
$$x = \frac{1}{2 \text{ x } 2}\left(--7 \pm \sqrt{(-7^2 - 4 \text{ x } 2 \text{ x } 6)}\right)$$

Completing the square method

$$2x^2 - 7x + 6 = 0$$
$$x^2 - 3.5x + 6 = 0$$
$$\left(x - \frac{3.5}{2}\right)^2 = \left(\frac{3.5}{2}\right)^2 - 3$$
$$\left(x - \frac{3.5}{2}\right)^2 = 0.0625$$

Divide each term by 2. (Since $a \neq 1$)

Rearrange $x^2 - 3.5x + 6 = 0$ in the form

$\left(x + \frac{b}{2}\right)^2 = \left(\frac{b}{2}\right)^2 - c$.

Solve the equation for x.

Don't do this. $x - 1.75 = \sqrt{0.0625}$

$x - 1.75 = \pm\sqrt{0.0625}$

$x = 1.75 \pm 0.25$

$x = 1.75 + 0.25$ or $1.75 - 0.25 = 2$ or 1.5

Practise exercise 13

1 Solve the following quadratic equations.
(a) $x^2 = 16$ (b) $x^2 - 49 = 0$ (c) $3x^2 = 27$ (d) $5a^2 - 12 = 0$
(e) $x^2 - 6x = 0$ (f) $x^2 = -11x$ (g) $(x-1)^2 = 9$ (h) $(p-8)^2 - 7 = 0$

2 Solve the quadratic equations using factorization method.
(a) $(x-2)(x-5) = 0$ (b) $(a+7)(a-3) = 0$ (c) $2x^2 + 7x + 6 = 0$
(d) $3x^2 - 7x + 4 = 0$ (e) $x^2 - 2x - 3 = 0$ (f) $7x^2 - 10x + 3 = 0$
(g) $6x^2 + 100 = 50x$ (h) $x^2 - 4x = 5$ (i) $x^2 + 26x + 168 = 0$
(j) $x^2 = 9x - 8$ (k) $12 + x - x^2 = 0$ (l) $60 + 4x = x^2$

3 Solve the quadratic equations by direct formula or completing the square method.
(a) $x^2 + 4x = 6$ (b) $x^2 - 6x = 1$ (c) $4x^2 - 16x + 3 = 0$
(d) $x^2 + 6x + 9 = 0$ (e) $2x^2 + 4x - 1 = 0$ (f) $3x^2 + 15x - 2 = 0$
(g) $5x^2 - 8x + 3 = 0$ (h) $-2x^2 + 7x - 5 = 0$ (i) $9x^2 = 20x - 9$
(j) $6a^2 - 11a - 5 = 0$ (k) $p^2 - p - 99 = 0$ (l) $15p^2 + 17p = 23$
(m) $x^2 - 3x - 4 = 0$ (n) $2x^2 - 11x + 12 = 0$ (o) $x^2 + 4x - 22 = 0$

4 (a) Show that the equation $\frac{2}{x-1} = \frac{x+3}{4}$ simplifies to the quadratic equation $x^2 + 2x - 11 = 0$.
(b) Solve the equation $x^2 + 2x - 11 = 0$.

5 Find the two consecutive integers whose product is 812.

6 Aleesa thought a number. When she subtracted 5 from it and then squares the result, it exceeds the original number by 267. What number did she think?

7 Given that x, (x + 1) and (x + 2) are three consecutive integers.
(a) Write down in terms of x,
(i) the square of the first integer x,
(ii) the product of the other two integers.
(b) The sum of the answers to parts (a)(i) and (ii) is 46. Write down a quadratic equation in its simplest form.
(c) Find the integer value of x.
(d) Write down the three consecutive integers.

8 A shopkeeper buys x article for a total cost of $ 640.
(a) Write down the cost of an article in terms of x.
(b) If the cost per article were $ 4 less, the number of articles that were bought for $ 640 would have been 8 more.
Show that this information forms the equation $x^2 + 8x - 1280 = 0$.
(c) Solve the equation $x^2 + 8x - 1280 = 0$.
(d) How many articles does the shopkeeper buy?

9

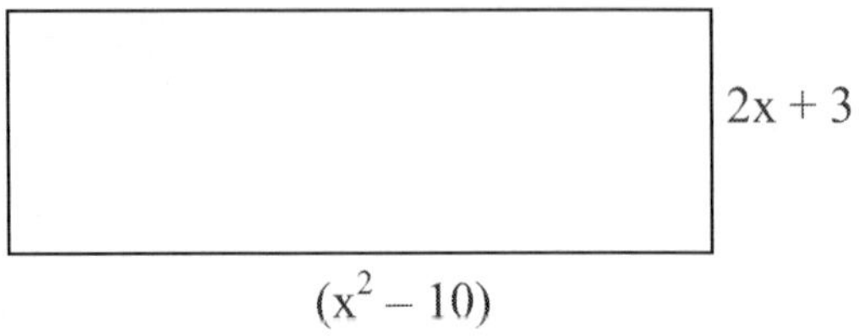

The diagram shows a rectangle. The lengths of its sides are given in terms of x.
The perimeter of the rectangle is 108 metres.
(a) Construct a quadratic equation to show this information.
(b) Solve the equation and write down the possible value of x.
(c) Write down (i) the length of the rectangle, (ii) the breadth of the rectangle.

10

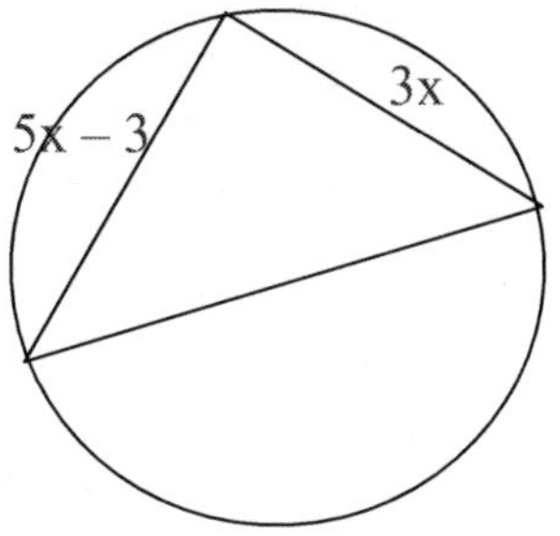

The figure shows a right angled triangle inscribed in a circle.
The lengths of its two sides are given in terms of x.
The diameter of the circle is 41 cm.
(a) Write down an equation in terms of x.
(b) Solve the equation for x.
(c) Write down the lengths of the sides of the triangle.
(d) Calculate the area of the triangle.

2.05 Linear Inequalities

EXTENDED CURRICULUM

Remember

1. The symbols for strict inequalities are > **(is greater than) and < (is less than).**
2. The symbols for weak inequalities are **≥ (is greater than or equal to)** and **≤ (is less than or equal to).**
3. The number at the vertex of an inequality is always smaller.
   E.g.(1) $2 < 5$ (2 is the vertex number) (2) $6 > -10$ (–10 is the vertex number)
4. Inequalities can be solved like solving simple equations except two things.

First thing: When you multiply or divide by a negative number on both sides, just reverse the inequality symbol.

Second thing: When you take the reciprocal of all the quantities on both sides, just reverse the inequality symbol.

Example Solve the inequalities.
(a) $2x + 5 > 3x - 1$

Working: $2x > 3x - 1 - 5$
$2x - 3x > -6$
$-1x > -6$
$x < 6$

Subtract 5 from both sides.
Subtract 3x from both sides.
Divide both sides by –1 and reverse the inequality symbol.

Don't do this. Last answer as $x > 6$

(b) $-10 < 4x - 5 \leq 7$
Working: $-10 + 5 < 4x \leq 7 + 5$

Add 5 with all the three expressions.

Divide by –4 with all the three expressions.

$\frac{-5}{4} < x \leq \frac{12}{4}$ or $-1.25 < x \leq 3$

(c) $\frac{1}{x+1} \leq \frac{1}{5-2x}$

Working: $x + 1 \geq 5 - 2x$ — Take the reciprocal on both sides and reverse the inequality symbol.

$x + 2x \geq 5 - 1$ — Add 2x on both sides. Subtract 1 from both sides.

$3x \geq 4$

$x \geq \frac{4}{3}$ — Divide both sides by 3.

Practise exercise 14

1 Solve the following inequalities.

(a) $x - 2 > 5$ (b) $x + 3 < 1$ (c) $-4x \leq 20$
(d) $3x - 5 \leq 10$ (e) $9 - 12x > 0$ (f) $-45 \geq 13 + 8t$
(g) $2x - 3 < 4 + 5x$ (h) $7 - 10x \geq 7x - 1$ (i) $3(2 - x) > 2(1 - x)$
(j) $9 - 8x > 5(7 - 2x)$ (k) $2 \leq 3x + 1 < 10$ (l) $2x - 3 < x + 2 \leq 3x + 5$
(m) $0.96y - 1.8 \geq 1.2(0.7 - 3.6y)$ (n) $120 > 4x + 16 > -24$ (o) $1 < 2 - 3x < 5$

2 Solve the inequalities.

(a) $\frac{-x}{7} \geq -4$ (b) $\frac{1}{2x} \leq -12$ (c) $-x + \frac{1}{2} < 8$
(d) $\frac{4x-9}{4} < -2$ (e) $11a - 6 \leq \frac{10a-3}{20}$ (f) $\frac{x}{2} + \frac{x}{3} < 1 - \frac{x}{4}$
(g) $\frac{2x}{5} - 5 \leq 3 - \frac{2x}{3}$ (h) $3 + \frac{5x}{3} < 2x + \frac{7}{2}$ (i) $-(x + 0.1) < 6\left(\frac{1}{2} - x\right)$
(j) $4 > \frac{5x}{6} - 6 \geq -6$ (k) $\frac{2}{5} < \frac{1}{2x} - 4 < \frac{3}{4}$ (l) $2x - 5 < \frac{9-3x}{2}$
(m) $4 - 5x < 2(x + 4)$ (n) $\frac{2-3x}{5} < x + 2$ (o) $3 < 2x - 5 < 7$

3 (a) Find the greatest integer value of x for which $4 - \frac{3x}{2} \geq 8x + 63$.

(b) Find the least integer value of x for which $\frac{5x}{3} + 3 < \frac{7}{2} + 2x$.

2.06 Linear Programming

EXTENDED CURRICULUM

Graphical representation of inequalities

Working rule

To represent inequalities on a square grid,
1) change the inequalities into equations,
2) then draw a line for each equation,
3) shade the unwanted region to get the not shaded solution region.

Remember

1. Draw a broken line for a strict inequality(i.e. > or <).
2. Draw a continuous line for a weak inequality(i.e. ≥ or ≤).
3. Use reverse inequality to shade unwanted region.
4. The line x = a is parallel to y-axis. 5. The line y =a is parallel to x-axis.

Example

On a grid draw the following inequalities and shade the unwanted region.

$x \geq 1, \quad y < 2, \quad 2x + 3y \leq 12, \quad y \geq 0.$

Working: $x \geq 1$

Change each inequality in to equality.

$x = 1$

$y < 2$

$y = 2$

$2x + 3y \leq 12$

$2x + 3y = 12$

x	0	6
y	4	0

Put x = 0 to get y = 4.

Put y = 0 to get x = 6.

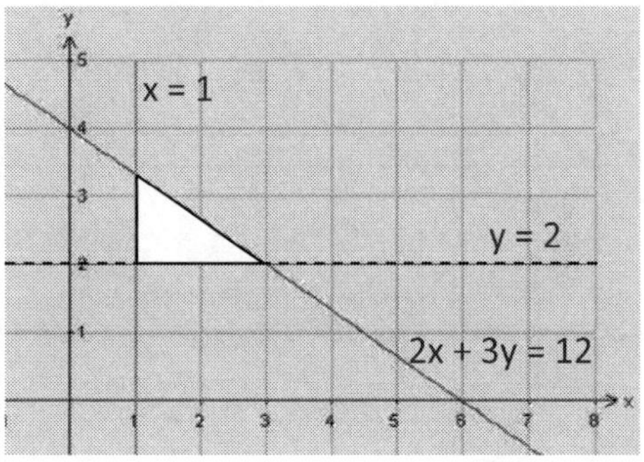

Solving linear programming problems

Remember

Inequality symbol	Phrases
>	Is greater than / is more than / must exceed
<	Is less than / is fewer than
≥	Is greater than or equal to / is not less than / is at least / minimum / Is not fewer than / must not be less than
≤	Is less than or equal to / is at most / must not exceed / is not greater than / is not more than / maximum

Working rule

In general, to solve a linear programming problem,
1) form inequalities using the information given in the question,
2) then represent the inequalities on a grid,
3) mark integer valued points in the solution region (except the points that are on the dotted line) and use them to get the possible solution (i.e. max value or min value).

Example

Hassan buys x text books and y note books.

(a) The total number of text books and note books must not be less than 5.
Write down an inequality to show this information.

(b) He buys at least twice the number of note books as text books.
The number of text books must be less than 5.
The number of note books must not exceed the number of text books.
Write down three inequalities using this information.

(c) (i) On a grid, represent the inequalities in parts (a) and (b). Shade the unwanted region.
(ii) Find all the possible integer values of x and y that satisfy the four inequalities.

Working: (a) $x + y \geq 5$

(b) $2y \geq x$

$x < 5$

$y \leq x$

Replace 'must not be less than' by $\geq$.
Replace 'at least' by $\geq$.
Replace 'less than' by $<$.
Replace 'must not exceed' by $\leq$.

(c) (i) $x + y \geq 5$

$x + y = 5$

Change each inequality in to equality.

x	0	5
y	5	0

Put $x = 0$ to get $y = 5$.
Put $y = 0$ to get $x = 5$.

$2y \geq x$

$2y = x$

x	4	3
y	2	6

Put $x = 4$ to get $y = 2$.
Put $y = 3$ to get $x = 6$.

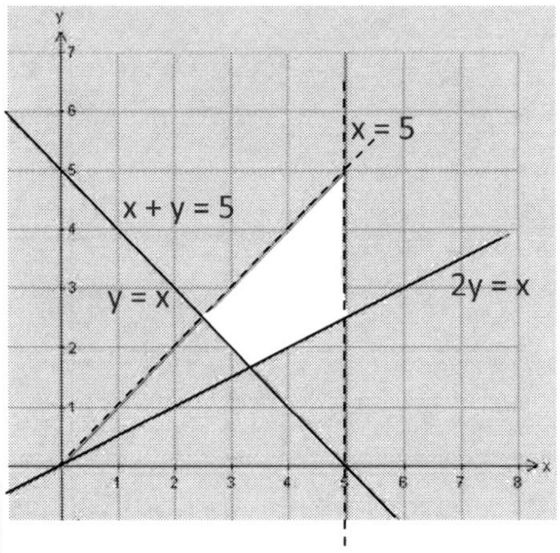

$x < 5$

So $x = 5$

$y < x$

$y = x$

x	4	5
y	4	5

Put $x = 4$ to get $y = 4$.
Put $y = 5$ to get $x = 5$.

(ii) All the integer valued points that satisfy the inequalities are (3, 2), (4, 2) and (4, 3).

Practise exercise 15

1 On a grid draw each set of inequalities and shade the unwanted region.
(a) $x \geq 1$, $x < 3$, $x + y \leq 4$.
(b) $y > x$, $x \leq 6$, $2x + y < 8$.
(c) $x \leq 3$, $2x + 1 \geq y$, $1 - y \leq 2x$.
(d) $x + 2y < 12$, $3x + y \leq 12$, $x \geq 0, y \geq 0$.
(e) $3x + 5y \geq 15$, $5x + 7y \leq 35$, $1 \leq y < 2$.
(f) $x + y > 1$, $5x + 4y \leq 20$, $1 < x \leq 3$.
(g) $10x + 13y \leq 130$, $25x - 12y > 6$, $x \leq 4y$.

2 x and y are two integers. The minimum sum of x and y is 3.
The sum of x and y is less than 6. The number x is more than 2.
The number y is greater than or equal to 0.
(a) Write down four inequalities using this information.
(b) On a grid, draw the inequalities in part(a) and shade the unwanted region.
(c) Write down three possible pairs of values for x and y that satisfy all the inequalities.

3 The total number of girls(g) and boys(b) in a Grade is less than 120.
There are at least 48 girls in the Grade. There are not more than 60 boys in the Grade.
(a) Write down three inequalities using the given information.

(b) Using a scale of 2 cm to represent 20 children on each axis draw three lines on a grid to represent the inequalities in part(a). Also shade the region on the grid, except the solution region.
(c) State the possible combination of girls and boys that satisfy all the inequalities.

4 Aminaa participates in a quiz competition. She answers x multiple choice questions and y descriptive questions.
(a) The maximum time to answer each multiple choice question is 2 minutes and each descriptive question is 6 minutes.
The length of time for the competition is 1 hour.
Show that $x + 3y \leq 30$.
(b) The total number of questions is not more than 20.
Write down an inequality to show this.
(c) Represent the inequalities in parts(a) and (b) on a graph and shade the region that does not satisfy the inequalities.
(d) In the competition, each multiple choice question carries one point and each descriptive question carries five points.
(i) Find the maximum possible points she can score in the competition.
(ii) Write down the number of descriptive questions she must answer to score maximum points. (Assume that her all answers are correct.)

5 A manufacturer produces two types toys A and B on a particular day.
He produces at most 1000 toys of both types. The number of type-A toys must exceed 400.
The number of type-B toys is fewer than 350.
He produces x number of type- A toys and y number of type-B toys.
(a) Write down three inequalities to show the above information.
(b) Draw these inequalities on a graph shading the unwanted region.
(c) Mark with dots on the graph two possible combination of number of toys which satisfy all the inequalities.
(d) By selling the toys the manufacturer makes a profit of \$ 8 from type-A toys and \$15 from type-B toys.
Find the number of toys of each type the manufacturer must produce to make largest profit on that day.

6 A vegetable merchant sells vegetable packets of green peas and beans.
A small packet contains 200 grams of green peas and 300 grams of beans.
A big packet contains 400 grams of green peas and 500 grams of beans.
The total mass of green peas exceeds 12 kilograms.
The total mass of beans must not exceed 30 kilograms.
The merchant sells p packets of green peas and b packets of beans.
(a) Using the given information, write down two inequalities in terms of p and b.
(b) (i) Using a scale of 1 cm to represent 10 grams on each axis, draw a p-axis for $0 \leq p \leq 100$ and a b-axis for $0 \leq b \leq 60$.
(ii) Draw two lines on your graph to represent the inequalities in part(a).
Shade the unwanted parts of the graph.
(c) The merchant gets a profit of \$ 0.85 on a small packet and of \$ 1.45 on a big packet.
Find (i) the maximum profit, (ii) the minimum profit,
(iii) the difference between the maximum and the minimum profit.

2.07 Functions ⟵ EXTENDED CURRICULUM ⟶

- When each member of a set is related to exactly one member of another set, the first set is called a function to the second set.

E.g. The set of students in a class is a function to the set of chairs they sit in that class.

- **Notation of a function:** A function can be described using f(x) notation or mapping notation.

E.g. (a) $f(x) = 2x - 5$ [Read as 'f of x equals 2x – 5]

(b) $f : x \longrightarrow 2x - 5$ [Read as 'f maps x onto 2x – 5]

Finding the value of a function

Working rule

To evaluate a function for specific number or expression,
1) replace the variable(e.g. x) by the given number or exprcssion,
2) then simplify.

Example

If $f(x) = x^2 + 4$ and $g : x \longrightarrow \dfrac{2-x}{x}$, $(x \neq 0)$, find

(a) f(3), (b) f(2x), (c) g(–4), (d) $g\left(\dfrac{1}{8}\right)$.

Working: (a) $f(3) = 3^2 + 4 = 13$

(b) $f(2x) = (2x)^2 + 4 = 4x^2 + 4$

(c) $g(-4) = \dfrac{2--4}{-4} = -1.5$

(d) $g\left(\dfrac{1}{8}\right) = \dfrac{2-\dfrac{1}{8}}{\dfrac{1}{8}} = 15$

Replace x by 3 in f(x).

Replace x by 2x in f(x).

Replace x by –4 in g(x).

Replace x by $\dfrac{1}{8}$ in g(x).

Finding the value of x

Working rule

When two functions or a function and its functional value are given, to find the value of x,
1) set an equation using the condition, 2) then solve the resulting equation for x.

Remember

1. While solving equations, take care of negatives.
2. To solve a quadratic equation, factorization method is more efficient than quadratic formula.

Example

If $f(x) = x^2 - 7x + 8$ and $g(x) = 3x - 16$, then solve

(a) $f(x) = -2$, (b) $g(x) = 0$, (c) $f(x) = g(x)$.

Working: (a)

$$f(x) = -2$$
$$x^2 - 7x + 8 = -2$$
$$x^2 - 7x + 10 = 0$$
$$x^2 - 2x - 5x + 10 = 0$$
$$x(x-2) - 5(x-2) = 0$$
$$(x-2)(x-5) = 0$$

Replace f(x) by $x^2 - 7x + 8$ to set an equation.

Use factorization method to solve the quadratic equation.

Don't do this. Finding f(–2) for f(x) = –2

$(x - 2) = 0$ or $(x - 5) = 0$
$x = 2$ or $x = 5$

(b) $g(x) = 0$
$3x - 16 = 0$
$3x = 16$
$x = 5\frac{1}{3}$

Replace g(x) by 3x – 16 to set an equation.

Solve the simple equation.

(c) $f(x) = g(x)$
$x^2 - 7x + 8 = 3x - 16$
$- 10x + 24 = 0$
$x - 6x + 24 = 0$
$(x - 4) - 6(x - 4) = 0$
$(x - 4)(x - 6) = 0$
$4) = 0$ or $(x - 6) = 0$
$x = 4$ or $x = 6$

Replace f(x) by $x^2 - 7x + 8$ and g(x) by 3x – 16 to set an equation.

Use factorization method to solve the quadratic equation.

Practise exercise 16

1 If $f(x) = 6x - 7$, find the following.

(a) $f(0)$ (b) $f(2)$ (c) $f(-5)$ (d) $f\left(\frac{1}{3}\right)$ (e) $f(4) - f(0.4)$ (f) $\frac{f(-1)+f(1)}{f(0)}$

2 Given $g(x) = \frac{x}{x+1}$, (Where $x \neq -1$). Find the following.

(a) $g(-x)$ (b) $g(2x)$ (c) $g\left(\frac{x}{3}\right)$ (d) $g(x^2)$ (e) $g(x + 1)$ (f) $g(2 - x)$

3 If $f : x \longrightarrow x^2 + 1$ and $g : x \longrightarrow \frac{2-3x}{3}$, find the values of

(a) $f(-5) \times g(5)$, (b) $f(\sqrt{3}) - 2g\left(-\frac{1}{3}\right)$, (c) $\frac{f(-7)}{f(-2)}$.

4 If $h(x) = x(x - 2)$ then show that $h(-1) = h(3)$.

5 If $f(x) = x^2 - 5$ and $g(x) = x + 1$, then solve the following.

(a) $f(x) = 4$ (b) $g(x) = -1$ (c) $f(x) = g(x)$
(d) $f(x) = g(619)$ (e) $f(x) = 7 - 4x$ (f) $g(x) = x^2 - 19$

6 A function k is given by the formula $k(x) = 400 - 64x^2$.
Find the values of x, such that (a) $k(x) = 0$, (b) $k(x) = -41$.

7 $g: x \longrightarrow \frac{(x+1)(x+2)}{x-2}$, for $x \neq 2$ and $h: x \longrightarrow x + 2$. Solve the following.

(i) $g(x) = 0$ (ii) $h(x) = x^2 - 10$ (iii) $g(x) = h(x)$

8 $f: x \longrightarrow \frac{24-10x-x^2}{8x}$, for $x \neq 0$.

(a) Evaluate (i) $f\left(\frac{1}{2}\right)$, (ii) $f\left(-\frac{1}{4}\right)$, (iii) $f\left(1\frac{1}{4}\right)$, (iv) $f(-0.4)$.

(b) Explain what will happen when $x = 0$? (c) Solve (i) $f(x) = \frac{1}{8}$, (ii) $f(x) = -1$.

9 $f(x) = 3^x$ and $g(x) = x^x$.
(a) Calculate
(i) f(0), (ii) f(7), (iii) g(6), (iv) g(–2), (v) f(–5) as a fraction.
(b) Find the value of x, for which
(i) g(x) = f(x), (ii) f(x) = 1, (iii) g(x) = 256, (iv) f(x) = 729.
(c) (i) Explain why g(– 0.5) does not give a real value?
(ii) Explain why g(0) cannot be found out?

10 $f(x) = \frac{x+3}{x}$ $x \neq 0$. (a) Calculate $f\left(\frac{1}{8}\right)$. (a) Solve $f(x) = \frac{1}{8}$.

Inverse function

- The reverse effect of a function is called its inverse.

Working rule

When f(x) is given, to find $f^{-1}(x)$, 1) first put y = f(x),
2) then rearrange the resulting equation for x,
3) at last replace x by $f^{-1}(x)$ and replace y by x.

Example
Let f(x) = 2 – 7x. Find (a) $f^{-1}(x)$, (b) $f^{-1}(-3x)$, (c) $f^{-1}(-5)$.
Working: (a) f(x) = 2 – 7x

y = 2 – 7x

$7x = 2 - y$ or $x = \frac{2-y}{7}$

$f^{-1}(x) = \frac{2-x}{7}$

Put y = f(x).
Rearrange for x.

Replace x by $f^{-1}(x)$ and replace y by x.

(b) From part(a), we have

$f^{-1}(x) = \frac{2-x}{7}$

Replace x by –3x.

$f^{-1}(-3x) = \frac{2--3x}{7} = \frac{2+3x}{7}$

(c) From part(a), we have

$f^{-1}(x) = \frac{2-x}{7}$ or $f^{-1}(-5) = \frac{2--5}{7} = 1$

Replace x by –5.

Alternate method f(x) = 2 – 7x
–5 = 2 – 7x
7x = 2 + 5
x = 1
So $f^{-1}(-5) = 1$

Replace f(x) by –5.
Solve the resulting equation for x.
Replace x by $f^{-1}(-5)$.

Composite function

- A composite function can be formed by combining two or more given functions.
- **Notation of a composite function:** If f(x) and g(x) are two functions, then fg(x) and gf(x) are two composite functions.

Remember

1. $fg(x) = f(g(x))$ and $gf(x) = g(f(x))$. 2. $fg(x) \neq f(x)\,g(x)$ and $gf(x) \neq g(x)\,f(x)$.
3. In general, $fg(x) \neq gf(x)$.

Example

Given $f(x) = 3x + 1$ and $g(x) = x^2 - 2$. Find (a) $fg(x)$, (b) $gf(x)$, (c) $gf(-4)$, (d) $(fg)^{-1}(x)$.

Working: (a) $fg(x) = f(g(x))$
$= f(x^2 - 2)$
$= 3(x^2 - 2) + 1$
$= 3x^2 - 6 + 1$
$= 3x^2 - 5$

Replace $g(x)$ by $x^2 - 2$ and then x in $f(x)$ by $x^2 - 2$. Simplify the expression.

Don't do this. Expanding bracket as $3x^2 - 2 + 1$

(b) $gf(x) = g(f(x))$
$= g(3x + 1)$
$= (3x + 1)^2 - 2$
$= 9x^2 + 6x + 1 - 2$
$= 9x^2 + 6x - 1$

Replace $f(x)$ by $3x + 1$ and then x in $g(x)$ by $3x + 1$. Simplify the expression.

Don't do this. Expanding $(3x + 1)^2$ as $9x^2 + 1$

(c) $gf(-4) = g(f(-4))$
$= g(3 \times -4 + 1)$
$= g(-11)$
$= (-11)^2 - 2$
$= 119$

Replace x in $f(x)$ by -4 and then x in $g(x)$ by -11. Simplify the expression.

Don't do this. Writing $(-11)^2 = -121$

(d) From part(a), we have
$fg(x) = 3x^2 - 5$
$y = 3x^2 - 5$
$3x^2 = y + 5$
$x = \sqrt{\dfrac{y+5}{3}}$

So $(fg)^{-1}(x) = \sqrt{\dfrac{y+5}{3}}$

Put $y = fg(x)$.
Rearrange the resulting equation for x.
Replace x by $(fg)^{-1}(x)$ and replace y by x.

Don't do this. $(fg)^{-1}(x) = \dfrac{\sqrt{y+5}}{3}$

Practise exercise 17

1 Let $f(x) = 4x - 15$. Find (a) $f^{-1}(x)$, (b) $f^{-1}(9)$, (c) $f^{-1}(-3)$,
(d) $f^{-1}(2a)$, (e) $f^{-1}\left(\dfrac{x}{4}\right)$, (f) $f^{-1}(x - 7)$.

2 For each of the following functions, evaluate (i) $g^{-1}(x)$, (ii) $g^{-1}(8)$.
(a) $g(x) = x + 4$ (b) $g(x) = 2 - x$ (c) $g(x) = 3x - \dfrac{1}{2}$
(d) $g(x) = 10 - 5x$ (e) $g : x \rightarrow \dfrac{x+6}{3}$ (f) $g : x \rightarrow \dfrac{x-7}{x}$, $x \neq 0$
(g) $g : x \rightarrow 12(1 - 3x)$ (h) $g : x \rightarrow x^2 - 9$ (i) $g(x) = x^4$

3 Given $f(x) = 5x - 1$ and $g(x) = x + 3$.
Find (a) $fg(x)$, (b) $gf(x)$, (c) $fg(0)$, (d) $gf(-2)$.

4 Let $h(x) = 6 - x$ and $k(x) = x^2$. Find (a) $hk(x)$, (b) $hk(-1.5)$, (c) $kh(x)$, (d) $kh\left(\frac{1}{2}\right)$.

5 $g: x \longrightarrow 1 - 2x$ and $h: x \longrightarrow x + 5$. Find (a) $g^{-1}(x)$, (b) $g^{-1}h(x)$, (c) $g(h^{-1}(x))$, (d) $(gh)^{-1}(x)$, (e) the value of x, when $g(x) = h^{-1}(x)$.

6 Let $f: x \longrightarrow 8x - 25$.

(a) Find (i) $f^{-1}(x)$, (ii) $f(f(x))$, (iii) $f(f(0))$, (iv) $f^{-1}\left(\frac{3}{8}\right)$.

(b) Solve $ff(x) = -33$. (c) Show that $f^{-1}f(x) = ff^{-1}(x)$.

7 $f(x) = x + 3$ and $g(x) = x - 2$. (a) Find (i) $fg(9)$, (ii) $gf(-7)$.

(b) Solve $f^{-1}(x) = \frac{-x^2}{4}$. (c) Show that $(gf)^{-1} = g^{-1}f^{-1}(x)$.

8 Let $p(x) = 3x - 7$ and $q(x) = x^2 - 3x + 2$.
(a) Find $p^{-1}(x)$. (b) Solve $p(x) = q(x)$. (c) Find the value of $pq(-2) - qp(-3)$.

9 The function f(x) is given by $f(x) = 9x - 5$. Find in its simplest form
(a) $f^{-1}f(x)$, (b) $ff(x)$.

10 $f(x) = 3x + 1$, $g(x) = \frac{2}{x} + 1$, $h(x) = 2^x$.

(a) Find the value of $fg(4)$. (b) Find $g^{-1}(2)$.

(c) Find $hh(3)$. (d) Find x when $f(x) = g\left(-\frac{4}{7}\right)$.

EXTENDED CURRICULUM

2.08 Sets

- A set can be described in word form or in list form or in rule form.
  1) In word form, a set is described in words.
     E.g. A = {The set of integers between –3 and 5}.
  2) In list form, the elements of a set are separated by commas and listed within curly brackets. E.g. A = {–2, –1, 0, 1, 2, 3, 4}.
  3) In rule form, the elements of a set are stated within curly bracket according to their common property. E.g. A = {x : x is an integer and $-3 < x < 5$}.

Notation and language of sets

- **Elements:**

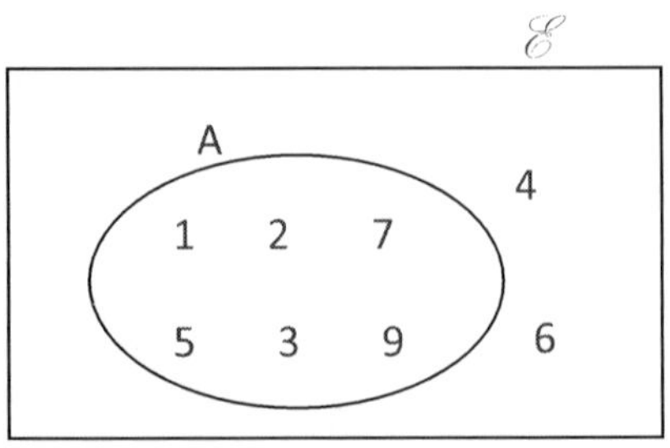

Diagram 1

In the adjacent diagram 1,
1) 7 is an element of set A.
   We say that '7 belongs to A'
   and write as $7 \in A$.
2) 4 is not an element of set A.
   We say that '4 does not belong to set A'
   and write as $4 \notin A$.
3) The total number of elements in set A is 6. We write this as n(A) = 6.
4) **Universal set,** $\mathscr{E}$ = {1, 2, 3, 4, 5, 6, 7, 9}

- **Relationship between two sets:** Let A and B are two sets.

1) **Subset ($A \subseteq B$): $A \subseteq B$** means all the elements of A are in B and all the elements of B may be in A.

Here **$A \subseteq B$.**

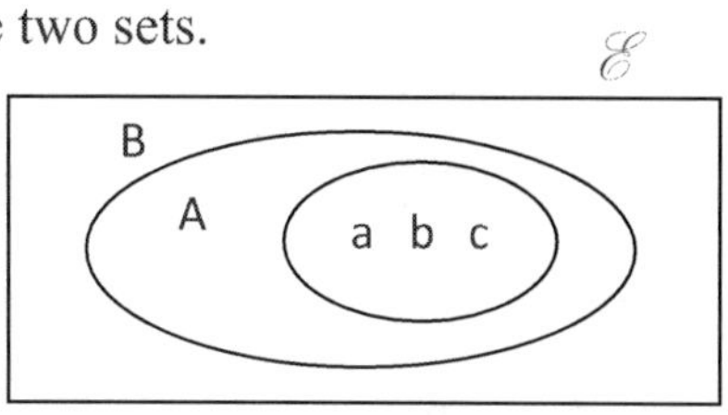

2) **Not a subset($A \not\subseteq B$): $A \not\subseteq B$** means at least one element of A is not in B. That is A and B are either intersecting or disjoint.

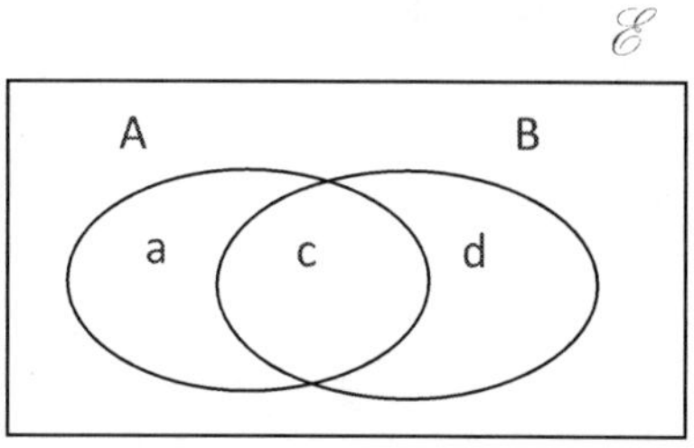

3) **Proper subset($A \subset B$): $A \subset B$** means all the elements of A are in B and at least one element of B must not be in A.

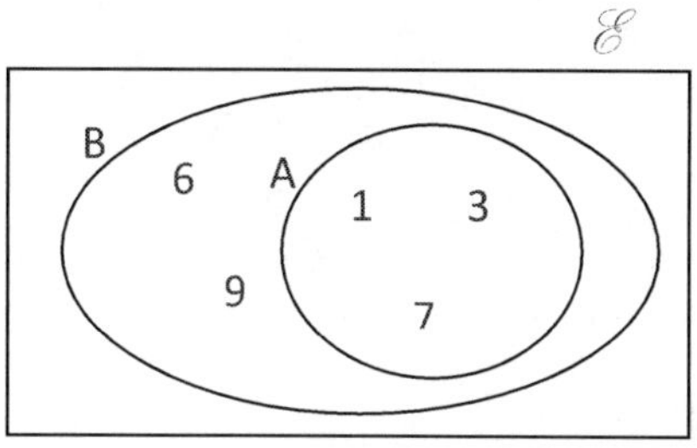

4) **Not a proper subset($A \not\subset B$):**
$A \not\subset B$ means at least one element of A is not in B or all the elements of B are in A.

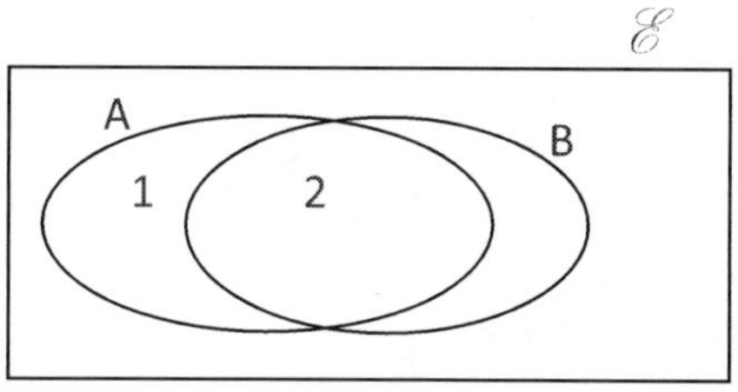

5) **Super set($B \supset A$): $B \supset A$** means **$A \subset B$.**

6) **Union of sets($A \cup B$):**
$A \cup B$ means the set of all the elements present in both A and B.
$A \cup B$ = {a, b, c, d, e} and n(**$A \cup B$**) = 5.

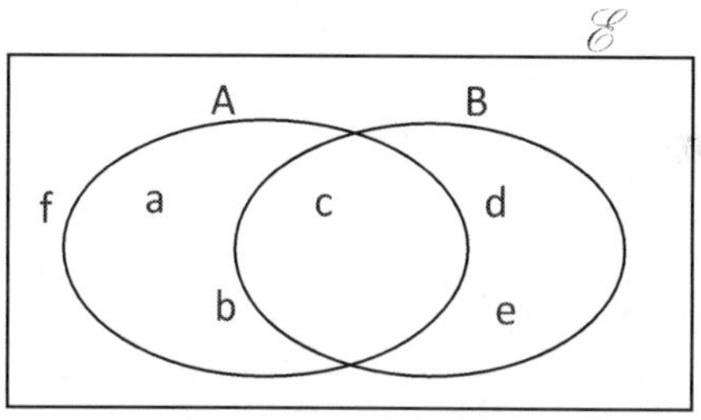

Diagram 2

7) **Intersection of sets($A \cap B$):**
$A \cap B$ is the set of common elements present in both A and B.
In diagram 2, **$A \cap B$** = {c} and n(**$A \cap B$**) = 1.

8) **Complement of a set(A^1):**
A^1 is the set of elements present in universal set, but not in A.
In diagram 2, **A^1** = {d, e, f} and n(**A^1**) = 3.
Also B^1 = {a, b, f} and n(B^1) = 3.

Remember

1. If $A \subseteq B$ and $B \subseteq A$, then A = B.	2. If $A \subset B$ and $B \subset C$, then $A \subset C$.	
3. $A \cup B = B \cup A$	4. $A \cap B = B \cap A$	5. $A \cup A^1 = \mathbf{E}$
6. $A \cap A^1 = \mathbf{Ø}$	7. $A \cup \mathbf{Ø} = A$	8. $A \cap \mathbf{Ø} = \mathbf{Ø}$
9. $A \cup \mathbf{E} = \mathbf{E}$	10. $A \cap \mathbf{E} = A$	11. $A \cup A = A$
12. $A \cap A = A$	13. $(A \cap B)^1 = A^1 \cup B^1$	14. $(A \cap B)^1 = A^1 \cap B^1$

Example

The Venn diagram shows the sets A, B and $\mathcal{E}$.

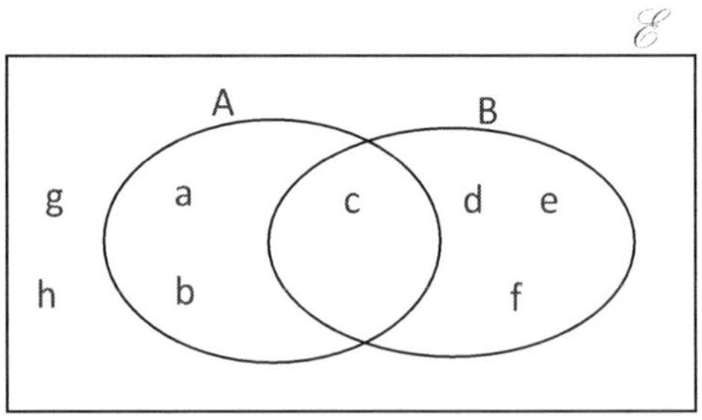

(a) List all the elements of

(i) A, (ii) B, (iii) $\mathcal{E}$,

(iv) $A \cup B$, (v) $A \cap B$, (vi) B^1,

(vii) $A \cap B^1$.

(b) Write down the value of

(i) n(A), (ii) n(B),

(v) $n(A^1 \cap B)$, (vi) $n(A \cup B)^1$,

(vii) $n(A \cap B)^1$.

Working: (a) (i) A = {a, b, c}

(ii) B = {c, d, e, f}

(iii) $\mathcal{E}$ = {a, b, c, d, e, f, g, h}

(iv) $A \cup B$ = {a, b, c, d, e, f}

(v) $A \cap B$ = {c}

(vi) B^1 = {a, b, g, h}

(vii) $A \cap B^1$ = {a, b}

All the elements present in A, B and outside of A and B.
All the elements present in A or B or both.
Common elements present in A and B.
Elements not present in B but in $\mathcal{E}$
Elements present in A but not in B.

(b) (i) n(A) = 3

(ii) n(B) = 4

(iii) n($\mathcal{E}$) = 8

(iv) $n(A^1) = 5$

(v) $n(A^1 \cap B) = 3$

(vi) $n(A \cup B)^1 = 2$

(vii) $n(A \cap B)^1 = 7$

The number of elements present in A is 3.
The number of elements present in B is 4.
The number of elements present in $\mathcal{E}$ is 8.
The number of elements not present in A is 5.
The number of elements not present in A but present in B is 3.
The number of elements not present in $A \cup B$ is 2.
The number of elements not present in $A \cap B$ is 7.

Practise exercise 18

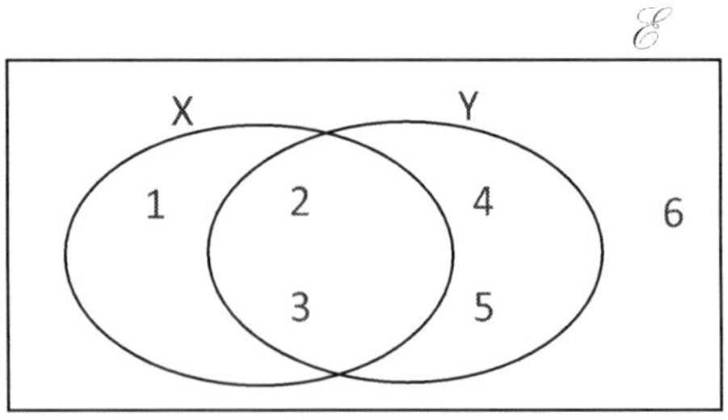

1 Using the appropriate set notation symbols $\in$, $\notin$, $\subset$ or $\not\subset$, fill in the blanks.

(a) 1 _______ X. (b) 2 ________ Y.

(c) 6 _______ X. (d) {1, 2} ______ Y.

(e) {1, 2, 3} ________ X

(f) Ø ________ Y. (g) 2 ________ $(X \cap Y)$.

2 $\mathcal{E}$ = { 1, 2, 3, 4, 5, 6, 7, 8, 9, 10, 11, 12, 13, 14, 15}

Write down the elements of the following sets.

(a) A = {Even numbers} (b) B = {Odd numbers}

(c) P = {Prime numbers} (d) M = {Multiples of 5}

(e) F = {Factors of 12}

3 $\mathcal{E}$ = {10, 11, 12, 13, 14, 15, 16, 17},

A = { 10, 12, 14}, B ={ 12, 13, 14, 15}

(a) Find (i) $A \cup B$, (ii) $A \cap B$, (iii) A^1, (iv) B^1,

(v) $A^1 \cup B$, (vi) $A \cap B^1$, (vii) $(A \cup B)^1$, (viii) $A^1 \cap B^1$.

(b) State the value of

(i) n(A), (ii) n(B), (iii) n($\mathcal{E}$), (iv) $n(A \cup B)$, (v) $n(A \cap B)$,

(vi) $n(A^1)$, (vii) $n(A^1 \cup B)$, (viii) $n(A \cap B^1)$, (ix) $n(A^1 \cup B^1)$.

(c) Show that $n(A) + n(B) - n(A \cap B) = n(A \cup B)$.

4 The Venn diagram shows the sets P, Q and $\mathscr{E}$.

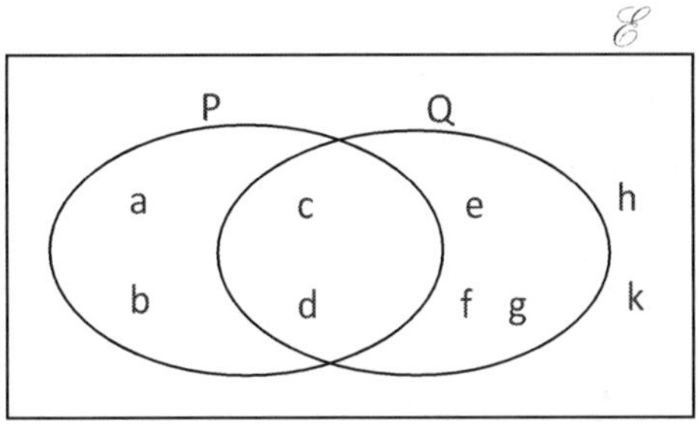

(a) List the members of each set.

(i) P	(ii) Q
(iii) $\mathscr{E}$	(iv) $P \cup Q$
(v) $P \cap Q$	(vi) P^1
(vii) Q^1	(viii) $P^1 \cup Q$
(ix) $P^1 \cup Q^1$	(x) $P^1 \cap Q^1$
(xi) $(P \cup Q)^1$	(xii) $(P \cap Q)^1$

(b) Write down the values of the following.

(i) n(P)	(ii) n(Q)	(iii) $n(\mathscr{E})$	(iv) $n(P \cup Q)$
(v) $n(P \cap Q)$	(vi) $n\left(P^1\right)$	(vii) $n(Q^1)$	(viii) $n(P^1 \cup Q)$
(ix) $n(P \cup Q^1)$	(x) $n(P^1 \cap Q)$	(xi) $n(P \cap Q^1)$	(xii) $n(P^1 \cup Q^1)$
(xiii) $n(P^1 \cap Q^1)$	(xiv) $n(P \cap Q)^1$	(xv) $n(P \cup Q)^1$	(xvi) $n(\mathscr{E}^1)$

5 A, B and C are three sets. Write the following sets in their simplest form.

(a) $A \cup \emptyset$	(b) $A \cap \emptyset$	(c) $A \cap A$
(d) $A \cup A$	(e) $(A^1)^1$	(f) $A \cup \mathscr{E}$
(g) $A \cap \mathscr{E}$	(h) $(A \cup B) \cap (A \cup C)$	(i) $(A \cap B) \cup (A \cap B)$
(j) $(A \cap B) \cup (A \cap B^1)$	(k) $(A \cap B) \cup (A^1 \cap B)$	(l) $(A \cap B) \cap (A^1 \cup B^1)$
(m) $A \cap A^1$	(n) $A \cup A^1$	

Venn diagrams

- Venn diagrams are used to represent sets in the form of pictures.

1) $A \subseteq B$ or $A \subset B$

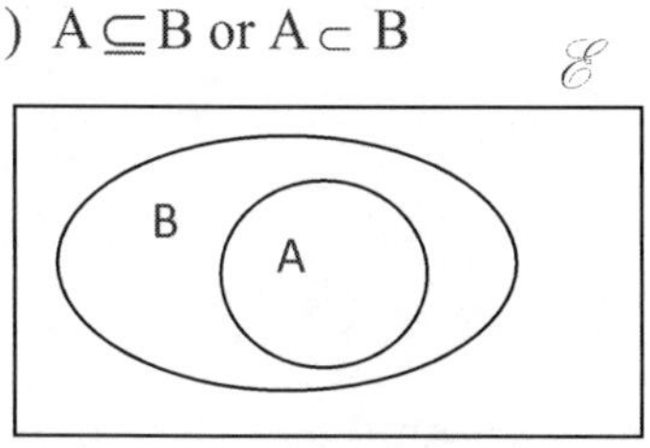

2) $A \cap B = \emptyset$

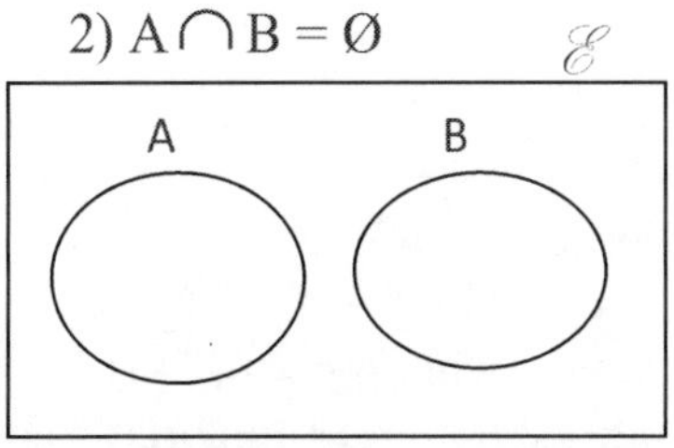

3) A^1

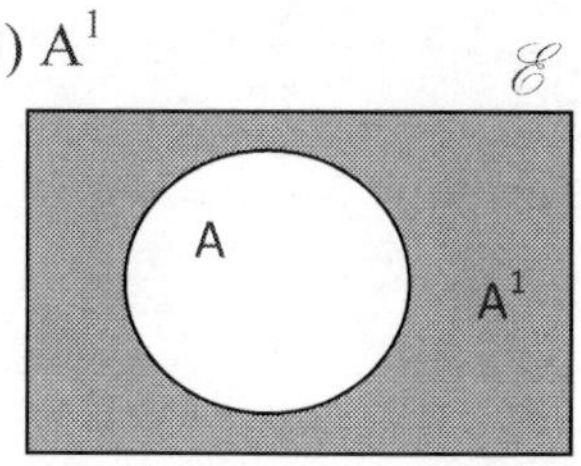

8) $A^1 \cup B$

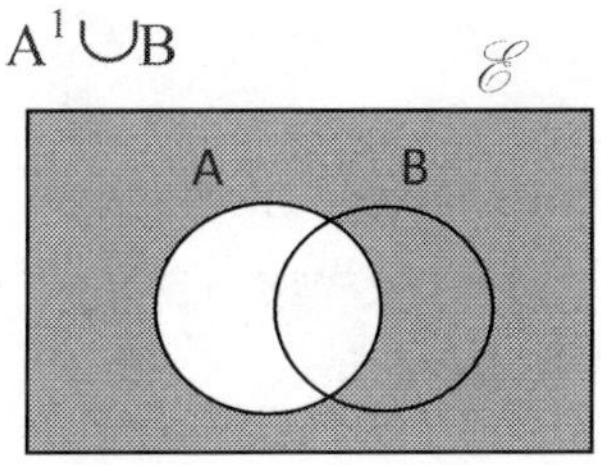

10) $A \cup A^1$

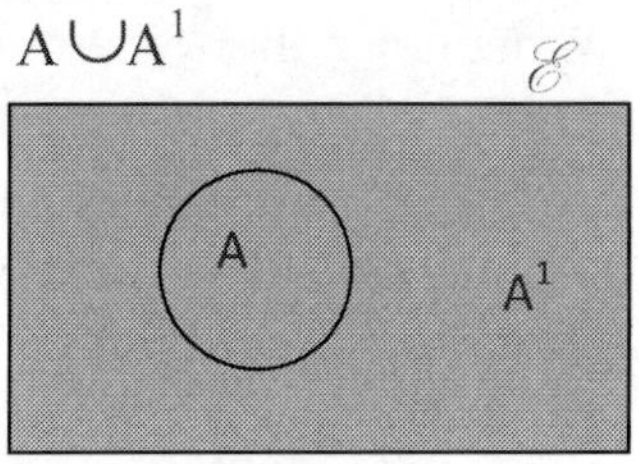

Application of sets

- The problems involving sets can be solved easily using Venn diagrams.

Remember

In a Venn diagram, the answer can be written in numeric form or algebraic form.

Example

In a group of 40 pupils, 24 like Mathematics (M), 18 like Science(S) and 7 like both subjects.

(a) Represent this information in a Venn diagram.

(b) Find the number of students who like

(i) only Science, (ii) neither of the subjects, (iii) Mathematics or Science or both.

(c) Write down the value of $n(M \cap S^1)$.

Working: (a)

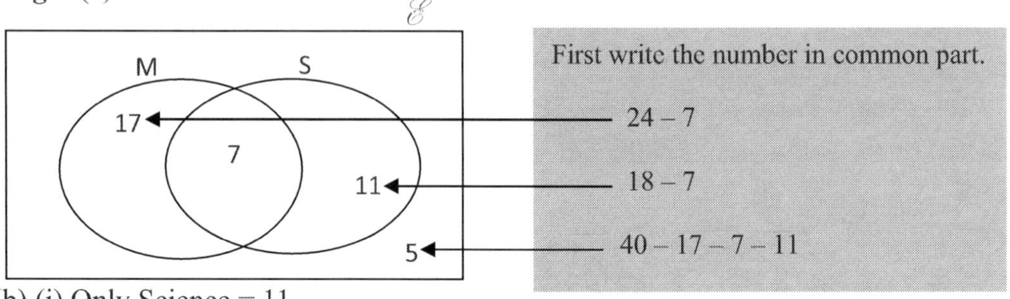

(b) (i) Only Science = 11

(ii) Neither of the subjects = 5

(iii) Mathematics or Science or both = 17 + 7 + 11 = 35

(c) $n(M \cap S^1) = 17$

$n(M \cap S^1)$ is the number of pupils who like M, but not S.

Practise exercise 19

1 In a class of 34 students, 19 like Physics, 18 like Chemistry and 6 like both subjects.

(a) Represent this information in a Venn diagram.

(b) Find the number of students who like

(i) only Physics, (ii) only Chemistry,

(iii) neither of the subjects, (iv) Physics or Chemistry or both.

2 In a group of 22 people, 13 take tea(T), 11 take coffee(C) and 3 take neither.

(a) Draw a Venn diagram to show this information.

(b) Find the number of people who take both tea and coffee.

(c) Write down the values of

(i) $n(T \cap C^1)$, (ii) $n(T^1 \cap C)$ (iii) $n(T \cap C)$.

3 There are 90 candidates appeared an examination. Exactly two-third of the candidates passed in Mathematics and one-third of the candidates passed in English.

17 candidates passed both in mathematics and English.

(a) Draw a Venn diagram to show this information.

(b) Find the number of candidates who passed

(i) just Mathematics, (ii) Just English, (iii) either Mathematics or English or both.

4 There are 200 people in a hostel. Exactly 40% of the people read news paper-A.

Exactly 35% of the people read news paper-B.

5% of the remaining people do not read either of the news papers.

(a) Find the number of people who do not read

(i) either of the news papers, (ii) both the news papers.

(b) Find the values of
(i) $n(A^1)$, (ii) $n(B^1)$, (iii) $n(A^1 \cap B)$, (iv) $n(A \cap B^1)$.

5 In a group of 181 children, 91 like apple, 83 like grape, 70 like orange.
Among them 28 like apple and orange, 30 like apple and grape, 26 like grape and orange.
(a) Draw a Venn diagram to show this information.
(b) Find the number of children who like all the three types of fruit.

2.09 Drawing graphs

Core Curriculum

Cartesian coordinate system

Remember

1. Each point on the coordinate system is an ordered pair.
2. In each ordered pair, the first element is x-coordinate and the second element is y-coordinate. E.g. In the ordered pair(2, 5), x-coordinate is 2 and y-coordinate is 5.
3. The line $y = 0$ is x-axis itself and the line $x = 0$ is y-axis itself.
4. Each axis may have different scales.

Practise exercise 20

1 Use suitable coordinate axes to plot each set of the following sets of points on a separate grid.
(a) (–2, 2), (–1, 1), (0, 0), (1, –1), (2, –2).
(b) (4.5, 2.1), (–3.7, 5.4), (6.2, –2.6), (–5.5, –4.9).
(c) (0, 7.25), (–3.75, 0), (4.37, 0.65), (–7.12, –3.46), (–5.15, 4.32).

2

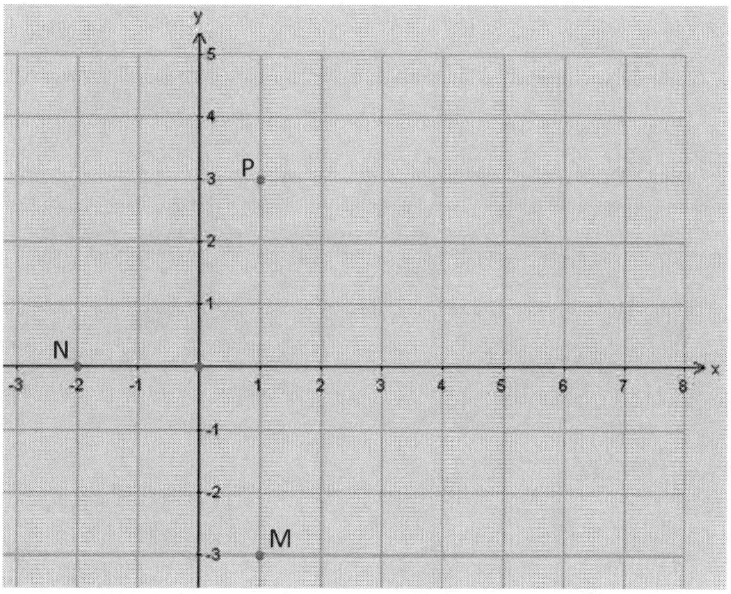

The points M, N and P are three vertices of a square MNPQ.
(a) Write down the coordinates of M, N and P.
(b) The fourth vertex of MNPQ is Q(4, 0). Mark the point Q on the same grid.
(c) Join the four points M, N, P and Q to form a square. Find the coordinates of the mid point of (i) MP, (ii) NP.

3 On a suitable grid, plot the points A(1, 2), B(–5, –4), C(–4, 0) and D(–1, –3).
Join the lines AB and CD.
Write down the coordinates of the point of intersection of both lines AB and CD.

4 (a) On a grid, plot the points A(0, 2), B(2, 2), C(4, 1) and D(–2, 1).
Use a scale of 1cm = 1 unit on both axes.
(b) Find the area of the quadrilateral ABCD.

Straight line graph: y = mx + c

- The graph of the linear equation 'y = mx + c' is always a straight line.

Remember

In the line y = mx + c,
'm' is the gradient and 'c' is the y-intercept(point at which the line touches y-axis).

Working rule

To draw the graph of an equation in the form y = mx + c,
1) Choose at least three values for x (if not given in the question).
2) Put the x-values ix the given equation to get respective y-values and draw up a table.
3) Plot the points on the grid and then join the points using a straight edge.

Example

Draw the graph of $y = -\frac{x}{4} + 2$.

Working: $y = -\frac{x}{4} + 2$

x	0	4	8
y	2	1	0

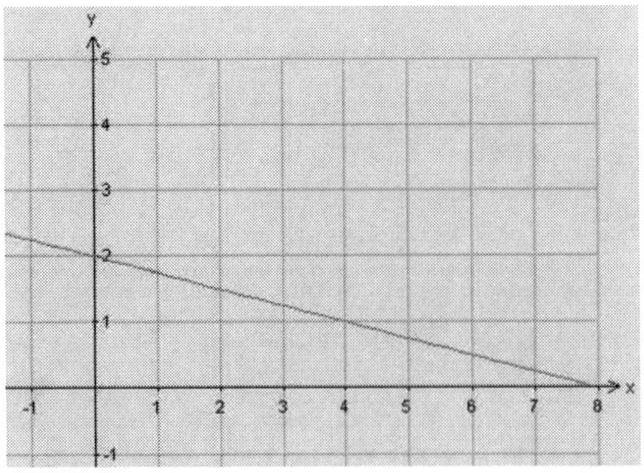

Choose x = 0, 4, 8.

Put the x-values in $y = -\frac{x}{4} + 2$ to get the respective y-values as 2, 1, 0.

Plot the points (0, 2), (4, 1) and (8, 0) and join the points straight way.

Practise exercise 21

1 Find the gradient and y-intercept of each line.

(a) $y = 2x + 3$ (b) $y = x - 1$ (c) $y = 6 - x$ (d) $2x = y - 7$

(e) $y = \frac{x}{4} + 8$ (f) $y = -7x + \frac{1}{2}$ (g) $\frac{y}{2} = 4 - 3x$ (h) $-y = 3x + 5$

(i) $y = 12 - 4x$ (j) $5y = 10 + 5x$ (k) $x = 5 - y$ (l) $9x = 3 - 27y$

2 Draw the graph of each line.

(a) $y = x$ (b) $y = -x$ (c) $y = 3x$ (d) $y = x - 4$
(e) $y = x + 7$ (f) $y = 5 - x$ (g) $y = 2 - 4x$ (h) $y = 3x + 2$
(i) $y = 12x - 5$ (j) $y = 100x - 25$ (k) $y = 1500 - 20x$ (l) $-y = 7 - 6x$

3 Plot the graph of each line.

(a) $4y = 2x + 3$ (b) $3x + 2y - 8 = 0$ (c) $6x = 18 - 6y$ (d) $2x + 4y - 6 = 0$
(e) $y = \frac{-x}{5} - 1$ (f) $y = 9 - \frac{2x}{3}$ (g) $\frac{y}{5} = 5 - 2x$ (h) $\frac{x}{8} - \frac{y}{4} = -2$

4 Draw a straight line that has the given gradient and y-intercept.

(i) Gradient = 3, y-intercept = 2. (ii) Gradient = $\frac{2}{5}$, y-intercept = – 3.

(iii) Gradient = $\frac{4}{3}$, y-intercept = (0, 3). (iv) Gradient = $-\frac{1}{2}$, y-intercept = (0, – 4).

Finding the gradient of a straight line

Working rule

To find the gradient from the graph of a given line,
1) Choose any two points on the line.
2) Then use the formula gradient, $m = \frac{\text{changein y - value}}{\text{changein x - value}}$.

Remember

1. The gradient of a line is positive, if it slopes foreward (like /).
2. The gradient of a line is negative, if it slopes backward (like \).
3. The gradient of x-axis(or any line parallel to x-axis) is zero.
4. The gradient of y-axis(or any line parallel to y-axis) is not defined.

Example

Find the gradient of line *l*.

Working: Gradient, $m = -\frac{2}{4}$

$= -0.5$

Use $m = \frac{\text{changein y - value}}{\text{changein x - value}}$.

Here sign of gradient is –, for back slope.

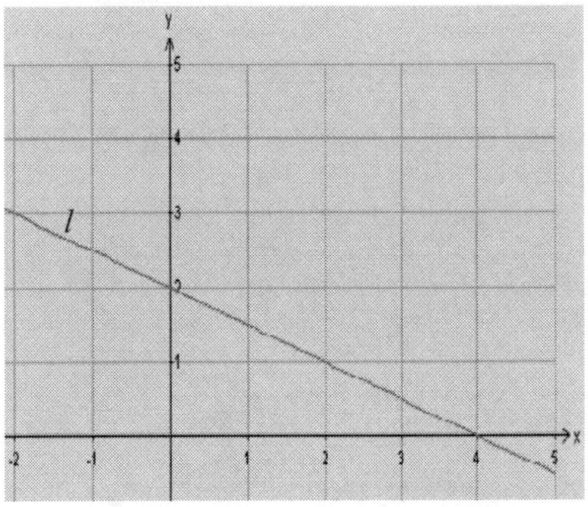

Finding its equation from a given line

Working rule

To find its equation from a given line,
1) first find the values of ‘c’(where the line touches the y-axis) and ‘m’ (gradient).
2) then substitute the values of c and m in the equation $y = mx + c$.

Example

Find the equation of the line shown on the grid.

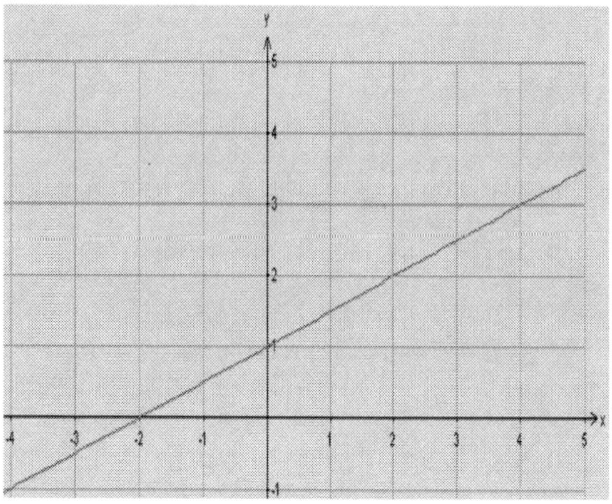

Working: Here c = 1

m = 1 ÷ 2 = 0.5

y = mx + c

y = 0.5x + 1

Use $m = \frac{\text{changein y - value}}{\text{changein x - value}}$.

Practise exercise 22

For each graph (a) find the gradient of each line, (b) find the equation of each line.

1

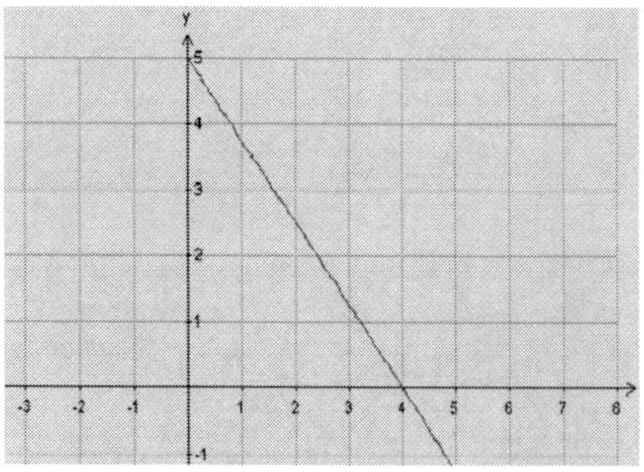

2

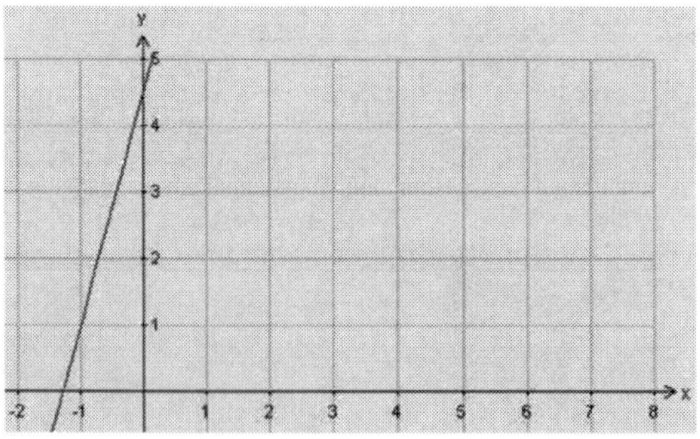

3

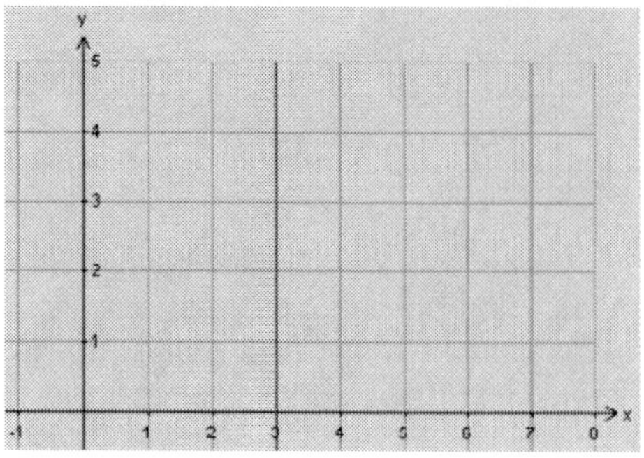

4

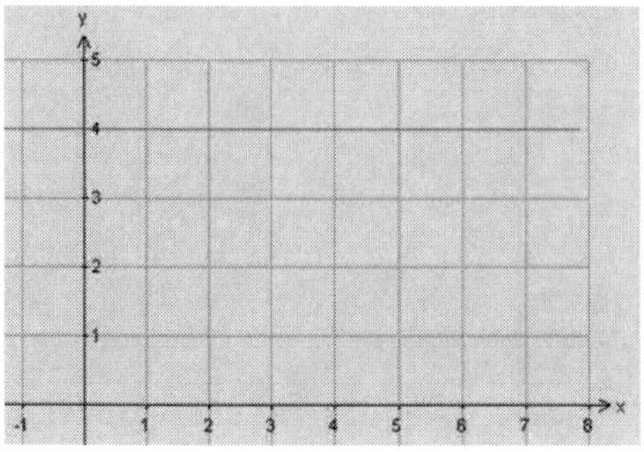

5

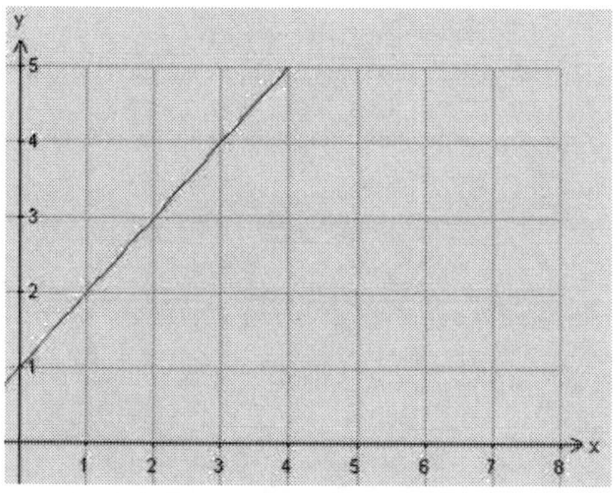

6

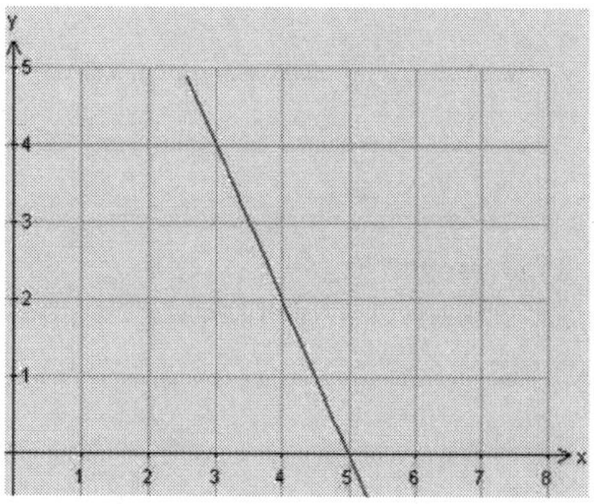

Finding the equation of a line parallel to a given line

► Parallel lines have same gradient.

Working rule

To find the equation of a line parallel to given line,
1) First find the gradient(m) form the given line.
2) Then use y = mx + c and the given things(y-intercept or any point) to obtain the required equation.

Example

Find the equation of the line parallel to y = 3x – 2 and passes through the point (1, 2).

Working: From y = 3x – 2, we have
gradient, m = 3.
y = mx + c passes through (1, 2).
2 = 3 x 1 + c
c = – 1

The required equation is y = 3x – 1.

Compare the gradients of y = 3x – 2 and y = mx + c.
Substitute x = 1, y = 2 and m = 3 in y = mx + c.
Solve the resulting equation for c.

Substitute m = 3 and c = – 1 in y = mx + c.

Practise exercise 23

1 Which of the following pairs of lines are parallel?
(a) y = 3x + 7, y = 3x – 9. (b) y = 5x – 10, y = 13 – 5x.
(b) 2x = y + 6, y = 15 + 6x. (d) y – 4x – 1= 0, 4x – 2y = 9.

2 Find the equation of the line with gradient 12 and passes through the point (6, 8).

3 Find the equation of a line parallel to y = 2x – 1 which passes through the point (2, 1).

4 Find the equation of a line parallel to 3x + 2y – 9 = 0 which passes through the point (–1, 5).

5 Find the equation of the line whose y-intercept is 5 and parallel to the line y = 6x – 2.

6 Find the equation of a line with y-intercept $\frac{-2}{3}$ and parallel to the line 8x = 3 – 4y.

7 Find the equation of the line parallel to x-axis which passes through the point (5, 2).

8 Find the equation of the line parallel to y = 10 – 5x and passes through the origin.

9 Find the equation of a line which passes through the pints (1, 3) and (5, 8).

10 Find the equation of a line which passes through the points (–3, 10) and (1, 2.5).

Solving a linear equation using graphs

- You can estimate the solution of a linear equation in the form $ax + b = c$ using the graph of $y = ax + b$.

Working rule

1) Draw the line $y = c$ on the given graph.
2) The x-value of the point of intersection of both the lines is the solution.

Example
On the grid, the graph shows $y = 2x + 1$. Using this graph, estimate the solution of $2x - 3 = 0$. Show all your working.

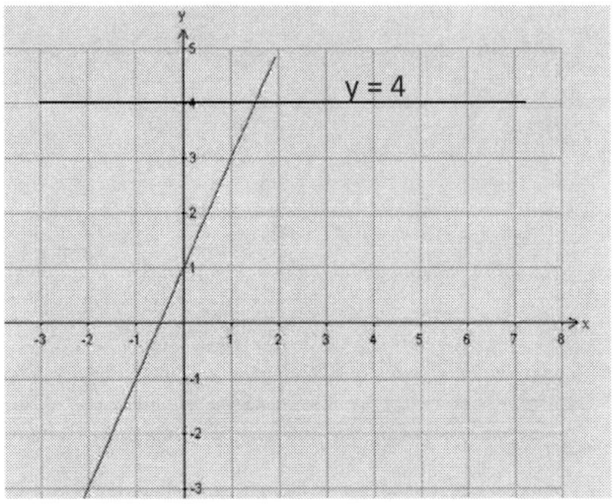

Working: $2x - 3 = 0$

$2x + 1 - 1 - 3 = 0$

$2x + 1 = 4$ or $y = 4$

Draw the line $y = 4$ on the same grid.

The point of intersection of both the lines is (1.5, 4).

So the estimated solution is $x = 1.5$.

Don't do this.
Lasr answer as $x = 1.5$ and $y = 4$.

Practise exercise 24

1 On the grid, the graph shows $y = 2x + 3$.

Using this graph, estimate the solution of $2x - 1= 0$. Show all your working.

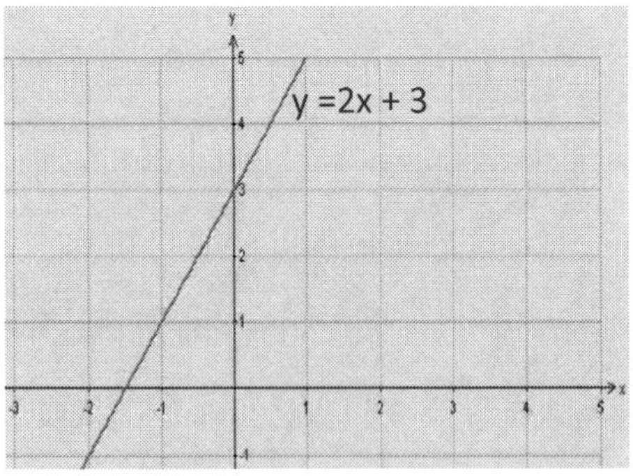

2 The graph shows the line $7y - 3 = 2x$.

Use this graph to estimate the solution of

$$\frac{2x}{7} - 2 = 0.$$

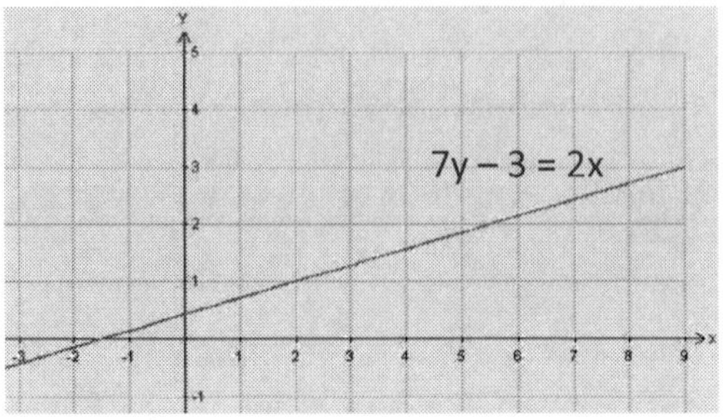

3 (a) Draw the graph of the equation y = 6 – x.
(b) Use your graph to solve (i) y = – 1, (ii) 3 – x = 0.

Solving simultaneous equations graphically

Working rule

To solve a pair of simultaneous equations by graphical method,
1) on the same grid, draw the graph of each of the equations,
2) look at the point of intersection of both the lines and write down as x = first coordinate, y = second coordinate.

Example

Use graphs to solve the simultaneous equations 3x + 2y = 9,
4x – y = 1.

Working: 3x + 2y = 9

X	0	1	2
Y	4.5	3	0

4x – y = 1

X	0	1	2
Y	–1	3	7

The point of intersection of both the lines is (1, 3).

So the solution is x = 1 and y = 3

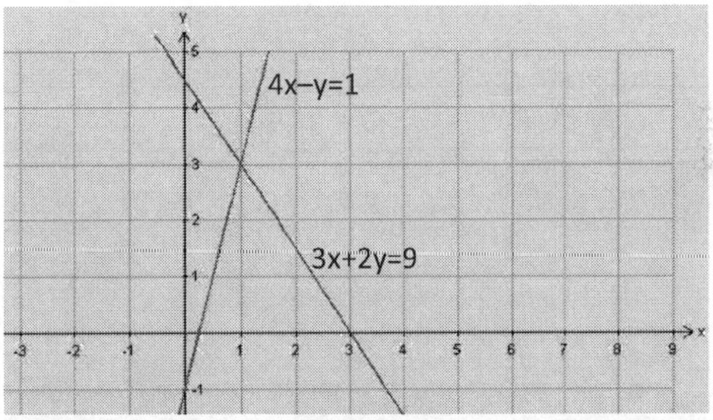

Draw the graphs of the lines 3x + 2y = 9 and 4x – y = 1 on the same grid.
Then locate the point of intersection of both the lines.

Practise exercise 25

1 Use graphs to solve the following simultaneous equations.
(a) y = x + 4, y = 3 – x.
(b) 3x – y = 3, 2x + y = 7.
(c) 4x – 3y = – 10, x + 2y = 3.
(d) x + y = 7, x – 2y = –2.
(e) 5x + y = – 15, x – 5y = 23.
(f) x – y = 1, 2x – 3y = 1.
(g) 4x + y = 54, 2x + 3y = 62.
(h) y = x + 1, $2x = y + \frac{1}{2}$.

2 (a) Draw the graph of y = x – 1.1 for the values of x from –2 to 2.
(b) On the same grid, draw the graph of y = 7.9 – x for the values of x from –2 to 2.
(c) Write down the point of intersection of both the graphs and the respective values of x and y.

3 There are x girls and y boys in a class. The number of boys is 4 more than the number of girls. Three times the number of girls minus the number of boys is 42.
(a) Write down two simultaneous equations in x and y to show the above information.
(b) On a suitable coordinate axes, draw the graphs of the equations in part(a).
(c) Use your graphs to find the solutions of the simultaneous equations.
(d) Use algebra to solve the simultaneous equations in part(a).

Quadratic graphs: $y = \pm x^2 + bx + c$

- The graph of a quadratic function $y = \pm x^2 + bx + c$ is a smooth curve(parabola).

Working rule

To draw the graph of the function $y = \pm x^2 + bx + c$,
1) choose at least five values for x (if not give in the question),
2) put the x-values in the given equation to get the respective y-values and draw up a table,
3) plot the points on the grid and then join the points with free hand (to get a smooth curve).

- An equation of the form $x^2 + bx + c = d$ can be solved using the graph of $y = x^2 + bx + c$.

Working rule

1) Draw the line $y = d$ on the same grid.
2) the x-values of the point of intersections of the line and the curve is the solution.

Remember

You must give the x-values of the point of intersection as solution only, not the coordinates of the intersections.

Example (a) Draw the graph of $y = 6 - 3x - x^2$ for $-5 \leq x \leq 2$.
Use 1 cm = 1 unit on x-axis and 1 cm = 2 units on y-axis.
(b) Using the graph of $y = 6 - 3x - x^2$, estimate the solution of the following equations.
(i) $6 - 3x - x^2 = 0$ (ii) $y = 4$ (iii) $6 - 5x - x^2 = 10$
(c) Draw the line of symmetry for the curve $y = 6 - 3x - x^2$.
Working: (a)

x	–5	–4	–3	–2	–1	0	1	2
y	–4	2	6	8	8	6	2	–4

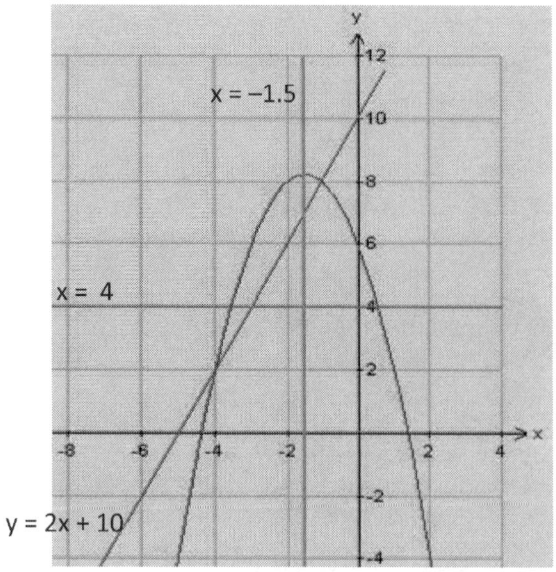

Plot all the points on the grid.

Draw a smooth single curve going through all the correct points with free hand.

Don't do this.
Joining the points in a series of straight lines instead of drawing a smooth curve.

(b) (i) $x = 1.5$ or $x = -4.5$ is the solution.

See the x-values where the curve crosses the x-axis. The line and the curve intersect at (0.5, 4) and (–3.5, 4).

(ii) Draw the line $y = 4$ on the same grid.
The solution is $x = 0.5$ or $x = -3.5$.

(iii) $6 - 5x - x^2 = 10$
$6 - 3x - 2x - x^2 = 10$
$6 - 3x - x^2 = 2x + 10$ or $y = 2x + 10$
Draw the line $y = 2x + 10$ on the same grid.
The solution is $x = -1$ or $x = -4$.

Rearrange $6 - 5x - x^2 = 10$ to get an equation in the form $y = d$.

The line and the curve intersect at (–1, 8) and (–4, 2).

(c) The line of symmetry for the curve is at $x = -1.5$.

Practise exercise 26

1 Draw the graph of each quadratic function.
(a) $y = x^2 + 1$ for $-4 \le x \le 4$
(b) $y = -x^2$ for $-4 \le x \le 4$
(c) $y = x^2 - 2x + 3$ for $-3 \le x \le 5$
(d) $f(x) = x^2 - 4x$ for $-1 \le x \le 5$
(e) $y = -x^2 + 4x - 7$ for $-1 \le x \le 5$
(f) $y = 4 + 3x - x^2$ for $-2 \le x \le 5$
(g) $y = 10 - 3x - x^2$ for $-6 \le x \le 3$
(h) $f(x) = 9 - x^2$ for $-4 \le x \le 4$

2 (a) On the same grid, draw the graphs of the functions $y = \frac{x^2}{2}$, $y = x^2$ and $y = 2x^2$.
(b) State the line of symmetry of all the three graphs.

3 (a) Draw the graph of $y = x^2 - 4x + 3$ for the values of x from – 1 to 5.
(b) Use your graph to solve the following equations.
(i) $x^2 - 4x + 3 = 0$
(ii) $y = 2$
(ii) $x^2 - 4x + 3 = 4.5$
(iv) $x^2 - 5x + 9 = 0$

4 (a) Complete the table of values for the function $y = 7 - 2x - x^2$.

x	–4	–3	–2	–1	0	1	2
y	–1					4	

(b) On a graph paper draw the graph of $y = 7 - 2x - x^2$.
(c) Use your graph to estimate the solution of the equation $y = 0$, correct to 1 decimal place.
(d) Complete the table below for $y = 3 - x$.

x	–5	0	5
y		3	

(e) By drawing the line $y = 3 - x$ on the same graph paper, find the approximate solution of both the equations.
(f) Find the gradient of the line $y = 3 - x$.

5 (a) On a suitable coordinate axes, draw the graphs of $y = x^2 - 10x + 26$ for $2 \le x \le 8$. and $y = x$ for $1 \le x \le 8$.
(b) Use your graphs to estimate, correct to 1 decimal place, the solution of $x^2 - 11x + 26 = 0$.
(c) On the same grid, draw the graph of $y = 5$.
(d) Write down the x-coordinates of the points of the intersections of the graphs of $y = x^2 - 10x + 26$ and $y = 5$.

6 (a) On the same grid, draw the curves $y = (1 - x)^2$ and $y = 1 + 2x - x^2$ for x-values from –2 to 4.

(b) Write down the points of intersections of both the curves.
(c) Draw the line $y = 2 - 2x$ for $-2 \le x \le 4$ on the same grid.
(i) Write down the exact solution of the curve $(1 + x)^2$ and the line $y = 2 - 2x$.
(ii) Write down the approximate solution of the curve $y = 1 + 2x - x^2$ and the line $y = 2 - 2x$, correct to one decimal place.
(d) Draw the line of symmetry of the curves. What can you say about the gradient of the line of symmetry?

Reciprocal graph: $y = \frac{a}{x}$, $(x \neq 0)$

- The graph of the reciprocal function $y = \frac{a}{x}$ has two curves (called hyperbola) lie on opposite quadrants.

Remember

1. The graph $y = \frac{a}{x}$ has two lines of symmetry with equations are $y = x$ and $y = -x$.
2. The graph has rotational symmetry of order 2 about the origin.
3. When $a > 0$, the two sections of the graph lie on first and third quadrant.
4. When $a < 0$, the two sections of the graph lie on second and fourth quadrant.
5. Do not join the two sections of the curve together through the axes.
6. Do not take $x = 0$ to evaluate y-value, since division by 0 is not defined.

Example

(a) Draw a table of values for the function $y = \frac{6}{x}$ for $-6 \le x \le 6$ except $x \neq 0$.

(b) Draw the graph of $y = \frac{6}{x}$ for $-6 \le x \le 6$.

(c) Use your graph to
(i) find the value of x when $y = 5$, (ii) solve the equation $5x + 72 = 0$.

(d) What are the equations of the two lines of symmetry of the graph $y = \frac{6}{x}$.

Working: (a)

x	–6	–5	–4	–3	–2	–1	1	2	3	4	5	6
y	–1	–1.2	–1.5	–2	–3	–6	6	3	2	1.5	1.2	1

(b)

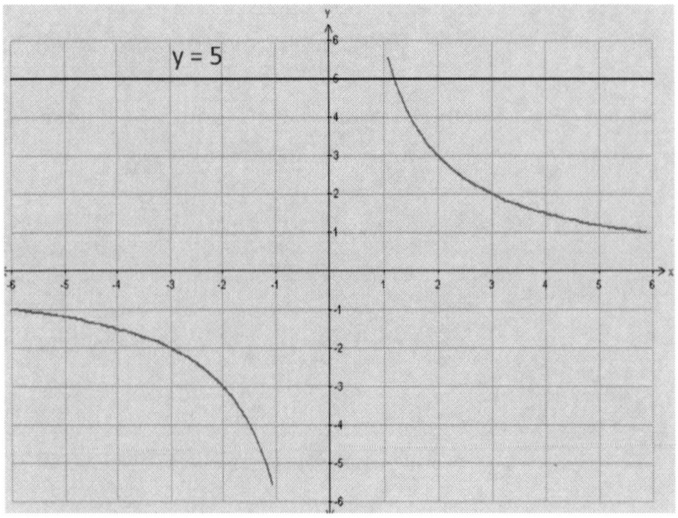

(c) (i) Draw the line y = 5 on the same grid.
So when y = 5 x = 1.2

The x-value of the point of intersection of the graphs of $y = \frac{6}{x}$ and y = 5 is 1.2.

(ii) 5x + 72 = 0
x = –14.4
$\frac{x}{6} = -2.4$ or $y = -2.4$
Draw the line y = – 2.4 on the same grid.
So the solution is x = 2.5.

Rearrange 5x + 72 = 0 to get an equation in the form y = constant.

The x-value of the point of intersection of the graphs of $y = \frac{6}{x}$ and y = – 2.4 is 2.5.

(d) The two lines of symmetry for the graphs are y = x and y = – x.

Practise exercise 27

1 Draw the graph of each reciprocal function.
(a) $y = \frac{3}{x}$ for $-6 \le x \le 6$ (b) $y = \frac{8}{x}$ for $10 \le x \le 20$ (c)
(c) $y = \frac{-6}{x}$ for $-6 \le x \le 1$ and $1 \le x \le 6$ (d) $y = \frac{21}{-x}$ for $-8 \le x \le -1$
(e) xy = 300 for x = –60, –50, –40, –30, –20, –10, 10, 20, 30, 40, 50, 60.

2 (a) Draw a table of values for the function $y = \frac{12}{x}$ for the values of x from –6 to 6.
(b) Draw the graph of $y = \frac{12}{x}$ for $-6 \le x \le 6$.
(c) Use your graph to
(i) find the value of x when y = 2.5, (ii) solve the equation 23x + 60 = 0.
(d) Write down the order of rotational symmetry about the origin of the graph.

3 (a) Draw the graph of the function xy = 60 for $-8 \le x \le 8$.
(b) On the same grid, draw the line y = x – 1 for $-2 \le x \le 4$.
(c) Write down the coordinates of the points of intersection of the two graphs.

4 (a) Complete the table of values for the function $y = \frac{-150}{x}$.

x	–90	–80	–70	–60	–50	–40	–30	–20
y	1.7		2.1		3			7.5

(b) Draw the graph of $y = \frac{-150}{x}$ for $-90 \le x \le -20$.
(c) Use your graph to find the value of x (i) when y = 1.6, (ii) when y = 3.2.
(d) Use your graph to solve the equation $\frac{-150}{x} = 2$.

5 (a) Draw the graph of the function $y = \frac{-2}{x}$ for the values x from –5 to 5.
(b) Draw the line of symmetry of the graph through the origin that intersects both sections of the graph.
(c) (i) State the equation of this line of symmetry.
(iii) Write down the coordinates of the points of intersection of the line of symmetry and the graph.

EXTENDED CURRICULUM

Exponential graphs: $y = a^x$, $(a = 1, 2, 3, \ldots\ldots)$

- The exponential graph is a curve that always goes through the point (0, 1) and lies above the x-axis.

Example

(a) The table below shows some corresponding values of x and y for the function $y = 3^x$.

x	–1.5	–1	–0.5	0	0.5	1	1.5
y	0.2	m	n	1	p	q	5.2

Write down the values of m, n, p and q.

(b) Draw the graph of $y = 3^x$ for $-1.5 \le x \le 1.5$.

(c) Use your graph to

(i) find y when $x = -0.8$, (ii) find x when $y = 3.8$, (iii) find x when $3^x = 2.2$.

Working: (a) $m = 0.3$, $n = 0.6$, $p = 1.7$ and $q = 3$.

(b)

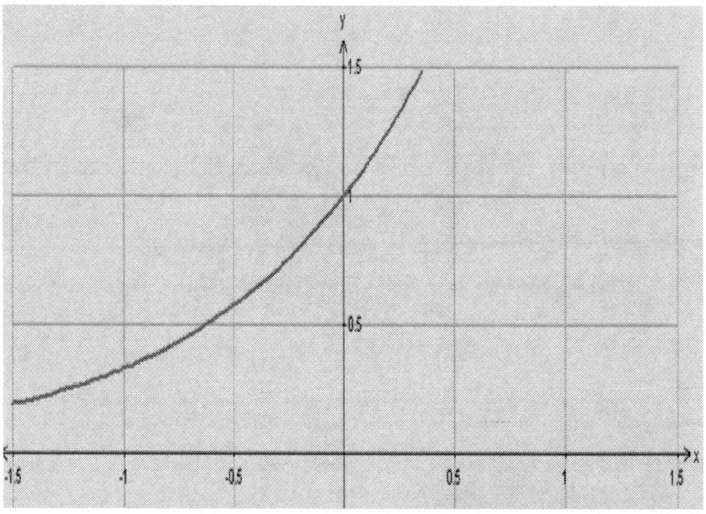

(c) (i) Draw the line $x = -0.8$ on the same grid.
So when $x = -0.8$ $y = 0.4$.

(ii) Draw the line $y = 3.6$ on the same grid.
So when $y = 3.6$ $x = 1.22$.

(iii) Draw the line $y = 2.2$ (since $3^x = 2.2$) on the same grid.
So when $3^x = 2.2$ $x = 0.72$.

The y-value of the point of intersection of the curve $y = 3^x$ and $x = -0.8$ is 0.4.
The x-value of the point of intersection of the curve $y = 3^x$ and $y = 3.8$ is 1.22.
The x-value of the point of intersection of the curve $y = 3^x$ and $y = 2.2$ is 0.72.

Practise exercise 28

1 (a) Draw the graph of $y = 2^x$ for $-3 \le x \le 3$.
(b) Use your graph to
(i) find y when $x = 2.5$, (ii) find x when $y = 0.7$, (iii) find x when $2^x = 5$.

2 (a) Draw the graph of $f(x) = 3^x$ for the values of x from –2 to 2.
(b) By drawing a suitable line on the same grid, solve the equation $f(x) = 6$.
(c) What happens to the shape of the graph as x goes towards smaller values?

3 (a) The below table shows x-values and some of their corresponding y-values for the function $y = 4^x$

x	–1.5	–1	–0.5	0	0.5	1	1.5	2
y	0.13	m	0.5	n	p	4	q	16

Write down the values of m, n, p and q.

(b) Using a scale of 1 cm = 2 units on both the axes, draw the graph of $y = 4^x$ for $-1.5 \leq x \leq 2$.

(c) Use your graph to find an approximate value of x when (i) y = 3, (ii) $4^x = 12$.

4 (a) Complete the table of values for the function $y = 5^x$

x	–0.8	–0.6	–0.4	–0.2	0.2	0.4	0.6	0.8
y		0.4		0.7		1.9		3.6

(b) Plot the graph of $y = 5^x$ for $-0.8 \leq x \leq 0.8$, using a scale of 2 cm = 1 unit on both the axes.

(c) Using the graph solve the equations

(i) $2 = 5^x$, (ii) $3 = 5^x$, (iii) $3 - 2x = 5^x$.

(d) Write down the y-intercept of the graph of $y = 5^x$.

5 (a) Plot the graph of the function $f(x) = 2^{x+1}$ for $-3 \leq x \leq 3$.

Use a scale of 1 cm = 1 unit on x- axis and 2 cm = 1 unit on y-axis.

(b) What happens to value of f(x) as x decreases?

(c) Use your graph to estimate the solution of the equation

(i) $f(x) = 1 - 2x$, (ii) $2^{x+1} = 2 - x$.

Graph of the function $y = ax^n$ (n = –2, –1, 0, 1, 2, 3)

Remember

1. When n = –2, the shape of the graph of $y = \frac{a}{x^2}$ has two sections and lie on adjacent quadrants.
2. When n = –1, the shape of the graph of $y = \frac{a}{x}$ has two sections and lie on opposite quadrants.
3. When n = 0, the shape of the graph of y = a is a straight line parallel to x-axis.
4. When n = 1, the shape of the graph of y = ax is a straight line with positive or negative gradient.
5. When n = 2, the shape of the graph of $y = ax^2$ is a quadratic curve.
6. When n = 3, the shape of the graph of $y = ax^3$ is a cubic curve.

Example

(a) Draw the graph of the function $y = 2x^3$ for $-3 \leq x \leq 3$.

(b) Use your graph to (i) find x when y = 10,

(ii) solve the equation $2x^3 = -25$, (iii) solve the equation $2x^3 + x = 0$.

Working: (a)

x	–3	–2	–1	0	1	2	3
y	–54	–16	–2	0	2	16	54

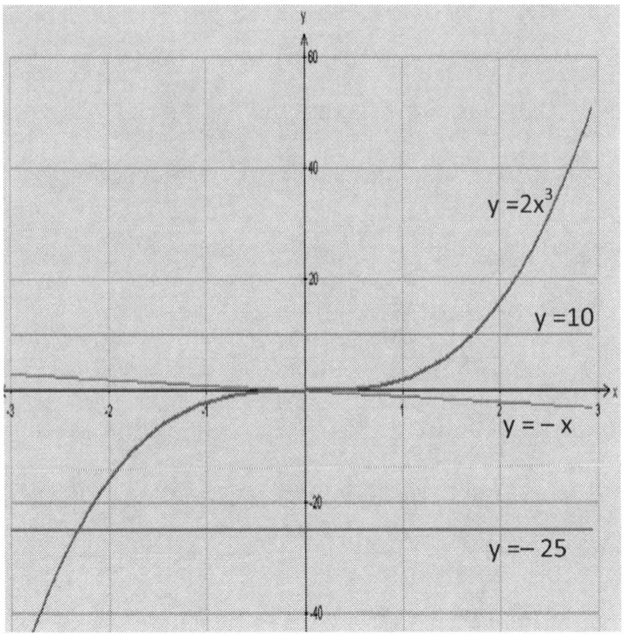

For solving equations using graphs, write down the x-value of the point of intersection only, but not both the x-value and y-value.

(b) (i) Draw the line $y = 10$ on the same grid.

So when $y = 10$, $x = 1.7$.

(ii) Draw the line $2x^3 = -25$ (or $y = -25$) on the same grid.
So $x = -2.32$ is the solution.

(iii) Draw the line $2x^3 = -x$ (or $y = -x$) on the same grid.
So $x = -2.32$ or 0 or 2.32.

The x-value of the point of intersection of the curve $y = 2x^3$ and $y = 10$ is 1.7.
The x-value of the point of intersection of the curve $y = 2x^3$ and $y = -25$ is -2.32.

Rearrange the equation $2x^3 + x = 0$ to get $2x^3 = -x$.
The x-values of the points of intersections of the curve $y = 2x^3$ and $y = -x$ are -2.32, 0 and 2.32.

Practise exercise 29

1 Draw the graph of each function.
(a) $y = x^3$ for $-4 \le x \le 4$.
(b) $y = -x^3$ for $-4 \le x \le 4$.
(c) $y = \dfrac{x^3}{2}$ for $-3 \le x \le 3$.
(d) $f(x) = \dfrac{5}{x^2}$ for $-6 \le x \le -1$ and $1 \le x \le 6$.
(e) $f(x) = \dfrac{-12}{x^2}$ for $-6 \le x \le -1$ and $1 \le x \le 6$.

2 (a) Draw the graph of $y = x^3 - 4x$ for the range of values of x from -4 to 4.
(b) Write down the three values of x when $y = 0$.
(c) Use your graph to estimate the solution of the equation $x^3 - 3x = 0$.

3 (a) Complete the table for the function $f(x) = 1 - x^3$.

x	–2	–1.5	–1	–0.5	0	0.5	1	1.5	2
f(x)	9			1.13				–2.38	

(b) Use 2 cm to 1 unit on x-axis and 1 cm to 1 unit on y-axis, draw a pair of coordinate axes.
(c) Plot the graph of $y = f(x)$ for $-2 \le x \le 2$ on the coordinate axes.
(d) Draw a table of values for the function $h(x) = x^2 - 2x$ for $-2 \le x \le 4$.
(e) On the same grid, plot the graph of $y = h(x)$ for $2 \le x \le 4$.

(f) Use your graph to solve the equation f(x) = h(x).

4 (a) Using a scale of 1cm = 1unit on x-axis and 2 cm = 1 unit on y-axis, draw a pair of Cartesian coordinate system.

(b) Draw the graph of $h(x) = \frac{4}{x^2}$ for $-5 \le x \le -1$ and $1 \le x \le 5$ on the coordinate system.

(c) Write down the equation of the line of symmetry of the graph in part(b).

(d) What value does h(x) approach as x increases or decreases?

(e) Write down a value of x for which h(x) does not exist.

(f) On the same grid, draw the line $k(x) = \frac{7 - x}{2}$, for $-3 \le x \le 3$.

(g) Using h(x) = k(x), show that $x^3 - 7x^2 + 8 = 0$.

(h) Use your graph to write down two values of x that satisfy the equation $x^3 - 7x^2 + 8 = 0$.

Finding the gradient of a line

Working rule

1) When two points (x_1, y_1) and (x_2, y_2) lie on a given line,

its gradient, $m = \frac{y_2 - y_1}{x_2 - x_1}$ or $\frac{y_1 - y_2}{x_1 - x_2}$.

2) When the angle made by a line with the positive x-axis is given as x, then its gradient, m = tan x.

Example Given (1, 4) and (–2, 7) are two pints on a line. Find its gradient.

Working: Gradient, $m = \frac{y_2 - y_1}{x_2 - x_1} = \frac{7 - 4}{-2 - 1} = -1$

Compare (1, 4) and (–2, 7) with (x_1, y_1) and (x_2, y_2).

Estimating the gradient of a line

Working rule

To estimate the gradient of a line

1) draw a tangent (a line that touches the curve exactly at one point) to the curve.

2) locate any two points on the tangent(Take care of scale).

3) then gradient, $m = \frac{\text{Changein y - value}}{\text{Changein x - value}}$.

Remember

The gradient of a curve at a given point = The gradient of the tangent to the curve at that point.

Example The curve shows the graph of the function $y = x^2 - 1$.

(a) Draw a tangent to the curve at the point $x = 1$.

(b) Estimate the gradient of the curve at $x = 1$.

(c) Write down the equation of the tangent.

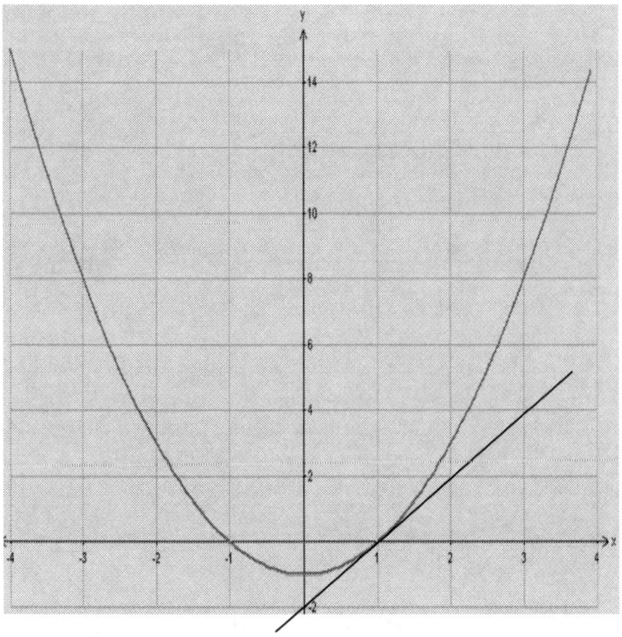

Working: (a) On the graph, t is the tangent.

(b) Gradient of the tangent,

$$m = \frac{\text{Change in y - value}}{\text{Change in x - value}}$$

$$= \frac{1}{1} = 1$$

Draw a line that touches the curve exactly at $x = 2$. Since the tangent slopes from bottom left to the top right, the sign of its gradient is positive.

(c) For the tangent, gradient, $m = 1$ and y-intercept, $c = -2$.
The equation of the tangent is $y = x - 2$.

Since the tangent is a straight line, use $y = mx + c$. ($m = 1$ and $c = -2$)

Practise exercise 30

1 Each pair of the following points lies on a line. Calculate the gradient of each line.

(a) (1, 2) and (3, 4)
(b) (2, 5) and (6, 1)
(c) (–3, 4) and (7, 3)
(d) (0, 0) and (–3, –5)
(e) (0, –1) and (–1, 0)
(f) (–2, –4) and (–6, –9)
(g) (10, 2.45) and (–3.25, 7)
(h) $\left(\frac{1}{2}, 4\right)$ and $\left(\frac{5}{3}, \frac{7}{10}\right)$

2 A(5, 8) and (a, 4) are two points on a line. The gradient of the line AB is –1. Calculate the value of a.

3 Points X(8, 3) and Y(a, b) are on a straight line. The value of a is twice the value of b. The gradient of the line XY is 1. Calculate the values of a and b.

4 Points P(4, 9) and Q(–2, –1) lie on a straight line. Another line MN is parallel to the line PQ. Find the gradient of the line MN.

5 The size of the angle made by a line with positive x-axis is 45°.

Find the gradient of this line.

6 The curve shows the graph of the function $y = x^2 + 1$.

(a) Draw a tangent to the curve at the point when $x = 1$.

(b) (i) Estimate the gradient of the curve at $x = 1$.

(ii) Explain why the gradient of the curve is an estimation.

(c) Write down the equation of the tangent.

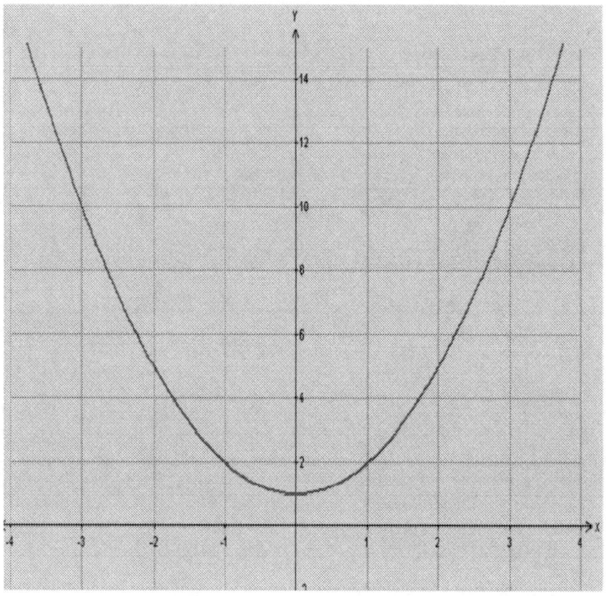

7 The curve shows the graph of the curve $y = \frac{-20}{x^2}$.

(a) (i) Draw the tangent t_1 to the curve at the point (–2, –5).

(ii) Draw another tangent t_2 to the curve at the point $x = 2.4$.

(b) Find the gradient of the curve

(i) at (–2, –5),

(ii) at 2.4.

(c) Write down the equation of

(i) the tangent t_1,

(ii) the tangent t_2.

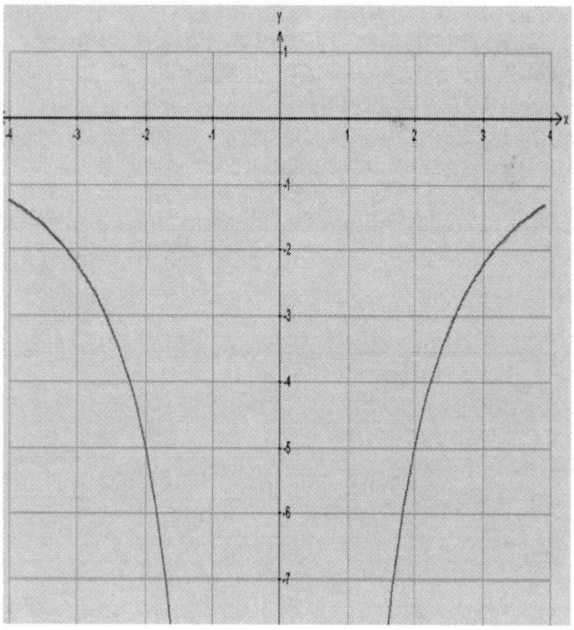

8 The curve shows the graph of the function $y = \frac{7}{x}$.

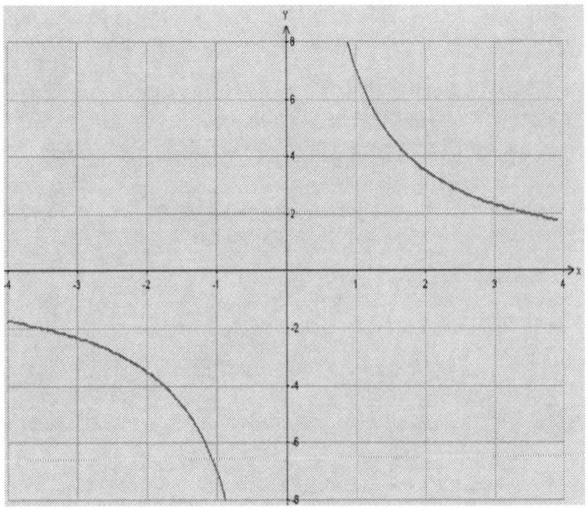

(a) Draw a tangent to the curve at the point when $y = 3$.

(b) Find the gradient of that tangent.

(c) Write down the equation of the tangent.

9 The graph shows the curve of the function $y = -x^3$.

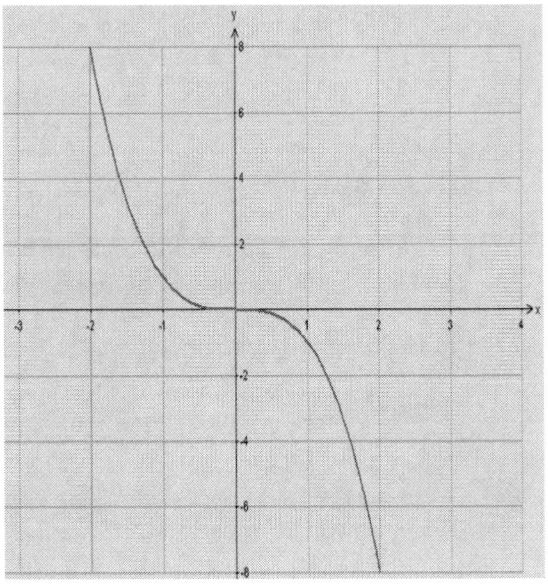

(a) Draw two tangents to the curve at $x = -1.5$ and at $x = 1.5$.

(b) Find the gradient of each tangent.

(c) What is the gradient of the curve at both the points?

(d) Are both the tangents are parallel to each other? Justify your answer.

Distance formula

Working rule

When the coordinates of the end points of a line segment are (x_1, y_1) and (x_2, y_2), the length of the line segment is given by $l = \sqrt{(x_1 - x_2)^2 + (y_1 - y_2)^2}$.

Example

The end points of a line segment are (–3, 3) and (5, 9). Calculate its length.

Working: $l = \sqrt{(x_1 - x_2)^2 + (y_1 - y_2)^2}$

$= \sqrt{(-3-5)^2 + (3-9)^2}$

$= 10$ units.

Compare (–3, 3) and (5, 9) with (x_1, y_1) and (x_2, y_2).

Practise exercise 31

1 Each pair of the following points represents the end points of a line segment. Calculate the length of each line segment.
(a) A(4, –3), and B(–2, 5) (b) C(2, 3) and D(1, 3)
(c) E(–5, 2) and F(–1, –4) (d) M(0, 0) and N(0, –8)
(e) P(0.6, –2.4) and (1.2, 0.7) (f) X(2, –2) and Y$\left(\frac{1}{2}, \frac{3}{4}\right)$

2 The points A(3, 5) and B(–3, 2) lie on a plane. Calculate the distance between the points A and B. Give your answer correct to one decimal place.

3 Triangle ABC is shown in the grid below.
(a) Write down the coordinates of the triangle ABC.
(b) Calculate the lengths of the side AB, BC and CA.
(c) Calculate the distance of
(i) the vertex A from the origin,
(ii) the vertex B from x-axis.

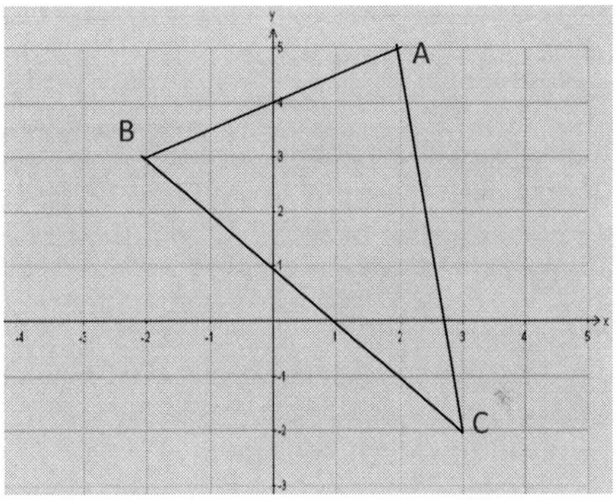

4 MNPQ is a quadrilateral whose vertices are M(–2, –2), N(1, –1), P(6, 8) and Q(3, 7). Show that MNPQ is a parallelogram.

5 The vertices of a right angled triangle are P(1, 4), Q(1, 1) and R(3, 1).
(a) Calculate the length of the sides
(i) PQ, (ii) QR, (iii) RP.
(b) Write down the length of the hypotenuse of the right angled triangle.

6 The length of a line segment joining the points (2, 0) and (a, 7) is 7 units. Find the value of a.

7 The points A(2, 3), B(8,11) and C(–1, –1) lie on the same plane. Show that the points A, B and C lie on the same line.

Mid-point formula

Working rule

When the coordinates of the end points of a line segment are (x_1, y_1) and (x_2, y_2), the coordinates of the mid-point(x, y) is given by $(x, y) = \left(\frac{x_1 + x_2}{2}, \frac{y_1 + y_2}{2}\right)$.

Example
The end points of a line segment AB are A(–1, 5) and B(3, 1).
Find the coordinates of the mid-point of AB.
Working: The coordinates of the mid-point,

$$(x, y) = \left(\frac{x_1+x_2}{2}, \frac{y_1+y_2}{2}\right) = \left(\frac{-1+3}{2}, \frac{5+1}{2}\right) = (1, 3)$$

Compare (–1, 5) and (3, 1) with (x_1, y_1) and (x_2, y_2).

Practise exercise 32

1 Each pair of the following points represents the end points of a line segment. Find the coordinates of the mid-point of each line segment.
(a) (3, 5) and (1, 7) (b) (2, –9) and (6, 15)
(c) (0, –4) and (8, 3) (d) (–12, –5) and (0, 0)
(e) (–1.2, –2.3) and (0.4, 1.5) (f) $\left(\frac{2}{3}, \frac{-4}{7}\right)$ and $\left(\frac{1}{2}, \frac{9}{11}\right)$

2 The vertices of a quadrilateral ABCD are A(2, 3), B(4, 8), C(–1, 5) and D(–2, 9). Find the coordinates of the mid-points of the sides
(i) AB, (ii) BC, (iii) CD, (iv) DA.

3 When the points (–4, 8) and (6, a) are joined in straight way, the coordinates of its middle point is (1, 6). Find the value of a.

4 The point C divides the line segment AB in the ratio 1:1. The respective coordinates of A and B are (3, –8) and (5, 2). Find the coordinates of the point C.

5 Find the coordinates of the point of bisection of the line segment MN, when M = (5, 11) and N = (–1, –3).

6 The point X is the mid-point of the line segment YZ. The coordinates of X and Y are (–2, 9) and (–6, 14) respectively. Find the coordinates of the point Z.

2.10 Graphs in practical situations

Core Curriculum

Conversion graph

▶ A conversion graph is used to convert between two quantities, when the rate is given.

Remember

A conversion graph is always a straight line.

Example
In January 2008, the exchange rate between US dollars($) and Indian rupees(Rs) was 1 $ = Rs 40.
(a) Draw a conversion graph using this exchange rate.
(b) Use your graph to change (i) Rs 110 into $, (ii) $ 5.90 into Rs.
(c) In January 2008, the cheapest car in the world costs Rs 100000.
Use your graph to estimate this cost in US dollars.
Working: (a)

$	1	2	3
Rs	40	80	120

Use the rate 1 $ = Rs 40 to draw up a table and then plot the graph.

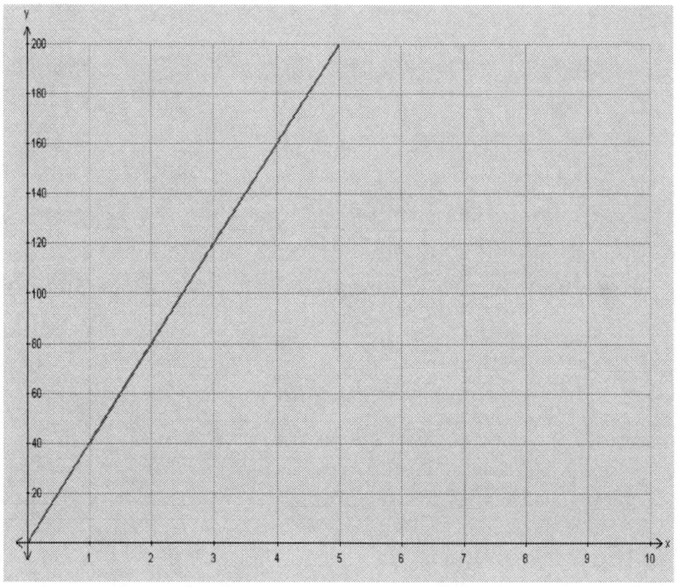

(b) (i) Rs 110 = $ 2.75 (ii) $ 5.90 = Rs 236

(c) From the graph, we have

Rs 100 = $ 2.50

So Rs 100 000 = $ 2.50 x 1000 = $ 2500

Choose 100 as a factor of 100 0000.

Practise exercise 33

1 The conversion between pounds(£) and euros(€) is 1 £ = 1.5 €.

(a) Draw a graph showing this conversion rate, using a scale of 1 cm = 1 unit on each axes. Take pounds on the horizontal axis and euros on the vertical axis.

(b) Use your graph to convert (i) £ 6.25 into euros, (ii) € 10 into pounds.

(c) The cost of a washing machine is € 270. What is the equivalent cost of this washing machine in pounds?

2 (a) The table below shows the approximate conversion between cubic inches(c. in.) and cubic centimetres(c.c.). Complete the table.

Cubic inches(c.in.)	1	2	3	4	5	6
Cubic centimeters(c.c.)	16	32		64		96

(b) Draw a graph showing the conversion between cubic inches and cubic centimetres, on a suitable grid.

(c) Using your graph, find the equivalent

(i) cubic centimetres for 3.5 cubic inches, (ii) cubic inches for 20 cubic centimetres.

(d) Water flows through a pipe at the rate of 80 c.c. per second for 10 seconds.

At the end of 10 seconds, calculate the quantity of water that flows through the pipe in cubic inches.

3 Alok says that 2 m/s = 7.2 km/h, 5 m/s = 18 km/h and 10 m/s = 36 km/h. This is true.

(a) On a suitable coordinate axes, draw the graph showing the conversion between m/s and km/h.

(b) Use your graph to change (i) 7 m/s to km/h, (ii) 14 km/h to m/s.

(c) A bus travels at an average speed of 64.8 km/h. Write down this average speed in m/s.

4 A shopkeeper give a discount of 10% on the marked price of each item for the annual stock clearance.

(a) Draw a conversion graph showing the marked price on the horizontal axis and the amount of discount on the vertical axis.

(b) Use your graph to estimate

(i) the amount of discount, when the marked price of a watch is \$ 150,

(ii) the marked price of a television, if the amount of discount is \$ 42.50.

5 A telephone company charges per call 60 cents plus 2 cents for each 10 seconds.

(a) Draw a graph showing the time on the horizontal axis and the call costs on the vertical axis.

(b) Using the graph, find the approximate

(i) cost of a call, that lasted for 1 minute and 15 seconds,

(ii) time for a call lasted, if the charge for the call is \$ 2.50.

Travel graph

- A travel graph is a distance-time graph which can be used to calculate the average speed of the whole or a part of the journey.
- **Illustration of a distance-time graph**

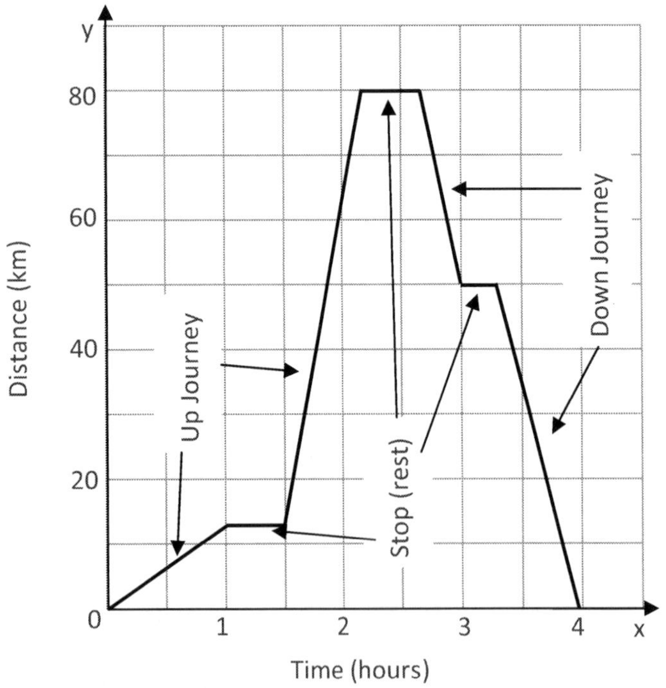

1) The horizontal line shows stop or rest.

2) From the starting point(origin),

(i) positive gradient(bottom left to the top right) shows increase in distance(up journey),

(ii) negative gradient(bottom right to the top left) shows decrease in distance(down journey).

Working rule

1. At a particular time, speed (velocity) = the gradient $\left(\text{i.e.} \dfrac{\text{Distance}}{\text{Time}}\right)$.
2. At a particular time interval, average speed $= \dfrac{\text{total distance travelled}}{\text{total time taken}}$.
3. Sharp slope shows the faster speed.

Example

The graph shows, Ahmed leaves his home at 0800 and drives to a theatre to watch a movie and returns back to home.

(a) How long does he stop at the traffic?

(b) Find his speed from the home to the traffic, in kilometres per hour.

(c) At what time does he reach the theatre?

(d) Find the length of the time he watches the movie.

(e) What is the distance between his home and theatre?

(f) Calculate the average speed of his journey from his home to the theatre,
(i) in kilometres per hour, (ii) in metres per second.

(g) Calculate his average speed for the whole journey, in kilometres per hour.

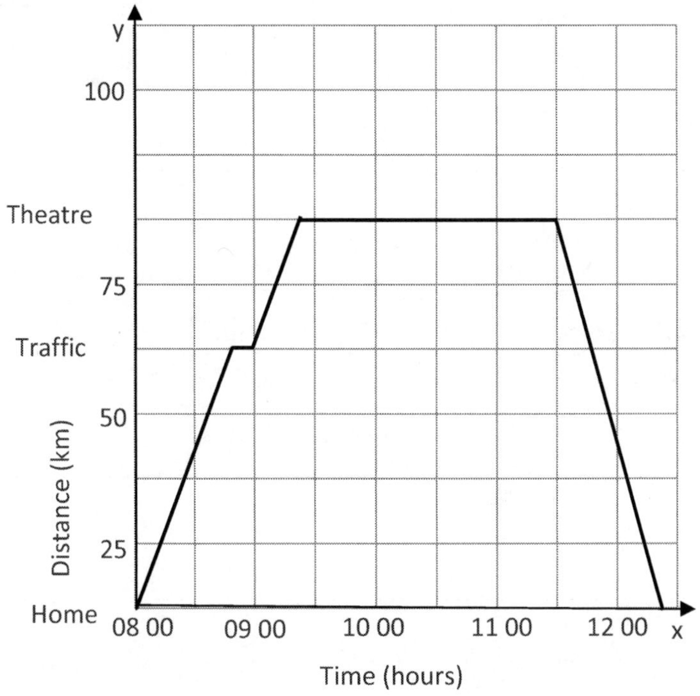

Working: (a) 12 minutes

(b) Speed = $\dfrac{50\text{ km}}{48\text{ min}} = \dfrac{50\text{ km}}{0.8\text{ h}} = 62.5$ km/h.

(c) 09 21 hours

(d) 2 hours 9 minutes

(e) 70 kilometres

(f) (i) Average speed = $\dfrac{70\text{ km}}{1.15\text{ h}} = 60.7$ km/h

(ii) 60.7 km/h = 16.9 m/s

(g) Average speed = $\dfrac{140\text{ km}}{2.05\text{ h}} = 68.3$ km/h

Use speed = gradient.
48 min = $\dfrac{48}{60}$ h = 0.8 h

Average speed = $\dfrac{\text{total distance travelled}}{\text{total time taken}}$
1 hour 9 min = 1.15 h.

To calculate the average speed for the whole journey, double the distance.
Total distance = 2 x 70 km = 140 km.
Total time = 1 h 9 min + 54 min = 2.05 h.

Practise exercise 34

1 Mourin cycles form her home to school. On the way she stops at a shop to purchase a book and then continued her journey. At the end of her classes, she cycles back from school to home without stop.

(a) At what time does she start her journey from home to school?

(b) How long does she stop at the shop?

(c) At what time does she reach the school?

(d) Find the length of time she stays in the school.

(e) What is the distance between her home and the school?

(f) Calculate the average speed of her journey from school to home,

(i) in kilometres per hour, (ii) in metres per second.

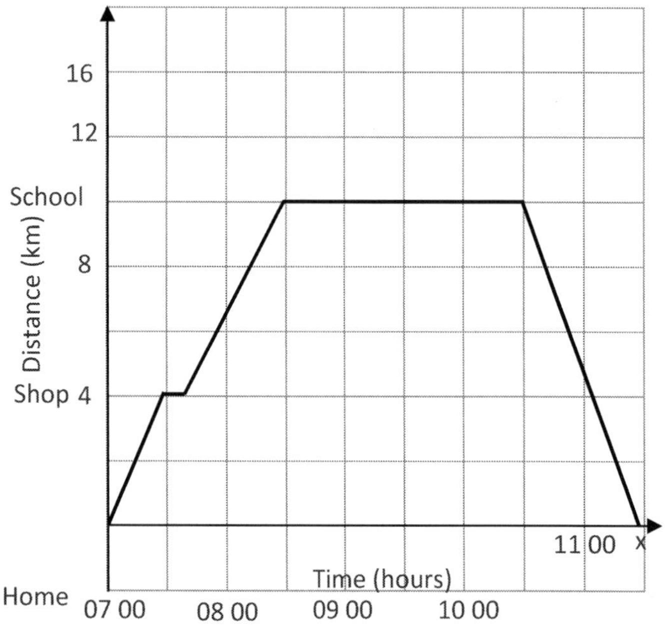

2

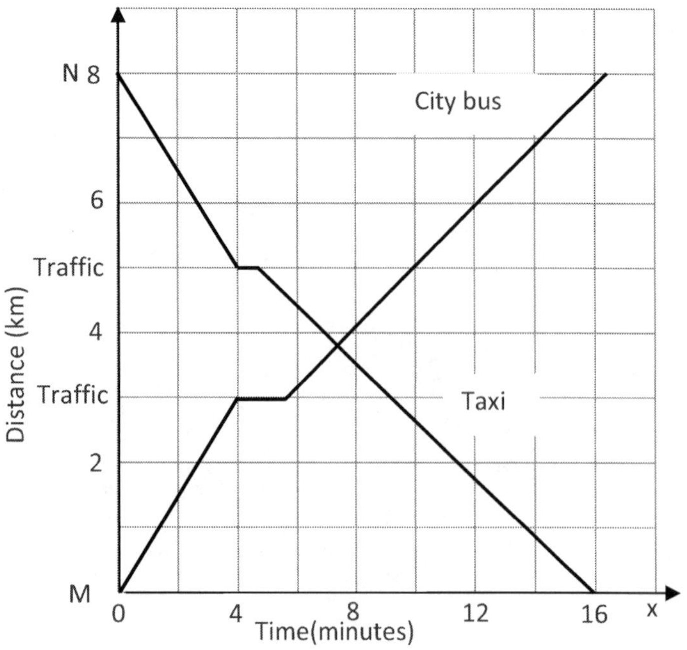

M and N are two locations in a City. A City bus leaves location M and travels to location N. At the same time a taxi leaves location N and travels to location M.

The graph above shows their journey.
(a) How long does the taxi stop at the traffic signal?
(b) For the first 4 minutes journey, calculate the average speed of the City bus,
(i) in kilometres per minute, (ii) in kilometres per hour, (iii) in metres per second.
(c) At what distance from location M, do they cross each other?
(d) How long does it take the City bus and the taxi to cross each other?
Give your answer in (i) minutes and seconds, (ii) minutes only,
(iii) hour only, correct to two decimal places.
(e) Which vehicle travels at a slower rate?

3 A motor cyclist travelled a distance of 500 metres in the first 60 seconds. He then stopped for 20 seconds at a traffic light. In the next 90 seconds, he travelled a distance of 800 metres and again stopped at a traffic light for 15 seconds. He covered the remaining 200 metres in 25 seconds to reach his office.
(a) Using a scale of 2 cm = 50 seconds on the horizontal axis and 2 cm = 400 metres on the vertical axis, draw a distance-time graph to show this information.
(b) Calculate (i) the total distance travelled, (ii) the total time taken,
(iii) the average speed for the whole journey.

4 A train departs at 08 00 hour from station S_1 and moves to station S_2 at steady speed of 60 km/h. The distance between the stations S_1 and S_2 is 585 kilometres.
Another train departs at 08 30 hour from station S_1 and moves to station S_2 at a steady speed of 90 km/h.
(a) On a suitable pair of axes, draw a distance-time graph showing the journey of both the trains.
(b) Use your graph to estimate
(i) the time at which the second train crosses the first train,
(ii) the distance at which the second train crosses the first train,
(iii) the difference in time at which both the trains reach at station S_2.

EXTENDED CURRICULUM

Speed-time graph

- A speed-time graph shows the increase in speed, constant speed and decrease in speed with respect to time.
- The total area under the speed-time graph is the total distance travelled.

▶ Illustration of a speed-time graph

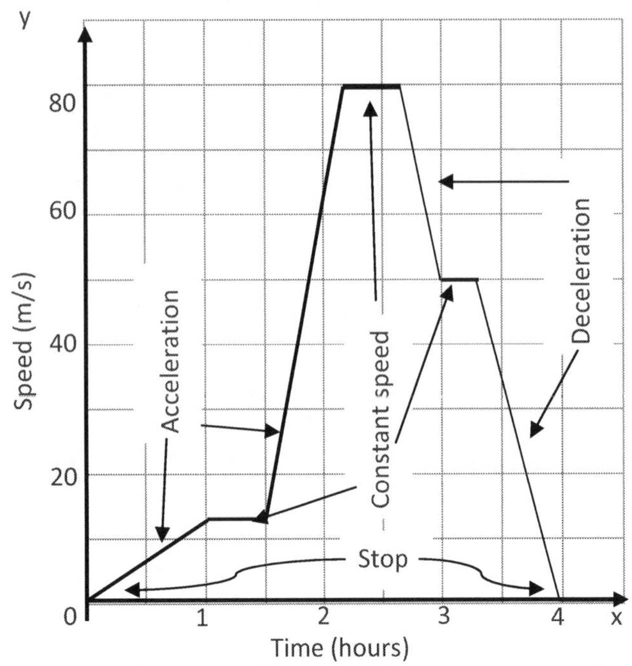

1) The horizontal line shows the constant speed.
2) From the starting point(origin),
(i) positive gradient(bottom left to the top right) shows increase in speed(acceleration),
(ii) negative gradient(bottom right to the top left) shows decrease in speed(deceleration).

Working rule

1) At a particular time, acceleration (or deceleration) = the gradient $\left(\text{i. e.} \dfrac{\text{speed}}{\text{time}}\right)$.
2) For the whole journey or part of a journey, distance = area under the curve.
3) At a particular time interval, average speed = $\dfrac{\text{total distance travelled}}{\text{total time taken}}$.

Example

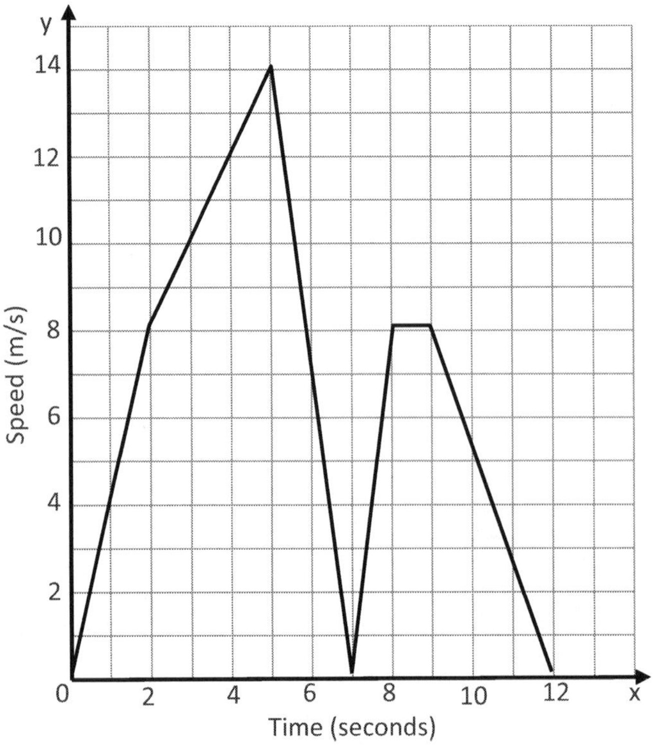

The speed-time graph shows the first 12 seconds of the journey of a car.

(a) Using the graph, describe the journey of the car between
(i) 0 and 2 seconds, (ii) 2 and 5 seconds, (iii) 5 and 7 seconds,
(iv) 7 and 8 seconds, (v) 8 and 9 seconds, (vi) 9 and 12 seconds.

(b) Calculate the acceleration of the car
(i) between 2 and 5 seconds, (ii) between 7 and 8 seconds.

(c) Calculate the deceleration of the car between 9 and 12 seconds.

(d) What is the maximum speed, in km/h of the car?

(e) Calculate the distance which the car travels in the first 7 seconds.

(f) What is the average speed, in metres per second of the car in the first 7 seconds?

Working: (a) (i) The car starts from rest and travels at an uniform acceleration to reach the speed of 8 m/s.

(ii) The car travels at a uniform acceleration to reach the speed of 14 m/s.

(iii) The car travels at a uniform deceleration to come to the first stop at the 7th second.

(iv) Again the car starts from rest and travels at a uniform acceleration to reach a speed of 8 m/s.

(v) The car travels at a constant speed of 8 m/s.

(vi) The car travels at a uniform deceleration to come to the second stop at 12th second.

(b) (i) Acceleration $= \frac{6\text{ m/s}}{3\text{ s}} = 2\text{ m/s}^2$

Acceleration (or deceleration) = the gradient.

(ii) Acceleration $= \frac{8\text{ m/s}}{1\text{ s}} = 8\text{ m/s}^2$

(c) Deceleration $= \frac{8\text{ m/s}}{3\text{ s}} = 2.67\text{ m/s}^2$

(d) Maximum speed = 14 km/h

Area of triangle $= \frac{1}{2} \text{ x b x h}$

(e) Area of triangle ABF $= \frac{1}{2} \text{ x } 2 \text{ x } 8 = 8$

Area of triangle CDE $= \frac{1}{2} \text{ x } 14 \text{ x } 2 = 14$

Area of trapezium $= \frac{1}{2} \text{ x } (a + b) \text{ x h}$

Area of trapezium BCEF $= \frac{1}{2} \text{ x } (8 + 14) \text{ x } 3 = 33$

Distance = Area under the curve

Distance = 8 + 33 + 14 = 55 m

(f) Average speed $= \frac{55\text{ m}}{7\text{ s}} = 7.86\text{ m/s}$

Average speed $= \frac{\text{total distance travelled}}{\text{total time taken}}$

Practise exercise 35

1 The speed-time graph shows the journey of a bus during the first 10 seconds.

(a) Using the graph, describe the journey of the bus between

(i) 1 and 3 seconds,

(ii) 7 and 10 seconds.

(b) Calculate the acceleration of the bus between first and third seconds.

(c) For how long the speed of the bus was 25 m/s?

(d) Calculate the deceleration of the bus between seventh and tenth seconds.

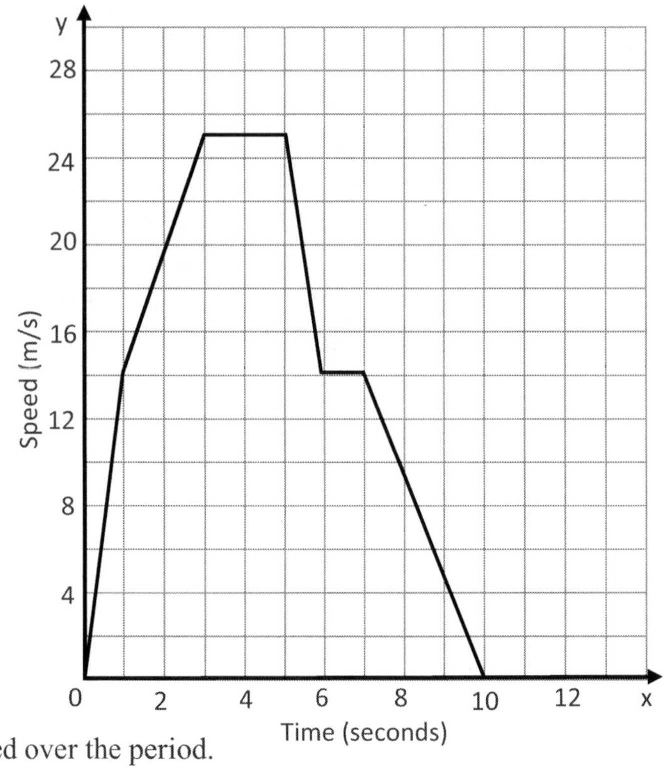

(e) Find the total distance travelled over the period.

(f) Find the average speed of the bus for the first 10 seconds.

2 A car starts from rest and accelerates to reach up to a speed of 10 m/s at the eighth second. It moves at this speed for 18 seconds and then decelerates to come to rest in a further 12 seconds.

(a) Draw a speed-time graph, using a scale of 1 cm as 4 seconds on the horizontal axis and 1 cm as 2 m/s on the vertical axis.

(b) Use your graph, to calculate

(i) the acceleration of the car, (ii) the deceleration of the car,

(iii) the distance travelled by the car in the whole period.

3 A motor bike starts from rest and reaches up to a speed of 18 m/s in 50 seconds. It continues at this speed for 30 seconds and then reduces its speed up to 12 m/s. After travelling at this speed for 15 seconds, it comes to rest at 120th second.

(a) Draw a speed-time graph, using a scale of 2 cm = 25 seconds on the horizontal axis and 2 cm = 5 m/s on the vertical axis.

(b) Using your graph, calculate

(i) the acceleration of the motor bike between t = 0 and t = 50,

(ii) the deceleration of the motor bike between t = 50 and t = 80,

(iii) the total distance travelled by the motor bike.

(c) What is the speed of the motor bike when t = 25?

4 The speed-time graph shows the speed of an air craft for a part of its journey.

(a) What is the speed of the air craft when t = 0?

(b) Calculate the acceleration of the air craft

(i) between t = 0 and t = 15,

(ii) between t = 15 and t = 20.

(c) What is the maximum speed of the air craft

(i) in metres per second,

(ii) in kilometres per hour.

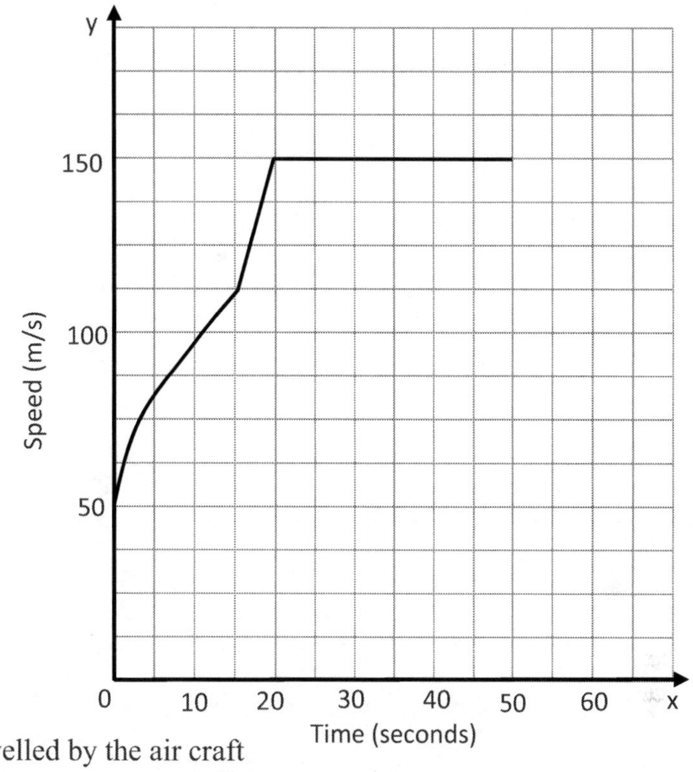

(d) Calculate the distance travelled by the air craft
(i) between t = 0 and t = 15, (ii) between t = 15 and t = 40.
(e) Calculate the average speed of the air craft between t = 15 and t = 40.

5 Starting from rest, a boat attains a speed of 20 m/s in 40 seconds.
It maintains this speed for 60 seconds and comes to rest in a further 15 seconds.
Then it increases the speed limit up to 30 m/s and continues at this speed for 50 seconds and comes to rest in another 18 seconds.
(a) Plot a speed-time graph to his information.
(b) Using your graph, calculate
(i) the acceleration of the boat for the first 40 seconds,
(ii) the deceleration of the boat for the last 18 seconds,
(iii) the distance travelled by the boat in the last 68 seconds.
(c) After how many seconds does the boat make
(i) the first stop, (ii) the second stop.

CHAPTER 3 Core curriculum Shape and Space

3.01 Geometrical terms and relationships

- The types of angles are given in the mind-map below.

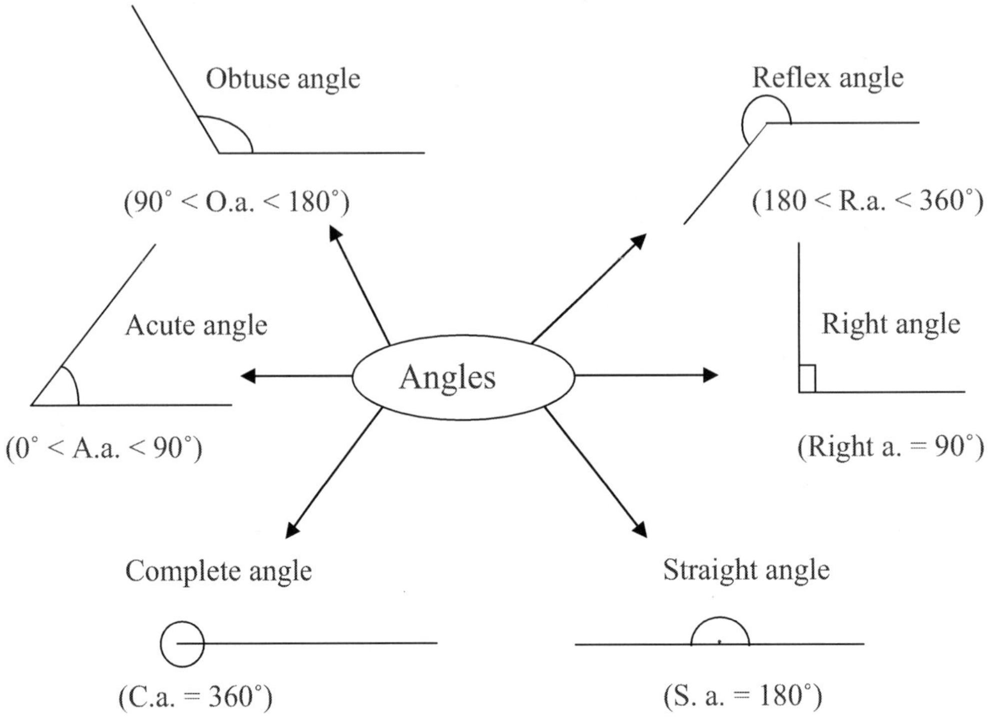

- The types of triangles are given in the mind-map.

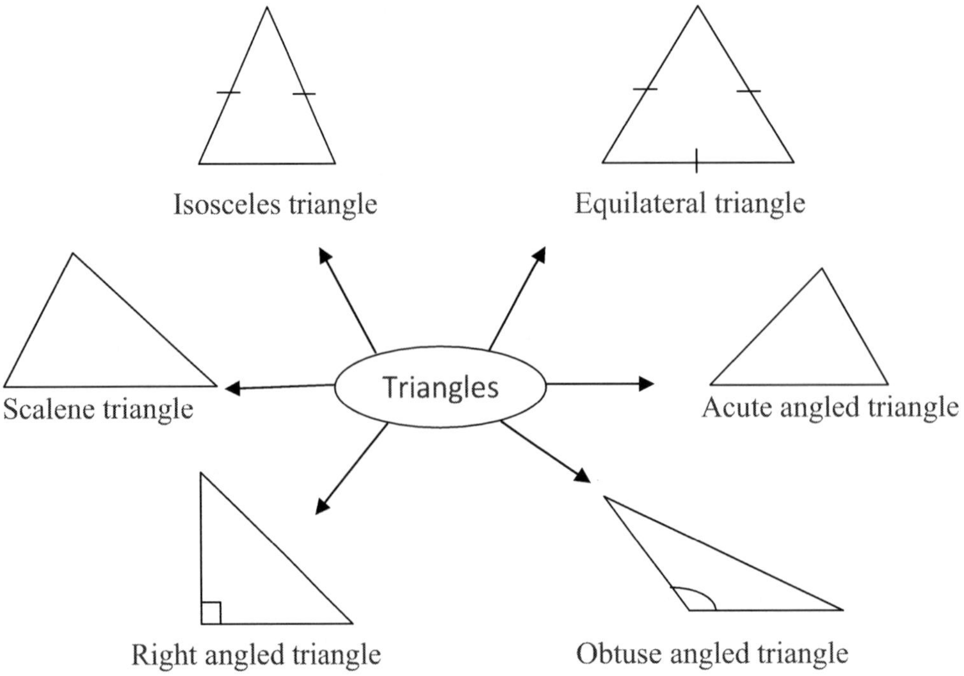

- The types of quadrilateral are given in the mind-map below.

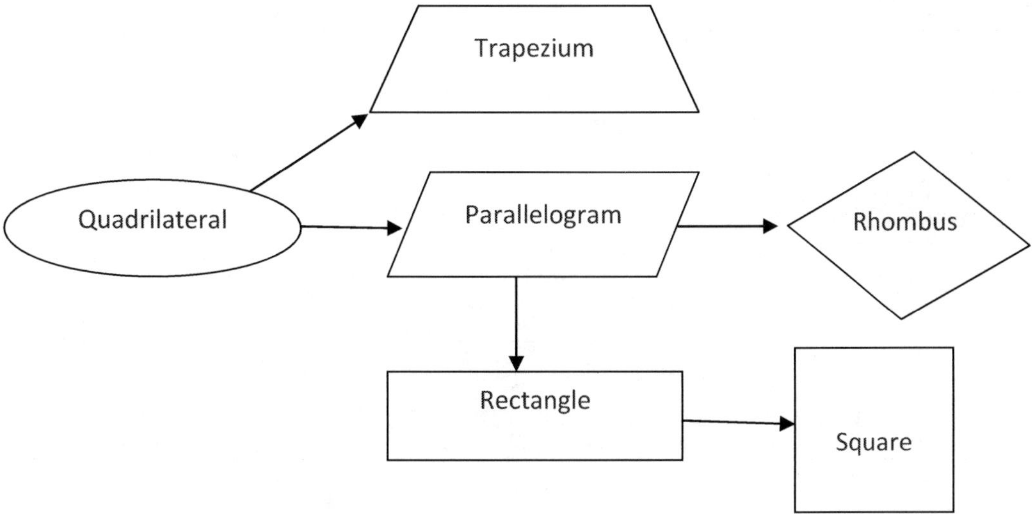

- Polygons are named according to their number of sides. The types of polygons are given in the table below.

Name of polygon	No. of sides	Name of polygon	No. of sides
Triangle	3	Octagon	8
Quadrilateral	4	Nonagon	9
Pentagon	5	Decagon	10
Hexagon	6	Dodecagon	12
Heptagon	7	Sexagon	60

Practise exercise 1

1 Write down the special geometrical term for each statement.
(a) Two lines with a common end point. (b) Two lines which intersect at 90°.
(c) The longest chord in a circle. (d) A polygon with five sides.
(e) A polygon with six sides. (f) A polygon with seven sides.
(g) A triangle whose all sides are equal. (h) A triangle with two sides equal.
(i) A triangle with a right angle. (j) A parallelogram with all sides equal.
(k) A quadrilateral with one pair of opposite sides parallel.
(l) A quadrilateral whose opposite pair of adjacent sides are equal.
(m) Two lines which maintain same distance at every point.

2 Acute angle Obtuse angle Reflex angle
Right angle Straight angle Complete angle
Use one of the above terms to describe each of the below angles.
(a) 1° (b) 77.5° (c) 90° (d) 179° (e) 180° (f) 180.4°
(g) 263° (h) 300° (i) 89.7° (j) 342.25° (k) 22.75° (l) 360°

3 State whether true or false.
(a) A right angle is 180°. (b) Every equilateral triangle is also isosceles.
(c) Every isosceles triangle is also equilateral. (d) A triangle can have two obtuse angles.
(e) A square is a special kind of rectangle. (f) A kite a quadrilateral.

(g) A parallelogram is a special kind of rhombus.
(h) An isosceles trapezium has two equal non-parallel sides.

- **Congruency and similarity**

- Figures with same shape and size are congruent (≅).
- Figures with same shape are similar (~).

Remember
In congruent figures, 1. respective angles are equal and sides are equal, 2. one figure is an exact copy of another. ***In similar figures***, 1. respective angles are equal and sides are in the same ratio, 2. one figure is an enlargement of another. All similar figures are congruent, but the reverse is not true.

Example 1

In the figure, triangle PQR is congruent to triangle STR.

Find the values of

(i) x, (ii) y,

(iii) a, (iv) b.

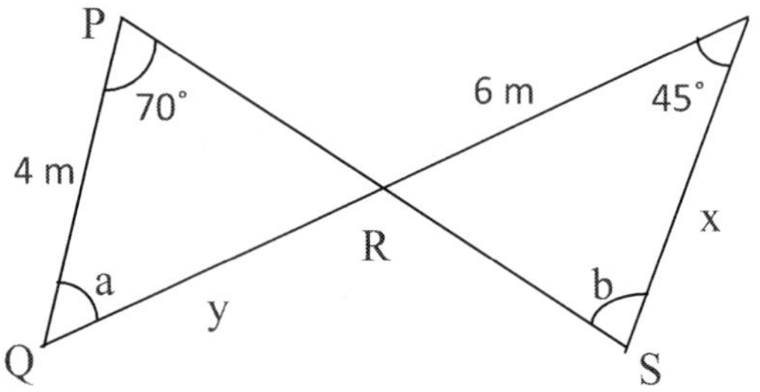

Working: (i) $x = 4$ m
(ii) $y = 6$ m
(iii) $a = 45°$
(iv) $b = 70°$

Since triangle PQR ≅ triangle STR,
PQ = ST (x = 4),
QR = TR (y = 6).
Angle PQR = Angle STR (a = 45°)
Angle RPQ = Angle RST (b = 70°)

Example 2

In the figure, triangle ABC is similar to triangle DEC.

Find the values of

(i) x, (ii) y, (iii) z.

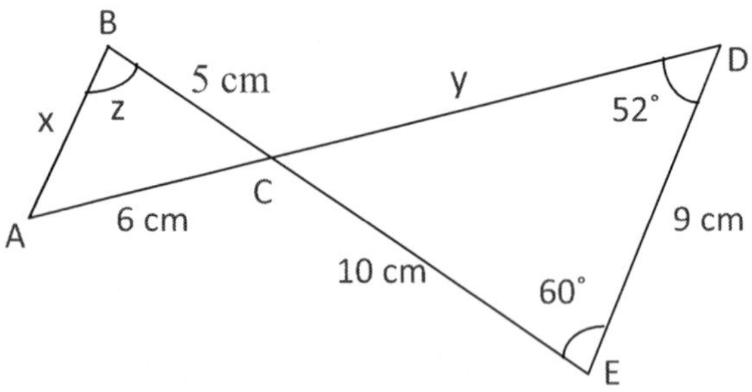

Working: (i) $\frac{x}{9} = \frac{5}{10}$
$x = 4.5$ cm
(ii) $\frac{y}{6} = \frac{10}{5}$
$y = 12$ cm
(iii) $z = 60°$

Since triangle ABC ~ triangle DEC, the respective sides are in the same ratio.
$\frac{AB}{DE} = \frac{BC}{EC}$
$\frac{CD}{CA} = \frac{EC}{BC}$
The respective angles are equal.

Practise exercise 2

1 Choose a pair of figures which is either congruent or similar.
All the dimensions are given in cm.

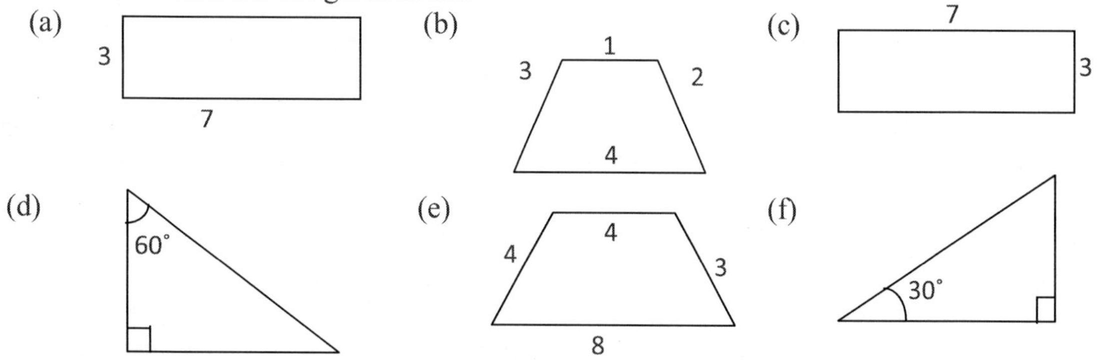

2

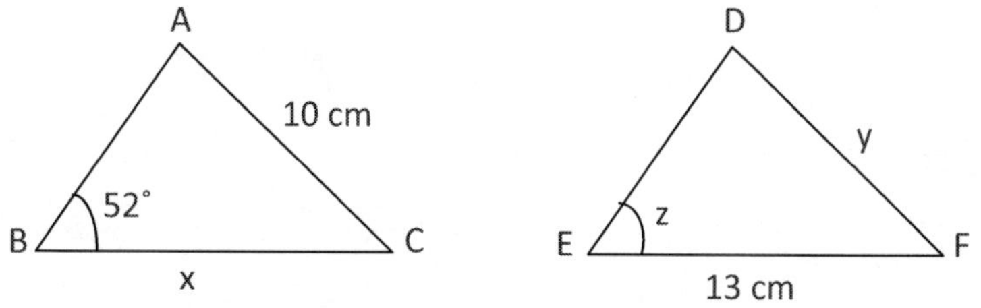

Triangles ABC and DEF are congruent to each other. Find the sizes of x, y and z.

3 ABCD is a parallelogram with AB = 20 mm and DA = 9 mm. Diagonal AC is joined.

(a) Is triangle ABC congruent to triangle CDA?

(b) Justify your answer.

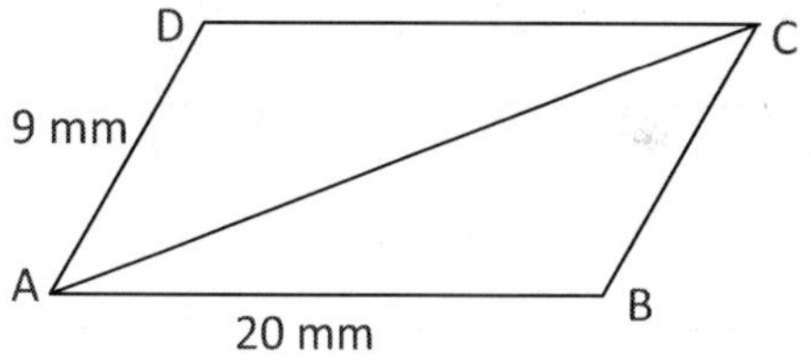

4 Triangle MNP and triangle QRS are similar.

Find the values of x and y.

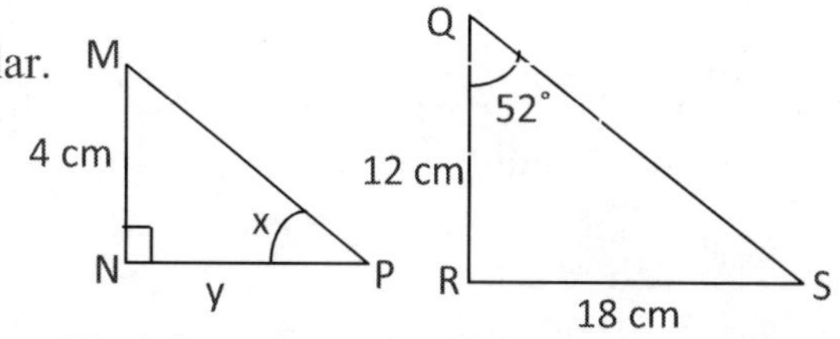

5 Triangle ADE and triangle ABC are similar. Find the values of x and y in each case.

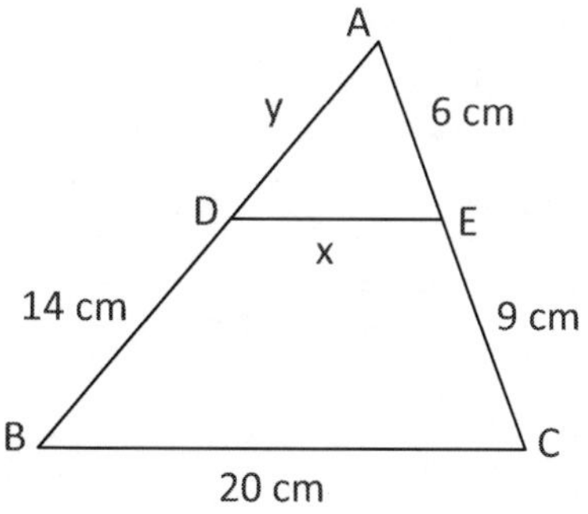

6 In the diagram AB is parallel to ED.

AD and BE intersect at C.

Find the values of

(i) x, (ii) y, (iii) z.

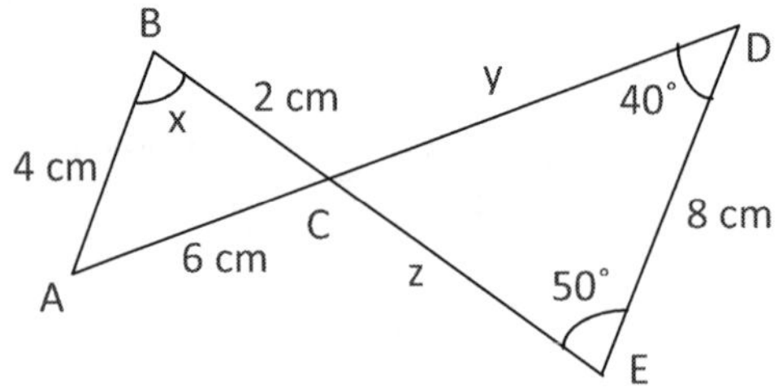

7 In the diagram BC is parallel to MN.

MN = x cm, BC = (2x+1) cm, AN =(x – 2) and NC = (x+1) cm.

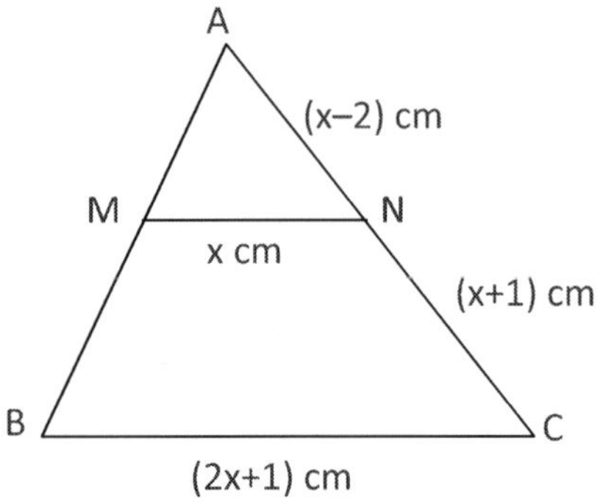

(a) Show that $x^2 - 4x - 2 = 0$.

(b) Find the lengths of (i) MN, (ii) AC.

8 In the diagram AB is parallel to CD.

AD and BS intersect at E.

AE = (2x – 1) cm, ED = (3x – 2) cm,

BE = (x + 2) cm and EC = (2x+1) cm.

(i) Show that $x^2 - 4x + 3 = 0$.

(ii) Find the two values of x which satisfy the equation in part(i).

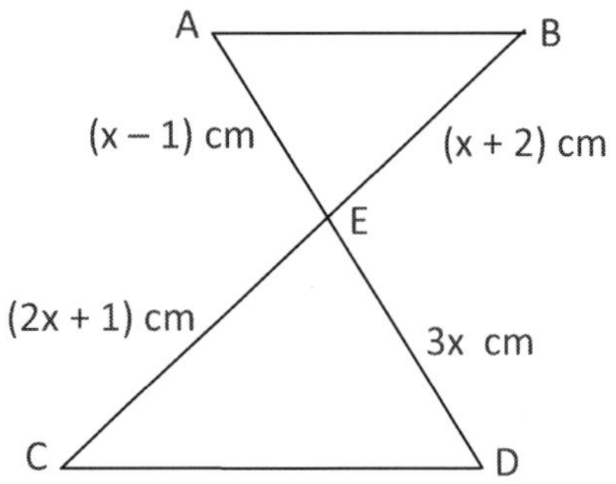

Area of similar shapes — Extended Curriculum

- A scale factor can be defined as the ratio between the corresponding sides or areas or volumes of similar shapes.

Remember

1. For two dimensional shapes,
   (a) the ratio between any pair of corresponding sides is a linear scale factor k,
   (b) the ratio between their areas is a quadratic scale factor k^2.
2. The ratio of the areas of two similar triangles is equal to the ratio of squares on
   (a) any two corresponding altitudes,
   (b) any two corresponding medians,
   (c) their perimeters.

Example

Triangle ABC is similar to triangle AED.

BC = 4 cm and ED = 12 cm.

The area of triangle ABC is 6 cm^2.

(a) Calculate the area of the triangle AED.

(b) Calculate the area of the trapezium BCDE.

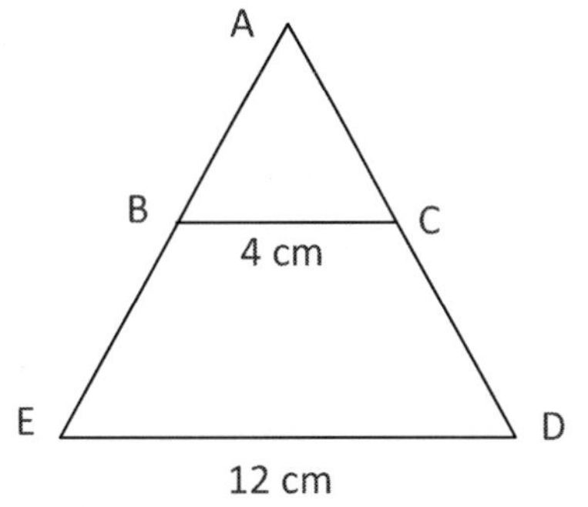

Working: Linear scale factor, $k = \frac{12}{4} = 3$

Quadratic scale factor, $k^2 = 3^2 = 9$

(a) Area of triangle AED = 9 x 6 = 54 cm^2

(b) Area of trapezium BCDE = 54 – 6 = 48 cm^2

k = ratio of sides
For area use quadratic scale factor k^2.

Practise exercise 3

1 Triangle MNP is similar to triangle QRS.

NP = 5 cm and RS = 10 cm.

The area of triangle MNP is 10 cm^2.

Calculate the area of the triangle QRS.

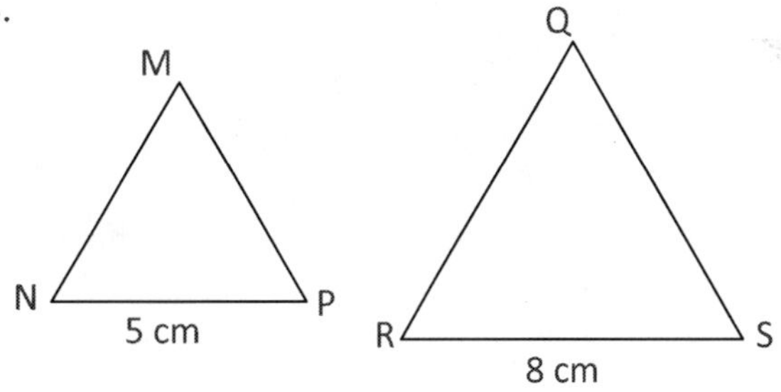

2 Triangle XYZ is similar to triangle XPQ.

YZ = 2 cm and PQ = 8 cm.

The area of triangle XPQ is 18 cm^2.

(a) Calculate the area of the triangle XYZ.

(b) Calculate the area of the trapezium YZQP.

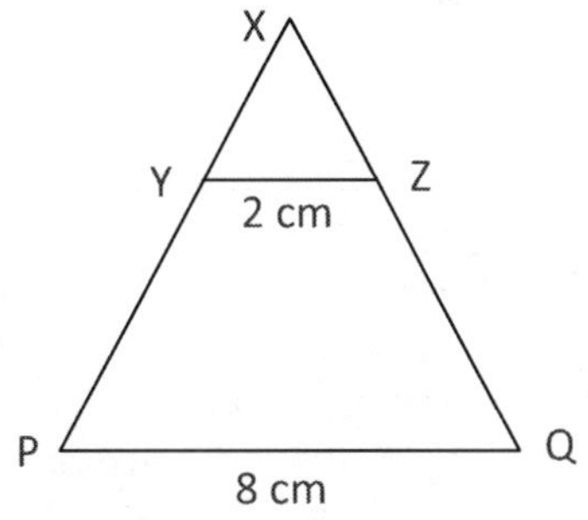

3 In triangle ABC, BC is parallel to ED.

Also BC = 30 cm, AD = 8 cm and CD = 16 cm.

(a) Find the length of the side ED.

(b) Find (area of triangle ABC) : (area of triangle AED).

(c) Find (area of trapezium BCDE) : (area of triangle AED).

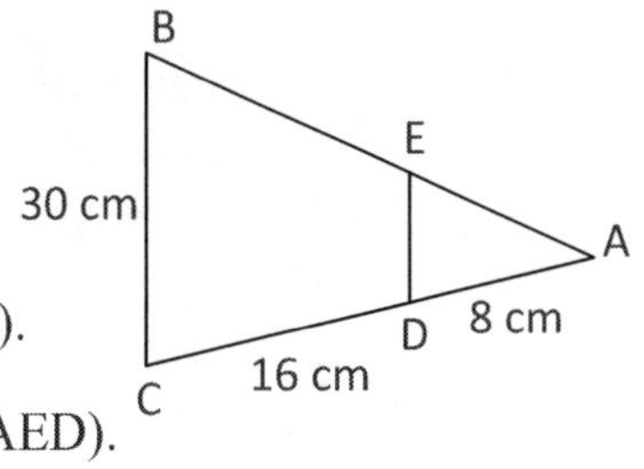

5 In the diagram MN is parallel to PQ and LP:LM = 1:4.

(a) What is the ratio of area of triangle LPQ and area of triangle LMN?

(b) The area of triangle LPQ is 20 cm^2. What is the area of triangle LMN?

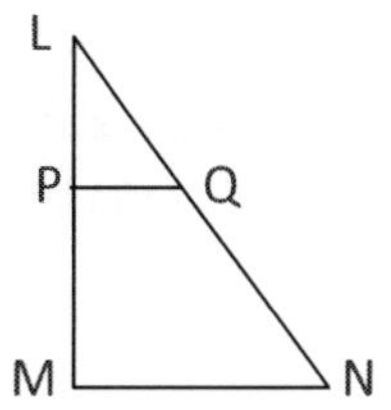

6 In the diagram AB is parallel to DE.

The area of triangle ABC is 14 cm^2.

The area of triangle EDC is 56 cm^2.

Also CD = 10 cm.

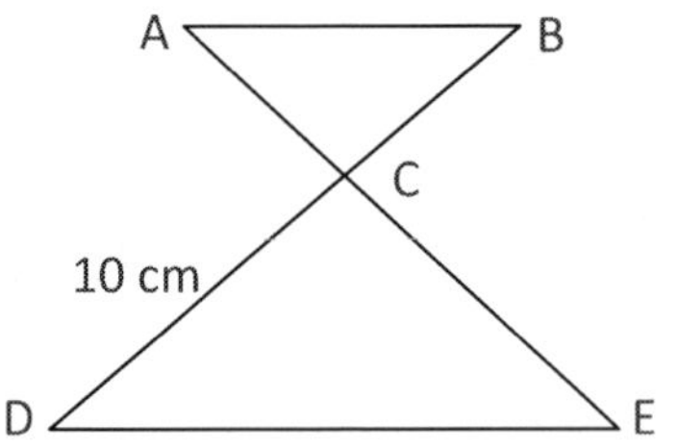

(a) Find the length of BC.

(b) If AB = y, write down an expression for DE in terms of y.

7 In the diagram EFG is a right angled triangle.

FH is perpendicular to GE.

FH = 6 cm and GH = 3 cm.

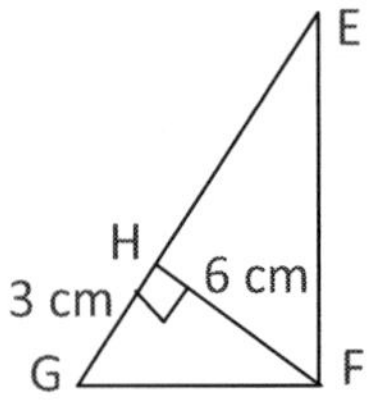

(a) Find (area of triangle FGH) : (area of triangle EGF).

(b) Find $\frac{\text{area of triangle EFH}}{\text{area of triangle EGF}}$.

(c) (i) Find the area of the triangle EGF.
(ii) Find the area of the triangle EFG.

8 Triangle ABC is similar to triangle DEF. Angle ABC equals angle DEF and angle BCA equals angle EFD. AB = 8 cm and DE = 24 cm.

Find $\frac{\text{area of triangle ABC}}{\text{area of traingle DEF}}$.

Volume of similar shapes

Remember

For three dimensional similar shapes,
(a) ratio between any pair of corresponding edges is a linear scale factor k,
(b) ratio between their volumes is a cubic scale factor k^3.

Example

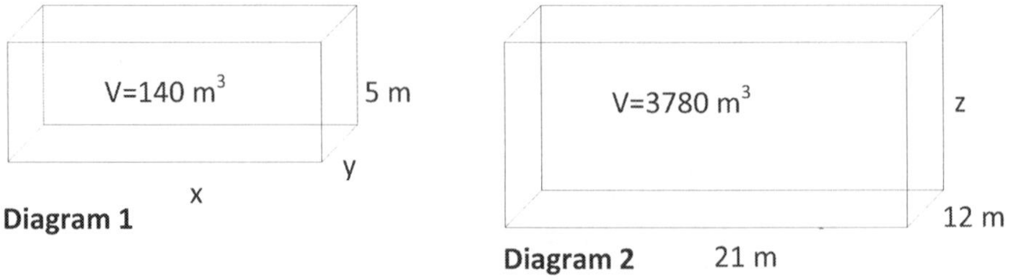

Diagram 1 shows a cuboid of volume 140 m^2 and height 5 m.
Diagram 2 shows a cuboid of volume is 3780 m^2, length 21 m and breadth 12 m.
Both the cuboids are similar. Find the values of x, y and z.

Working: Cubic scale factor $k^3 = \dfrac{3780}{140} = 27$

Linear scale factor $k = \sqrt[3]{27} = 3$

$kx = 21$ or $x = \dfrac{21}{3} = 7$ m

$ky = 12$ or $y = \dfrac{12}{3} = 4$ m

$k \times 5 = z$ or $z = 3 \times 5 = 15$ m

From the given volumes first find cubic scale factor and then linear scale factor.

Use linear scale factor to find the length of the edges.

Practise exercise 4

1

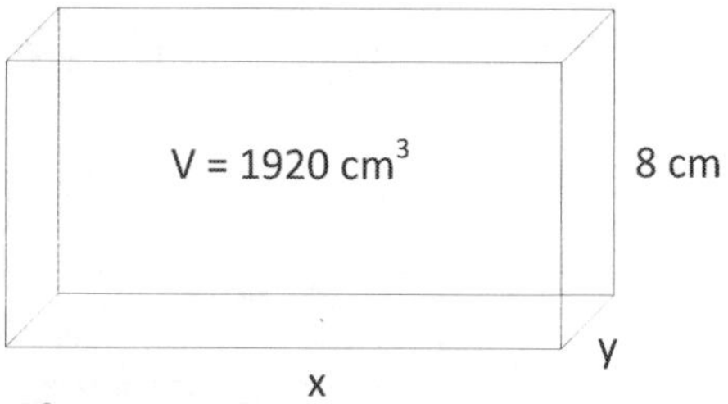

Diagram 1

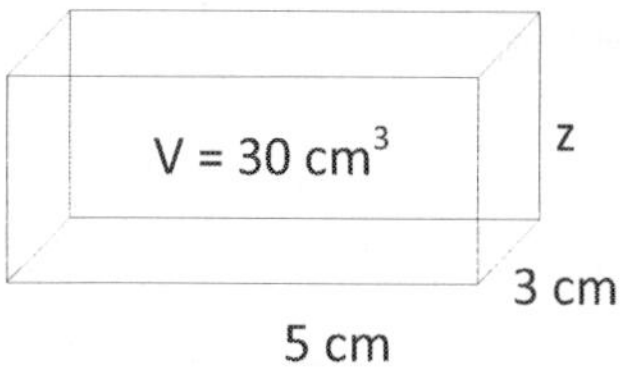

Diagram 2

Diagram 1 shows a cuboid of volume 1920 cm^3 and height 8 cm.
Diagram 2 shows a cuboid of volume 30 cm^3, length 5 cm and breadth 3 cm.
Both the cuboids are similar.
Find the values of x, y and z.

2 The ratio between the volumes of two similar cubes is 1:8. The volume of the larger cube cube is 1728 cm^3. (a) Calculate the volume of the smaller cube.
(b) Find the ratio between the total surface areas of the smaller cube to the larger cube.

3

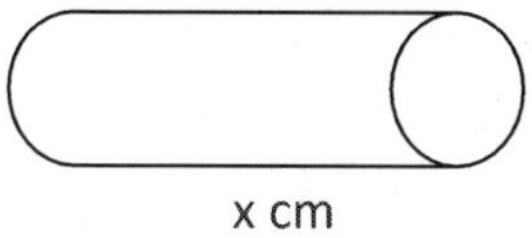

Diagram 1

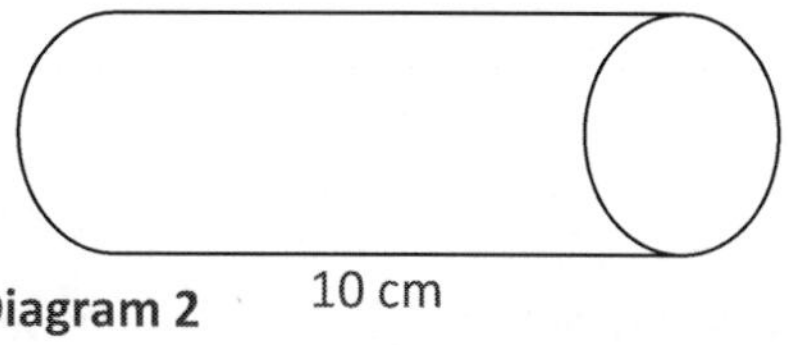

Diagram 2

In the above diagrams, both the cylinders are similar.
The volume of the cylinder in diagram 1 is 141 cm^3 and its length is x cm.
The volume of the cylinder in diagram 2 is 1131 cm^3 and its length is 10 cm.
(a) Calculate the value of x.
(b) Write down in its simplest form, the ratio between the areas of cross-sections of the first cylinder and the second cylinder.

4

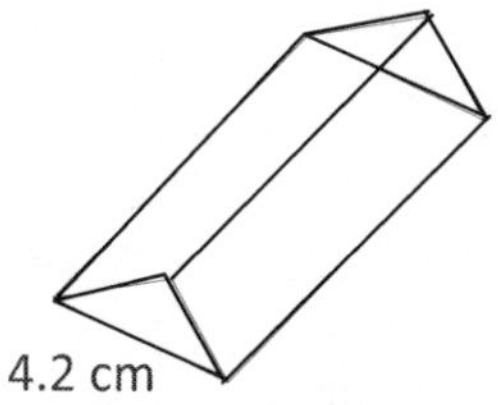

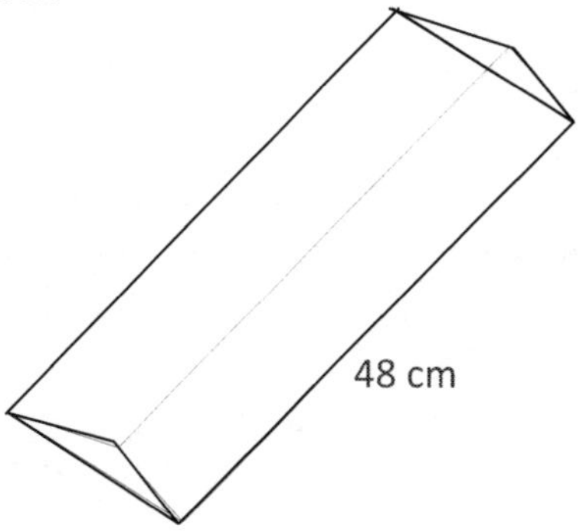

The triangular prisms in the above diagram similar to each other.
The area of cross-section of the small prism is 10.5 cm^2 and its base is 4.2 cm.
The area of cross-section of the large prism is 168 cm^2 and its length is 48 cm.
(a) Find the measure of the base of the cross-section of the large prism.
(b) Calculate the ratio between the volumes of the large prism to the small prism.
(c) Find the length of the small prism.

5 The volume of a cone is 29 cm^3 and its base radius is 2 cm.
The volume of another similar cone is 783 cm^3.
(a) Calculate the base radius of the second cone.
(b) Write down as a fraction, the ratio between the bases of first cone to the second cone.

3.02 Geometrical constructions

Core curriculum

Measuring lines and angles

Remember

1. For the angle ABC, the point B is the vertex.
2. To find the size of angle ABC, measure the angle at point B.
3. For the length of a line, give your answer correct to the nearest millimitre.

Practise exercise 5

1 In triangle MNP,
(a) find the sizes of
(i) angle MNP, (ii) angle NPM,
(iii) angle PMN,
(b) measure the length of the sides
(i) MN, (ii) NP, (iii) PM.

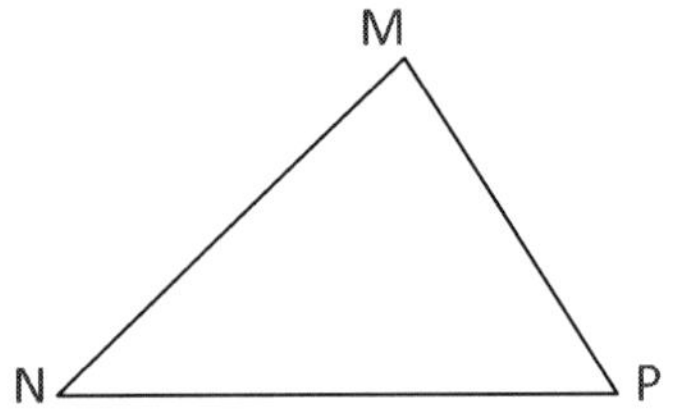

2 In the quadrilateral ABCD,
(a) find the measures of
(i) angle BCD, (ii) angle DAB,
(b) find the lengths of the sides (i) AB, (ii) DA.

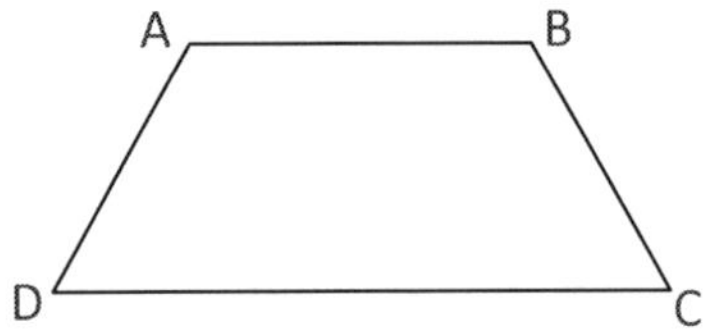

3 Draw angles with the following sizes.
(a) 27° (b) 64.5° (c) 90° (d) 140° (e) 100° (f) 180°

4 Using a ruler and compasses, draw angles with the following sizes.
(a) 60° (b) 30° (c) 90° (d) 45° (e) 120° (f) 150°

Angle bisector

Working rule

To bisect an angle using a straight edge and compasses,
1) with centre as the vertex and a constant radius, draw two arcs that cross each side of the angle,
2) then draw two intersecting arcs with centre as each of the crossing points on the sides,
3) join the vertex and the point of intersection of the arcs to get the angle bisector.

Remember

1. Your construction arcs must be clear and correct.
2. You should not erase the arcs after joining the bisector.
3. You must not locate a central point inside and join with the vertex to obtain the angle bisector.

Example

The diagram shows triangle ABC. Construct the bisector of angle CAB.

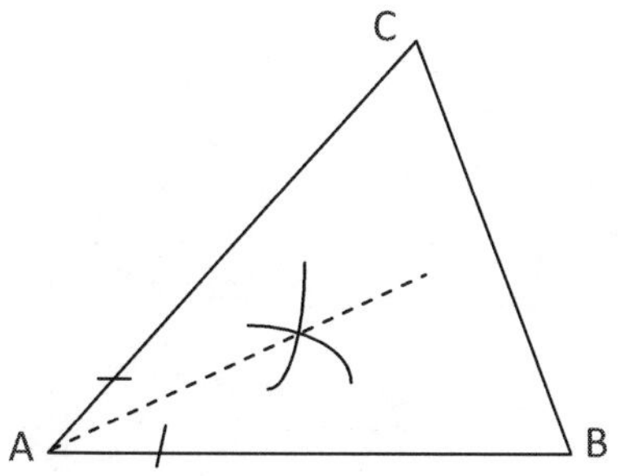

With centre as A and a constant radius draw two arcs.

With centre as each crossing point draw a pair of intersecting arcs.

Join A and the point of intersection.

- **Perpendicular bisector**

Working rule

1) With your compasses, take a radius of more than the given line.
2) With centre each end point of the line draw two pairs of intersecting arcs on either side of the line.
3) Join the points of intersections of both the pair of arcs to get the perpendicular bisector of the line.

Remember

You are not allowed to measure and find the mid-point of the line when constructing the perpendicular bisector.

Example

The diagram shows a parallelogram ABCD. Construct the perpendicular bisector of AB.

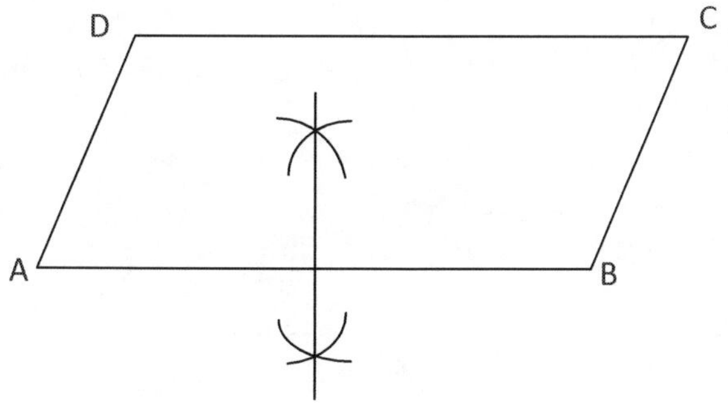

Take a radius just more than half of AB.

With A and B as centre draw two pairs of intersecting arcs.

Join the points of intersections.

Practise exercise 6

1 In triangle ABC, construct

(a) the bisector of angle ACB,

(b) the perpendicular bisector of side BC.

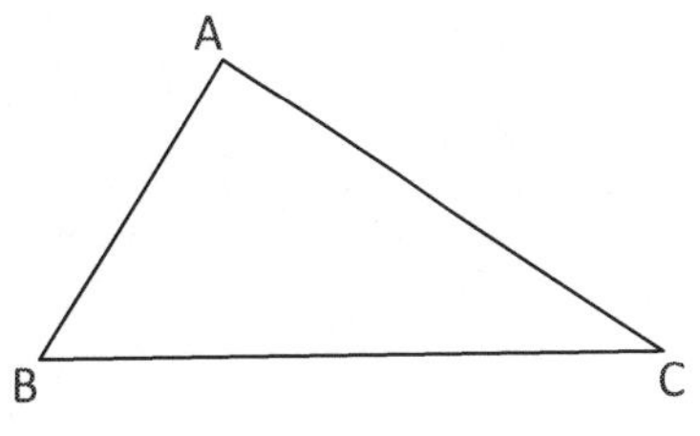

2 The diagram shows a kite LMNP.

(a) Construct the angle bisector of

(i) angle LMN,

(ii) angle PLM.

(b) What can you say about the angle bisectors of angle LMN and angle PLM?

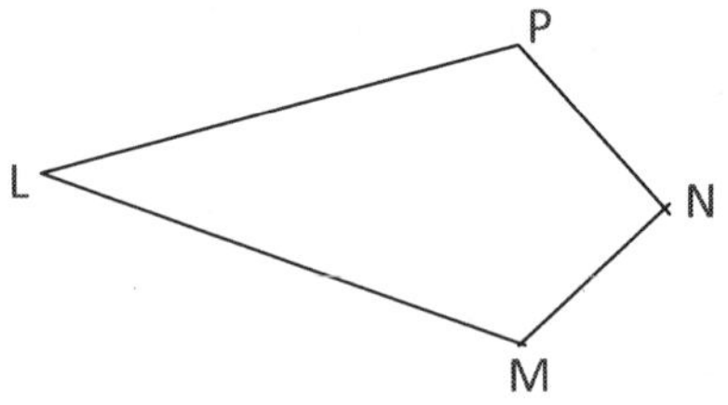

3 (a) Using a protractor, draw angle PQR = 74°. Construct an angle bisector of angle PQR.
(b) Draw a line segment AB= 8.4 cm. Construct the perpendicular bisector AB.

4 AB and CD are two chords of a circle.

(a) Construct the perpendicular bisector of

(i) AB, (ii) CD.

(b) Label the centre of the circle O.

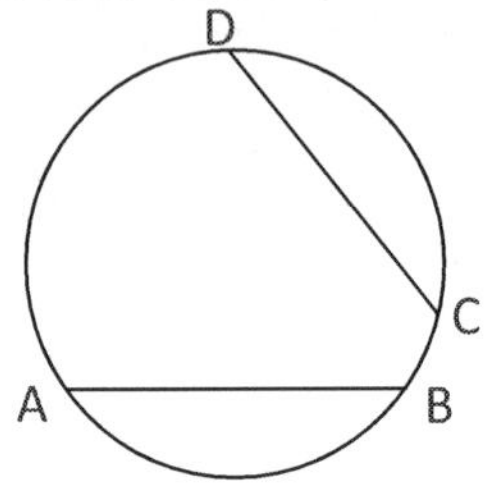

5 MN is a diameter and MP is a tangent to the circle.

(a) Find the centre of the circle and label it C.

(b) Construct an angle bisector of angle NMP.

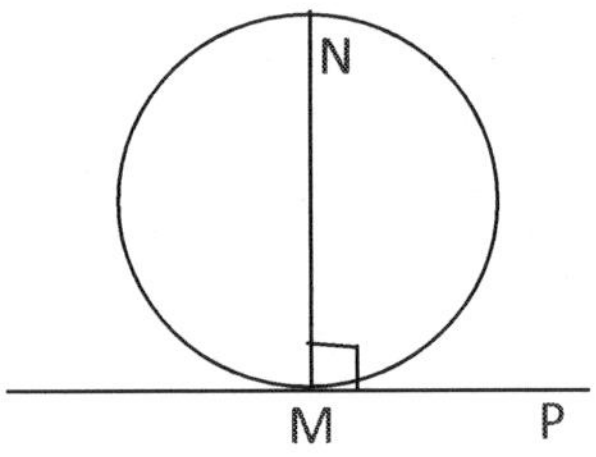

Locus

- Locus is a path consists of points whose location(position) is determined by a particular rule.

Working rule

1) To draw the locus of a point which is equidistant from a given point, construct a circle or an arc.
2) To draw the locus of a point which is equidistant from a given straight line, construct a pair of straight lines parallel to the given line and lying on both sides of it.
3) To draw the locus of a point which is equidistant from two given points, construct the perpendicular bisector of the line joining the two points.
4) To draw the locus of a point which is equidistant from two given intersecting lines, construct the angle bisectors of any two adjacent angles.
5) To draw the locus of a point which is equidistant from both sides of an angle, construct its angle bisector.

Example

Draw the locus of a point, above the line AB which is 2.4 cm away from it.

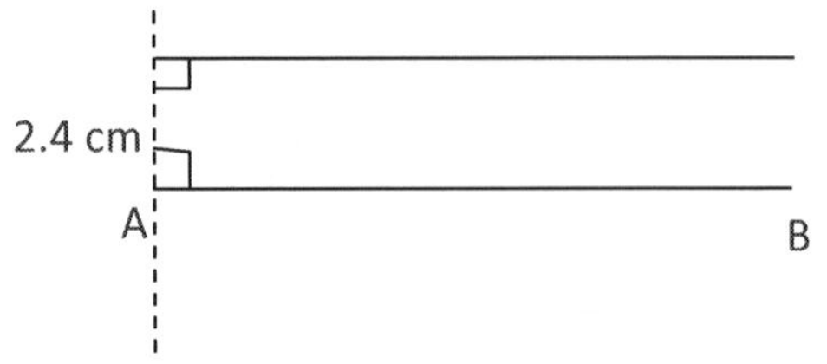

Use sets square to construct a line above AB at a distance of 2.4 cm.

Example

Two Islands C and D are in the sea. A fishing industry is equidistant from the point C and the point D. Construct the locus of possible positions on which the fishing industry lies.

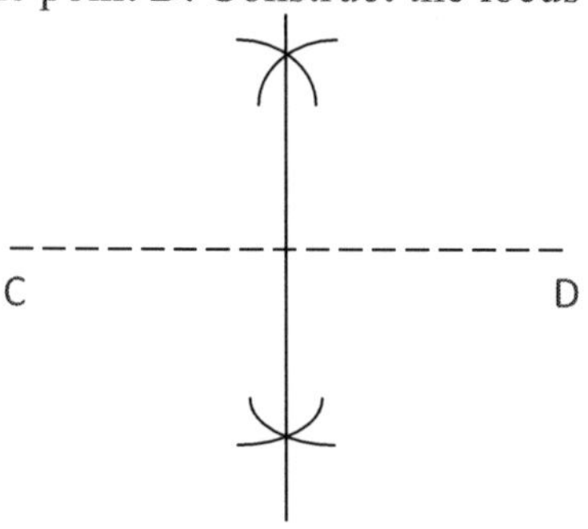

Draw the perpendicular bisector of the line joining CD.

Do not draw the locus by finding the mid-point of CD.

Don't erase the construction arcs.

Practise exercise 7

1 Draw the locus of a point, above the line which are 3.2 cm away from the line MN = 7.2 cm.

2 Draw the locus of a point, below the line which are 2 cm away from the line PQ = 7.2 cm.

3 Construct the locus of a point which is equidistant from a fixed point 4.3 cm away from it.

4 Two Villages A and B are in the same region. A market M is located equidistant from both the Villages. Construct the locus of possible position on which the market is lies.

5 Three Islands A, B, and C are in the same sea. An Air Port Island, P, is equidistant from the three Islands A, B, and C. Find the exact position of the air port island P.

6 Lines LM and NP intersect at the point Q.

Draw the locus of a point which is

equidistant from both lines.

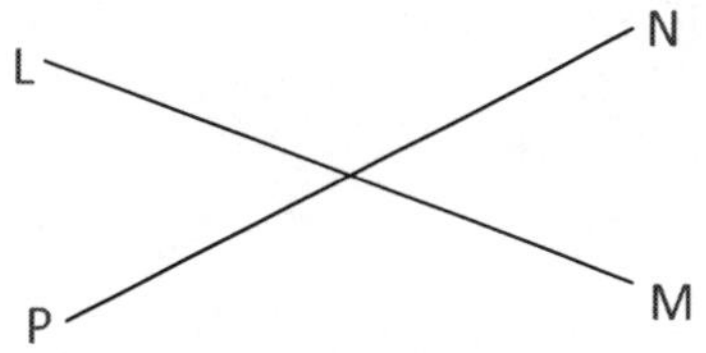

7 Draw a line XY = 6.4 cm. Construct the locus of a point, Z, above the line XY such that the area of triangle XYZ is 6.4 cm^2.

8 The diagram shows an angle ABC.

Draw the locus of a point which is

equidistant from both arms AB and BC.

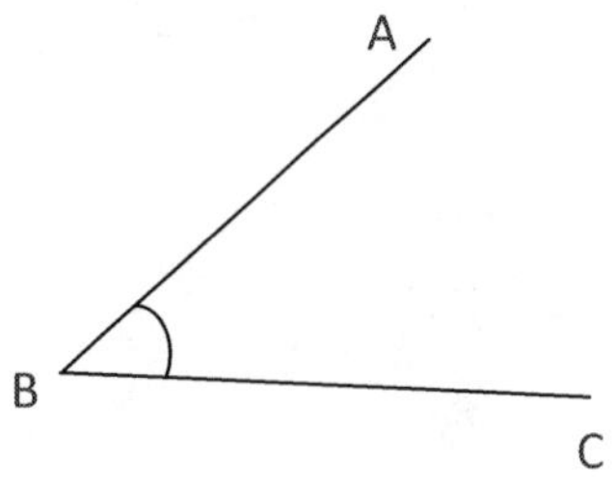

9 The diagram shows a quadrilateral MNPQ.

Construct the locus of a point which is equidistant from

(a) the points P and Q, (b) the sides MN and NP.

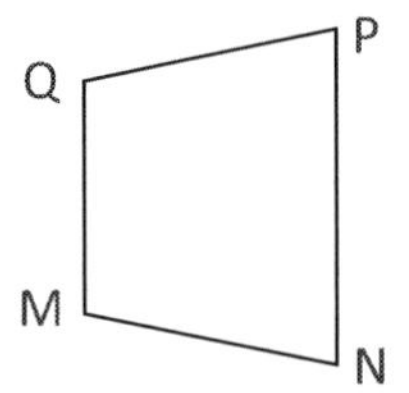

10 The figure shows a triangle STU.

Find a point, V, on SU so that TV = US.

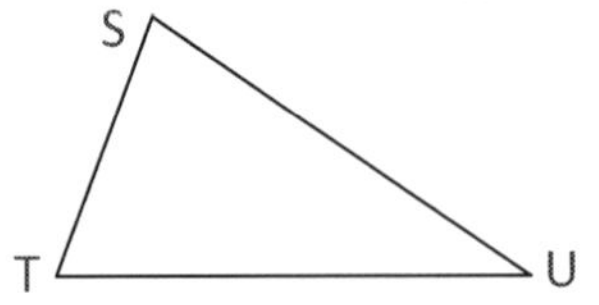

Construction of a triangle

Working rule

To construct a triangle using a straight edge and compasses,
1) draw the first side of given length,
2) with centre one end of the first side and radius the length of second side, draw an arc,
3) with centre as other end of first side and radius as length of third side, draw another arc so that it intersect the first arc,
4) join the point of intersection with both ends of first side.

Remember

1. Do not construct the given sketch of the triangle exactly.
2. Use accurate measurements labeled in the triangle to construct it.
3. DO not erase the construction arc.
4. Use protractor to measure the angles in a triangle.
5. You need at least the length of a side or the length of a diagonal to construct a square.
6. You need at least two parts to construct a rectangle or rhombus.
7. You need at least three parts to construct a construct a parallelogram or kite or trapezium.
8. You need the length of radius or diameter to construct a circle.

Example

Mark three points A, B and C which are not in straight line.
Construct a circle that passes through these points.

Working:

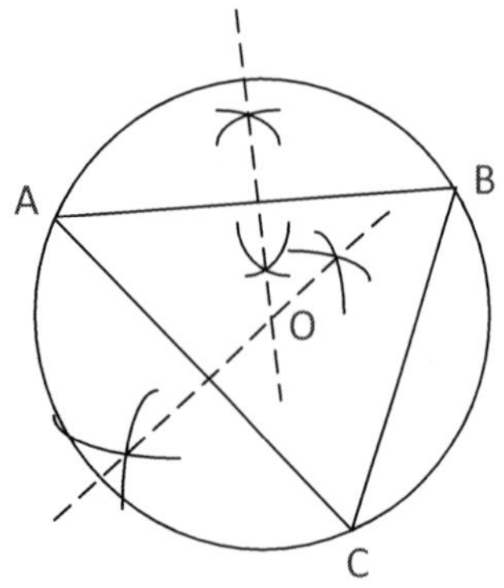

Mark the points A, B, C and join them to get a triangle.

Draw the perpendicular bisectors of AB and AC to get the point O.

With O as the centre and OA or OB as radius, draw a circle.

Practise exercise 8(A) **Core curriculum**

1 (a) Construct the adjacent triangle accurately, using a straight edge and compasses only.

(b) From your construction to part(a), write down the measure of (i) angle ABC, (ii) angle ACB.

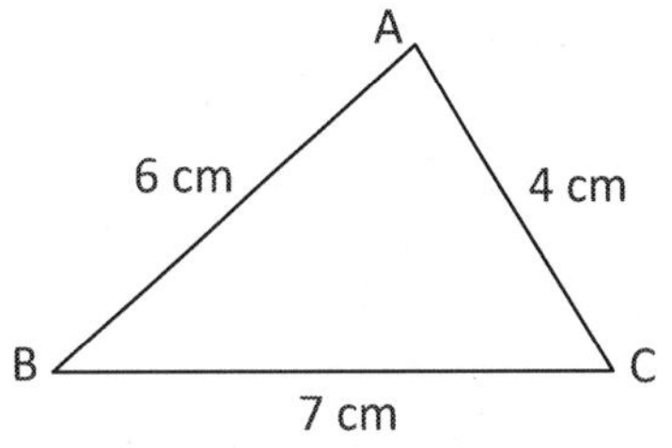

2 (a) Using a straight edge and compasses only, construct a triangle PQR in which PQ= 8.2 cm, QR = 4.8 cm and RP = 7 cm.
(b) On your diagram, draw the locus of a point which is equidistant from
(i) the points R and P, (ii) the sides PQ and QR.

3 (a) Using compasses and straight edge only, draw a triangle LMN in which LM = 5 cm, MN = 7.5 cm and NL = 8.4 cm.
(b) Draw the locus of a point inside the triangle which is 2.7 cm from M.
(c) (i) Construct a circle that passes through all the three points L, M and N.
(ii) Label the centre of the circle by the letter C.
(d) Write down the size of the angle LMN correct to the nearest degrees.
(e) Shade the region inside the triangle which is 2 cm more than L and closer to L than to N.

4 The points P, Q and R are non collinear. Construct a triangle which passes through all the three points P, Q and R.

5 (a) Construct a kite ABCD in which AB = 7.4 cm, BC = 4 cm and AC = 9 cm.
(b) On your diagram, (i) draw the locus of points that are 4.2 cm from the vertex A,
(ii) draw the perpendicular bisector of the diagonal AC.

6 (a) Construct a parallelogram EFGH in which EF = 8 cm, FH = 6.3 cm and HE = 5 cm, using a straight edge and compasses only.
(b) Shade the region inside the parallelogram EFGH which is
(i) nearer to F than E, (ii) nearer to EF than FG.

7 The diagram shows a semi circle. (a) Draw the tangent to the semicircle at C.
(b) Use a straight edge and compasses to locate the mid-point of AB.
(c) The point C is marked on the semi-circular part. Draw AC and BC.
(d) Shade the region inside the semicircle which is closer to A than to B and closer to AC than BC.

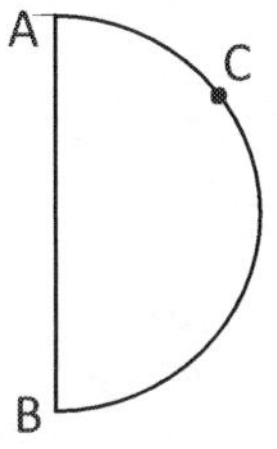

Practise exercise 8(B) **Extended curriculum**

1 Use a straight edge and compasses only in this question.
(a) Construct a circle with centre C and diameter 9 cm.
(b) Mark a point D such that CD = 8 cm.
(c) Construct two tangents DA and DB to the circle.
(d) (i) Complete the quadrilateral CADB.
(ii) What is the special name of the quadrilateral CADB?
(e) Shade the region inside the circle which is more than 1.3 cm from B and closer to B than to A.

2 (a) Construct a quadrilateral MNPQ in which MN = 7.9 cm, NP = 5.5 cm, PQ = 6.1 cm, QM = 4 cm and MP = 9.2 cm.
(b) Shade the region inside the quadrilateral which is 3 cm less than C, closer to BC than CD and closer to C than A.

3 Use straight edge and compasses only in this question.
(a) Construct an equilateral triangle XYZ in which XY = 6.7 cm.
(b) Mark, A, the mid-point of YZ through construction.
(c) Construct a circle that passes through A and touches ZX at X.
(d) Shade the region inside the circle which contains the points more than 2 cm from X and nearer to A than X.

Scale drawing

- The scale of drawing is usually written in the form 1: n.
- The scale 1: n means 1 cm on the diagram = n cm real length.

Working rule

To calculate
1) the scale of a drawing from a given length on the drawing, use direct proportion or simply divide,
2) the real length of the drawing from a given scale of drawing , use direct proportion or simply multiply.

Remember

For the real length, a measurement from the scale of a drawing is not acceptable.

Practise exercise 9(A) **Core curriculum**

1

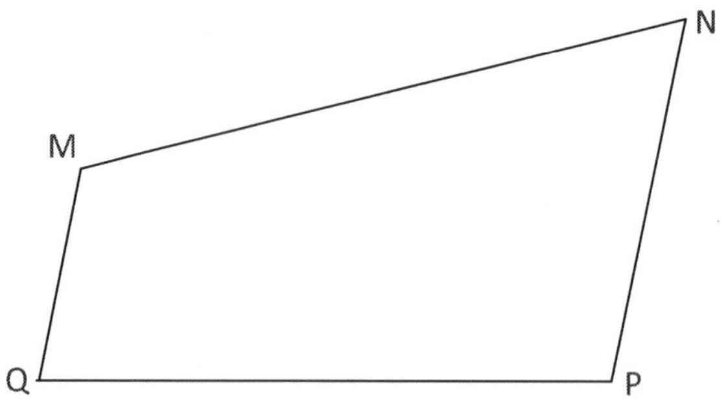

The diagram is a scale drawing of a field. The actual length of PQ is 900 m.
(a) Write down the scale of the drawing in the form 1: n.
(b) Calculate the actual length of (i) QM, (ii) MN, (iii) NP.
(c) (i) A tube well is equidistant from the point M and the point N. Construct the locus of the points on which the tube well lies, using a straight edge and compasses only.
(ii) The tube well is also same distant from the sides NP and PQ. Construct and mark, R, the position of the tube well.

2 The diagram is a scale drawing of a square shaped garden. The actual length of AB is 18 m.
(a) Write down the scale of the drawing in the form 1: n.

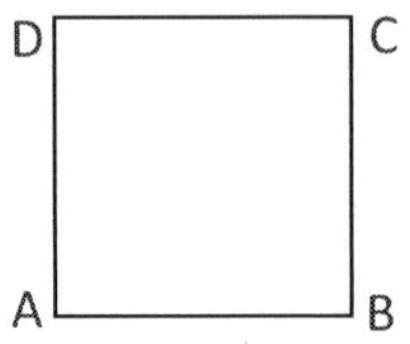

(b)(i) Using a straight edge and compasses only, locate the mid-point of the garden on the scale diagram. Label the mid-point M.

(ii) Calculate the actual distance on the garden between the point M and the point C.

(c)(i) A special kind of red flowers are planted on a curve of 2 cm equidistant from point A. Using compasses and straight edge only, on the diagram draw the curve on which the red flowers are planted.

(ii) Write down the actual distance of the red flowers from point A.

3 A plot of land WXYZ is represented on a map of scale 1: 200000.

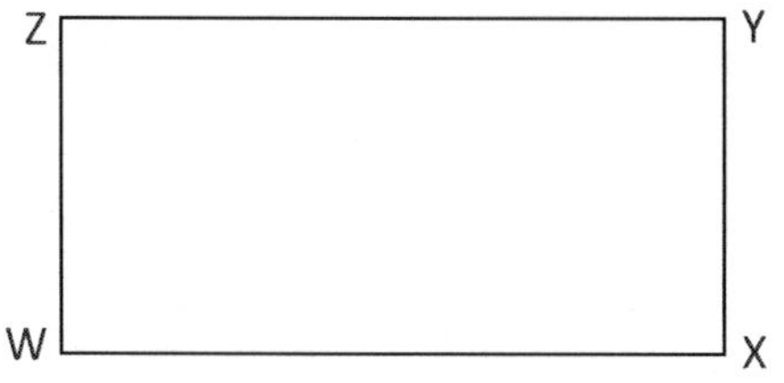

(a) Find the actual length of WX in kilometres.

(b) Find the actual length of XY in kilometres.

(c) Find the actual length of the diagonal

(i) in centimetres, (ii) in metres,

(iii) in kilometres.

4 The diagram shows the scale drawing of a plot of a house. In the diagram 1 cm represents 5.5 m.

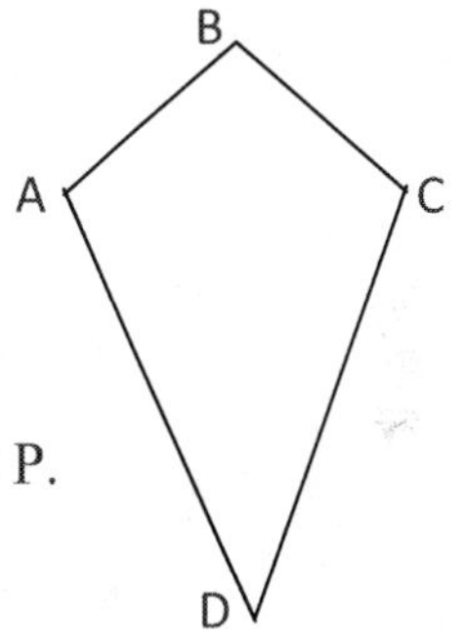

(a) A pillar is equal distances from the sides AB and BC. Construct the locus of points on which the pillar locates.

(b) The pillar is at same distances from the point C and the point D.

(i) By actual construction, find the position of the pillar and label it P.

(ii) What is the actual distance between the pillar and the point B.

(c) Find the actual perimeter of the plot of the house.

5 The perimeter of an equilateral triangular floor is 24 m.

(a) Using a scale of 1: 100, make a scale drawing of the floor and label it MNP.

(b) In your scale drawing

(i) draw accurately the mid-point of MN and label it Q,

(ii) find the scale length between P and Q,

(iii) find the real length between P and Q.

6

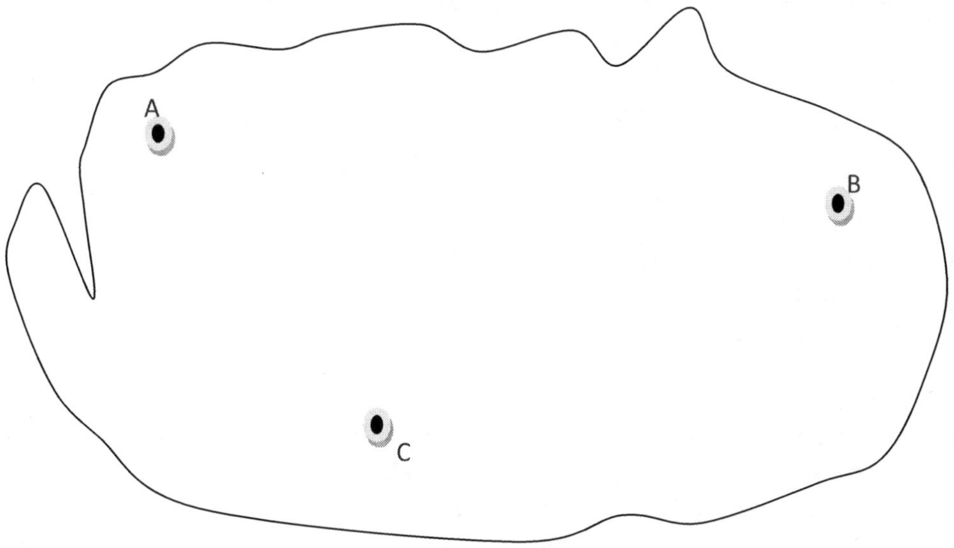

The diagram shows a City, drawn to a scale of 1 cm equals 10 km. Points A, B and C shows three densely populated areas in the City.

(a) Find the actual distance between the points A and C, in kilometres.

(b) On the diagram, construct the locus of points which are
(i) 40 kilometres from point B, (ii) 35 kilometres from point C.
(c) Label the point of intersection of both loci in part b(i) and (ii) as D.
(d)(i) An air port is situated exactly at the middle of the points A and D. Construct the position of the points on which the air port is situated.
(ii) The air port is also equidistant from the points A and C. Construct and label the position of the air port E.

Practise exercise 9(B) **Extended curriculum**

1 The diagram shows the rough plan of the floor of a room.

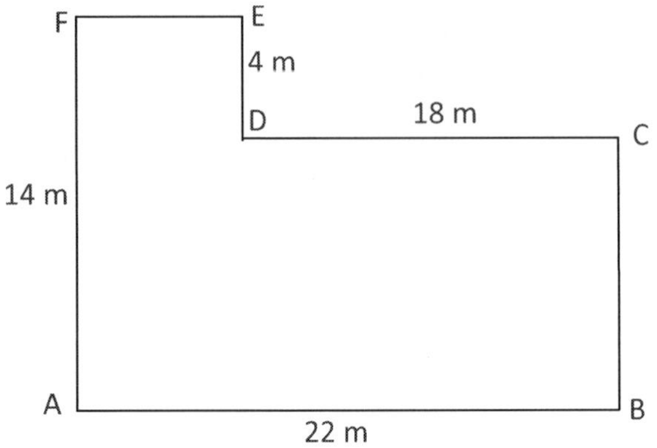

AB = 22 m, CD = 18 m, DE = 4 m and FA = 14 m.

(a) Using a scale of 1 cm to represent 2 m, draw an accurate plan of the room. Use a straight edge and sets squares only to draw the accurate plan of the floor.

(b)(i) What is the real distance between the points B and F.
(ii) Calculate the actual area of the triangular region ABF of the floor in square metres.
(iii) Carpet costing $2.25 per square metre is to be covered in the triangular region ABF. What is the total cost of the carpet to be covered?
(c) On your plan, draw the locus of points which are same distances
(i) from the points B and C, (ii) from the sides BC and CD,
(iii) mark G, the points of intersection of both the loci.

2 The diagram shows a rough plan of an isosceles trapezoidal play ground. PQ = 120 m, QR = 80 m, RS = 190 m and PR = 170 m.

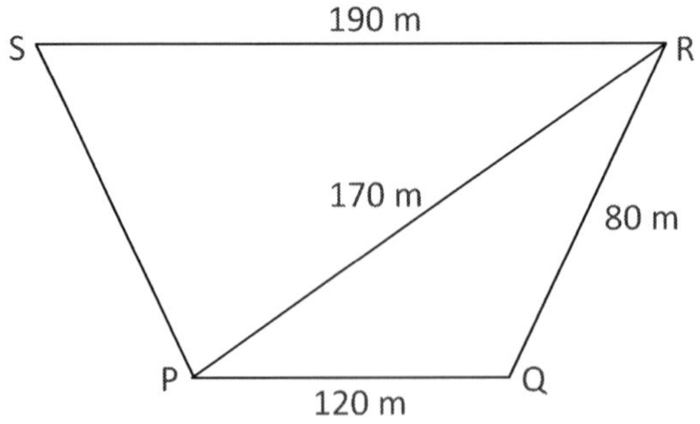

(a) Use a scale of 1 cm to represent 2000 cm, draw an accurate plan of the play ground.
(b) On your plan,
(i) construct the locus of points inside the play ground which are exactly 1.4 cm away from S,
(ii) construct the locus of points which are equidistant from Q and R,
(iii) a player is standing at the point of intersection of the curve to part(i) and the line to part (ii). Mark the position of the player P.
(c) An electric pole is on the fence and equidistant from R and S. On your plan, through accurate construction locate the position of the electric pole and label it as T.

3 The diagram shows the location of a radio mast R. Signals from the radio mast, R, can be received up a distance of 125 km.

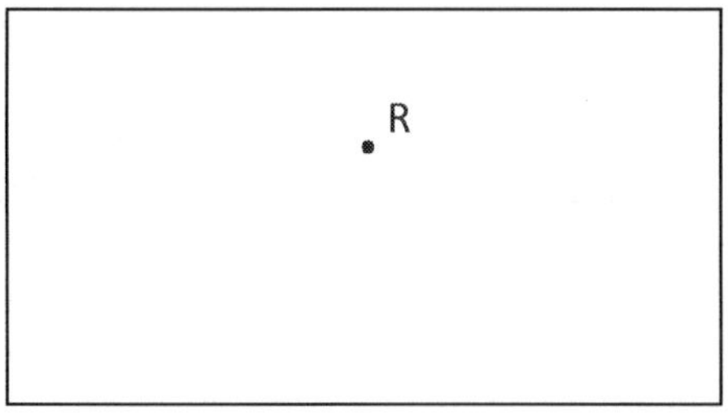

(a) Using a scale 1: 2500000, construct an outline of the region which can be received the signal.

(b)(i) On a diagram, shade the region which is 1.8 cm away from the mast.

(ii) Find the actual area, in km^2, of the shaded region.

4 The diagram shows the map of a rectangular plot of land ABCD. AB = 12 cm and BC = 8 cm. The scale of this map is 1: 3000.

(a) Calculate the actual dimension of the land ABCD.

(b) There is a path from the mid points of BC and DA. Using a straight edge and compasses, draw a line to show the path.

(c) A cow is fastened by a rope 40 m long at the point B.

(i) Shade the region that the cow can reach.

(ii) Calculate the area of the region that the cow can reach in square metres.

3.03 Angle properties

Core curriculum

Angles at a point/on a line/on intersecting lines

Remember

1. The angles around a point(AAP) always add up to 360°.
2. The angles on a line(AOL) always add up to 180°.
3. When two lines intersect, the vertically opposite angles(VOA) are equal.

Example 1

Write down an equation from the diagram and solve it for a.

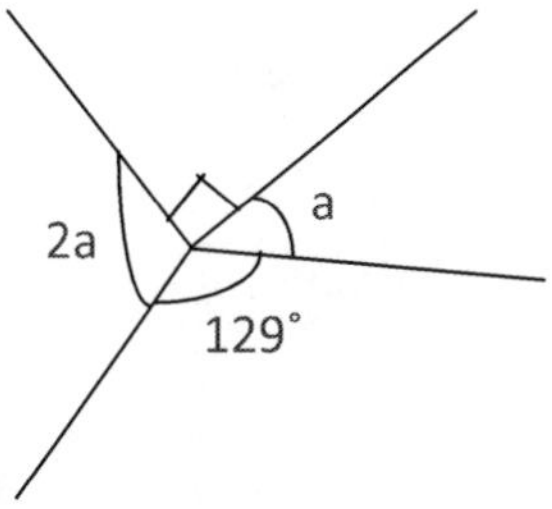

Working: $2a + 90° + a + 129° = 360°$

$3a = 360° - 219°$

$a = 47°$

Use AAP = 360° to form the equation.Use the rule 'side change, sign change' to solve the equation.

Example 2

Find the angles a, b and c.

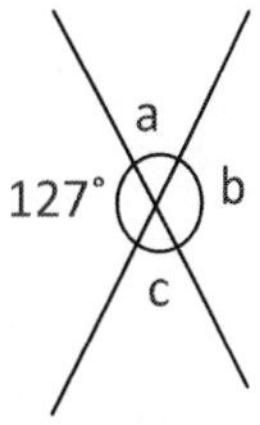

Working: $a + 127° = 180°$

$a = 53°$

$b = 127°$

$c = 53°$

Use AOL = 180°.

b and 127° are VOA.

c and a are VOA.

Practise exercise 10

1 Write down an equation from each diagram and solve it for the letter.

(a)

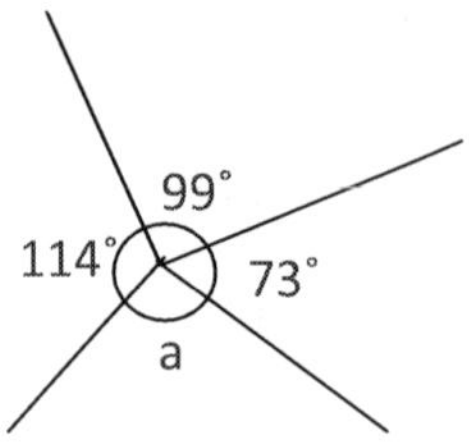

(b)

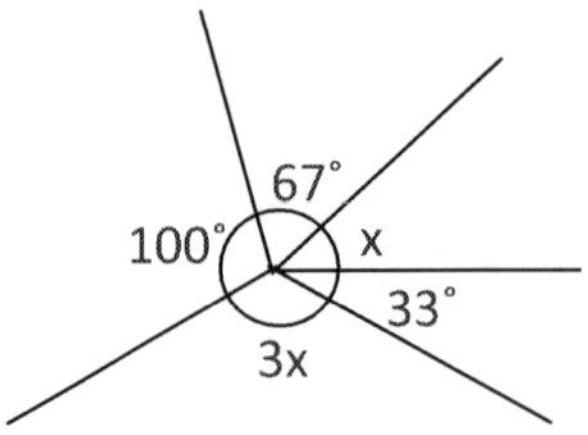

(c)

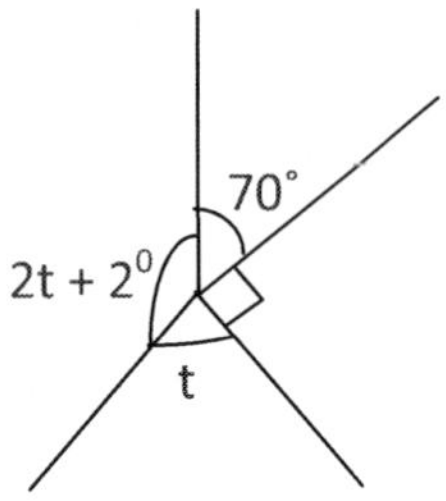

(d)

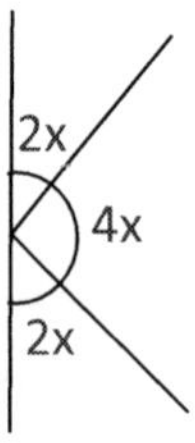

2 Find the value of each letter.

(a)

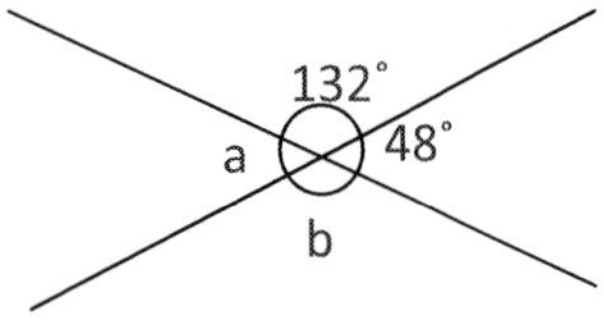

(b)

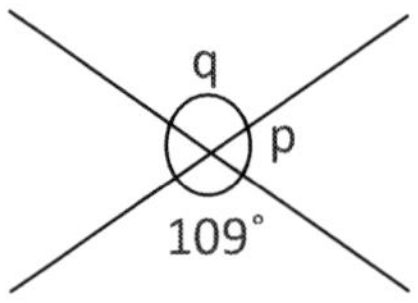

(c)

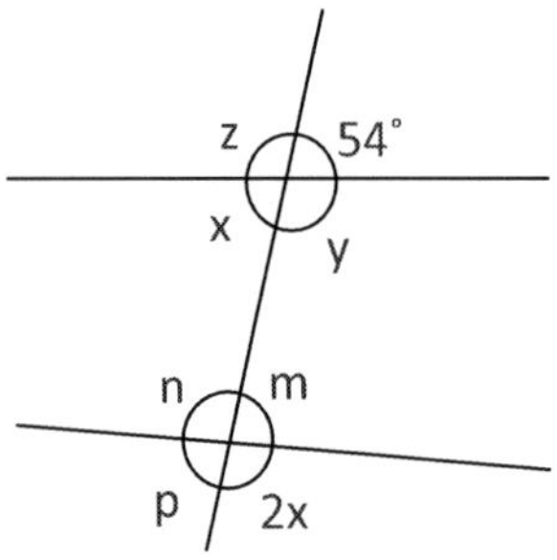

(d)

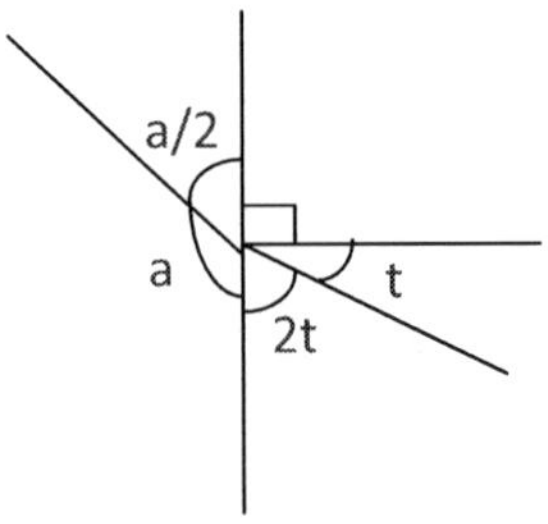

3 Find the values of l and m.

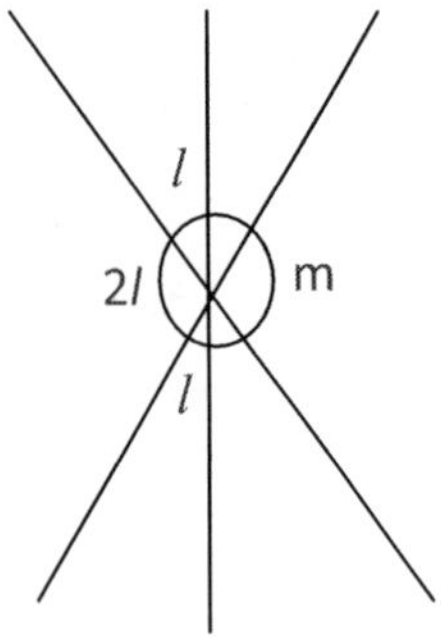

4 The diagram shows four angles around a point.
(i) Write down an equation in x and y.
(ii) Simplify your equation.
(iii) Find y when x = 52.

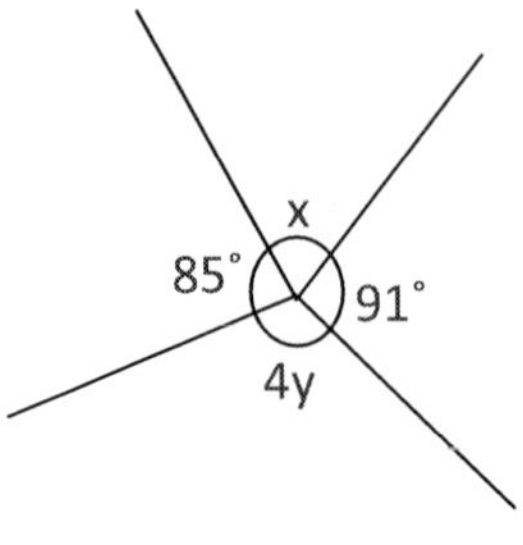

Angles in a triangle and quadrilaterals

Remember

1. The angles in a triangle(AIT) add up to 180°.
2. In an equilateral triangle each angles is 60°.
3. In an isosceles triangle the angles opposite to the equal sides are same.
4. Any exterior angle of a triangle is the sum of its interior opposite angles.
5. The angles in a quadrilateral(AIQ) add up to 360°.
6. In a square each angle is 90°.
7. In a parallelogram or rhombus each pair of opposite angles are equal.

Example 1

Find the angles x, y and z.

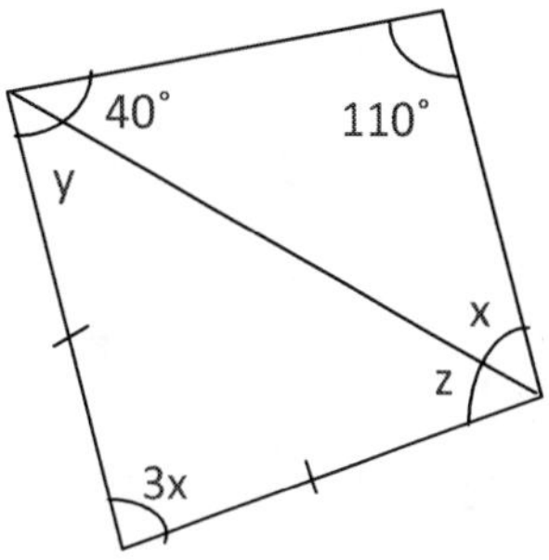

Working: In the top triangle,

$x + 40° + 110° = 180°$

$x = 30°$

In the bottom triangle,

$3x + y + z = 180°$
$90° + y + y = 180°$
$2y = 90°$
$y = 45°$ So $z = 45°$

Use AIT = 180° to form an equation and solve it.

Use AIT = 180° to form an equation.

Replace x by30° and z by y.

z = 45°, since z = y.

Example 2

Calculate the size of each angle in the adjacent quadrilateral.

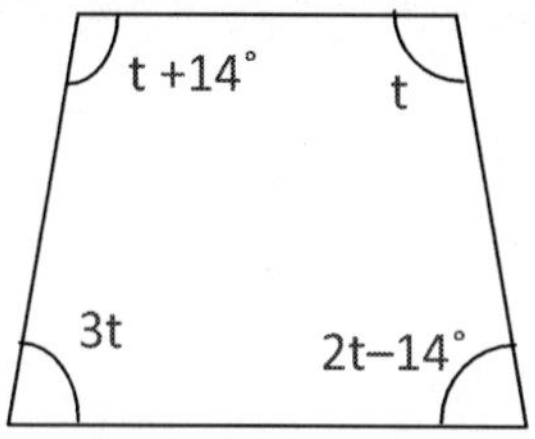

Working: $t + 2t - 14° + 3t + t + 14° = 360°$

$7t = 350°$ or $t = 50°$

So the other angles are $t = 50°$,
$2t - 4° = 2 \times 50° - 4° = 100°$,
$3t = 3 \times 50° = 150°$,
$t + 14° = 50° + 14° = 64°$.

Use AIQ = 360° to form an equation and solve it.

Replace t = 50° in other angles.

Practise exercise 11

1 Find the size of angles marked with letters in each diagram.

(a)

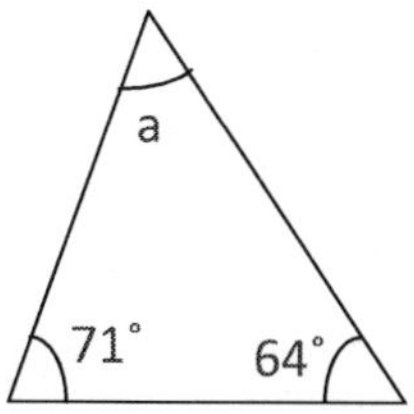

(b)

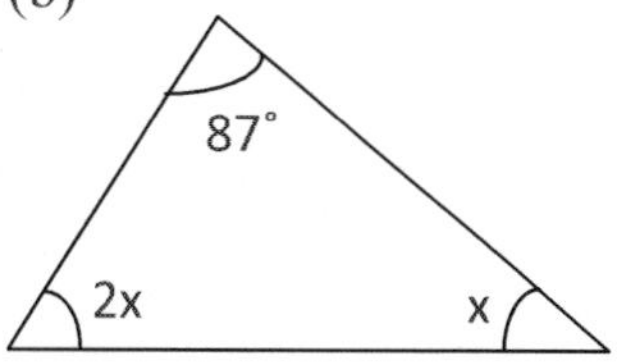

(c)

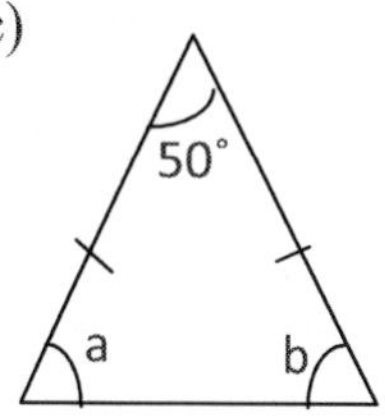

(d)

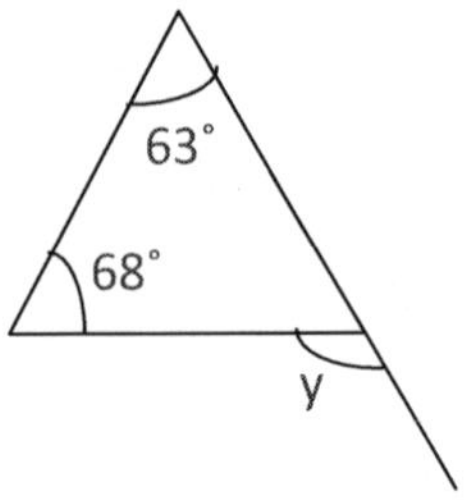

(e)

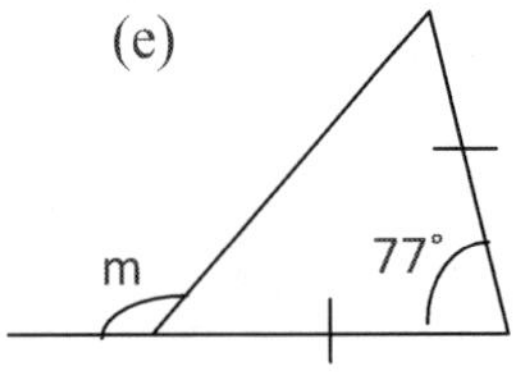

(f)

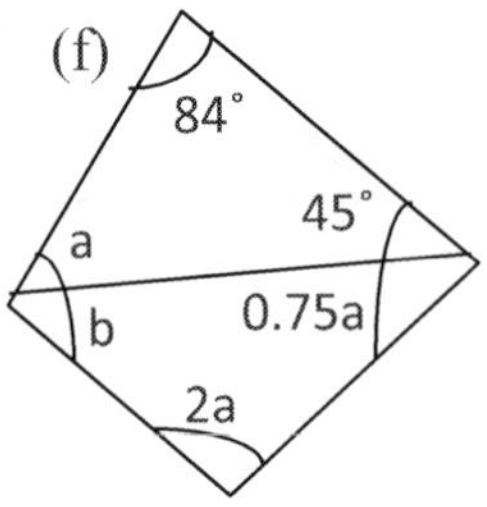

2 Find the values of the letters in each diagram.

(a)

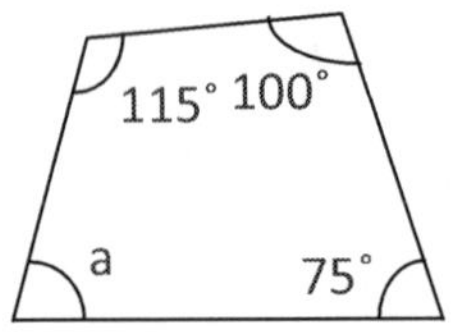

(b)

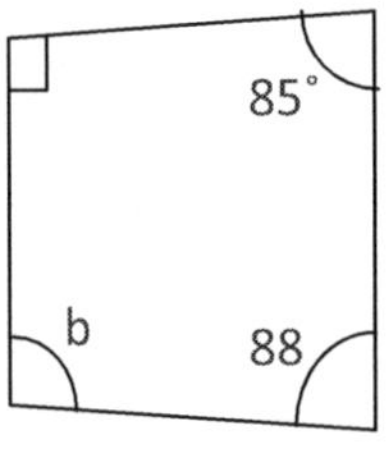

(c)

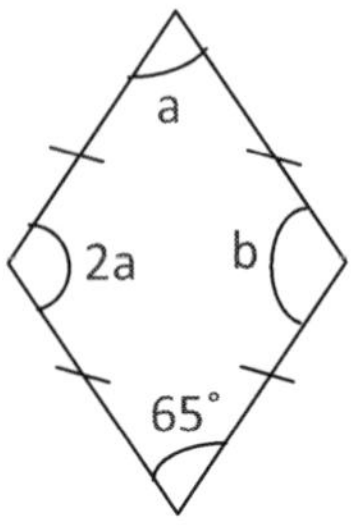

3 The diagram shows a kite ABCD.

Angle DAB = 32° and angle DBC = 55°.

Work out

(i) angle ADB, (ii) angle BCD.

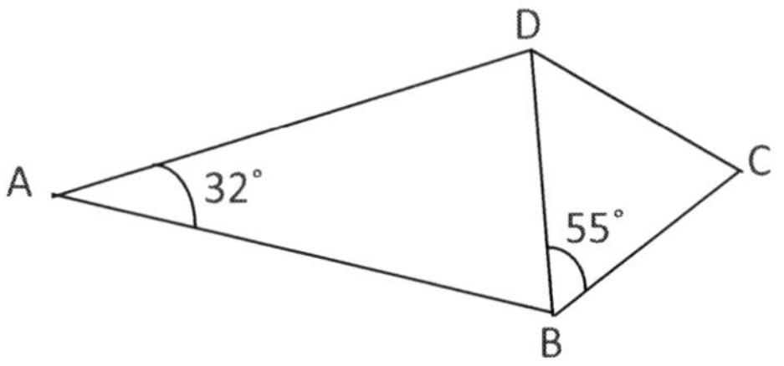

4 The diagram shows triangle NPR lie inside the rectangle.

R is the mid-pointof MQ. Angle MRN is 62^0.

(a)What is the special name of the triangle NPR?

(b)Find the size of (i) angle MNR, (ii) angle NRP, (iii) angle NRP.

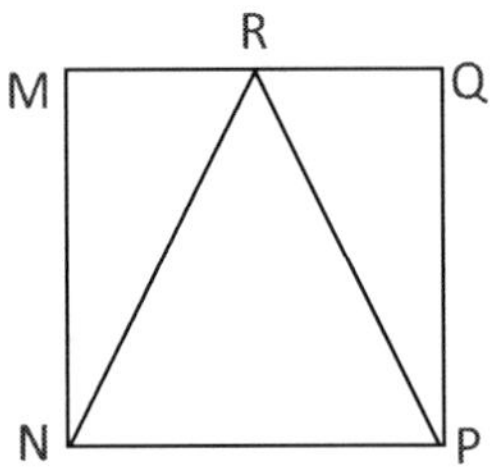

5 Find the angles e, f and g.

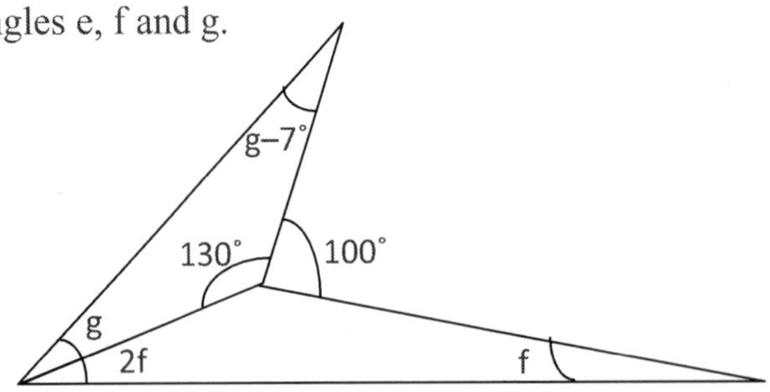

6 A triangle has angles of x°, 3x° and 2x°. Calculate the size of its
(a) smallest angle, (b) largest angle.

7 The interior angles of a quadrilateral are 2t+3°, 3t, 4t-24°, t+20°.
(a) Write down an equation in t. (b) Solve the equation for t.
(c) Write down all the angles of the quadrilateral.

8 (a)

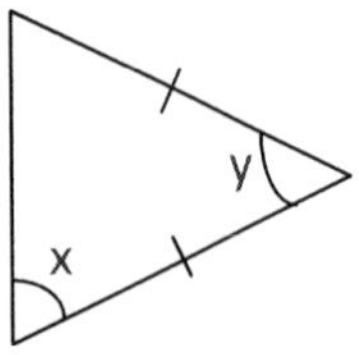

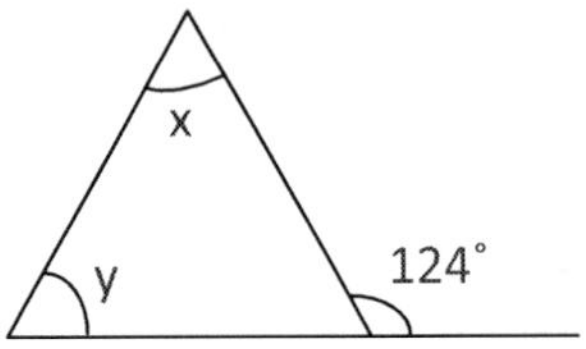

(a) Using each triangle write down an equation in x and y.
(b) Find the values of x and y by solving the simultaneous equations from part(a).

Angles formed within parallel lines

- In the diagram line L_1 is parallel to line L_2.
  The line t_1 is the transversal

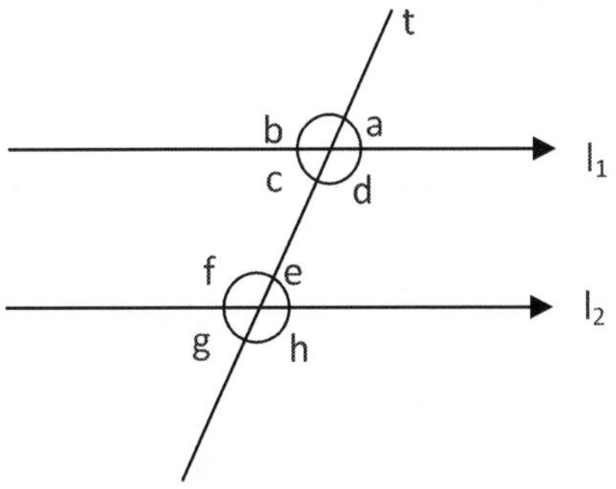

1) The four pairs of corresponding angles (C.A.) are equal

$a = e$, $b = f$, $c = g$ and $d = h$.

2) The four pairs of alternate angles(A.A.) are equal.

$a = g$, $b = h$, $c = e$ and $d = f$.

Remember

In short, odd angles are equal i.e. $a = c = e = g$ and even angles are equal, i.e. $b = d = f = h$.

3) The sum of the interior angles (or exterior angles) on the same side of the transversal is 180°. $c + f = 180°$, $d + e = 180°$, $a + h = 180°$ and $b + g = 180°$.

Example

Find the angles marked a, b, c, d, e, f and g in the diagram.

Working:

$f=50°$	f and 50 are V.O.A.
$b=50°$	b and 50 are A.A.
$d=50°$	d and 50 are C.A.
$g+50°=180°$	A.O.L is 180.
$g=130°$	
$e=130°$	e and g are V.O.A
$a=130°$	a and g are A.A.
$c=130°$.	c and g are C.A.

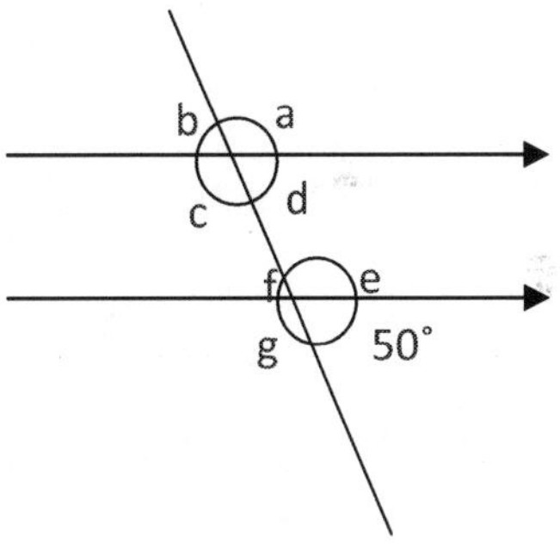

Practise exercise 12

1 Find the angles marked by letters in each diagram.

(a)

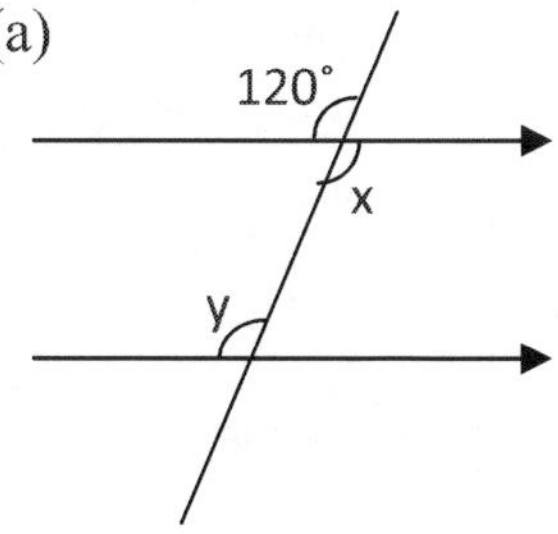

(b)

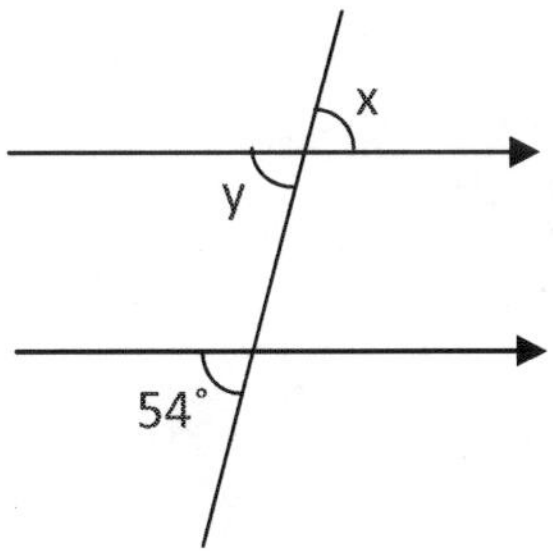

(c)

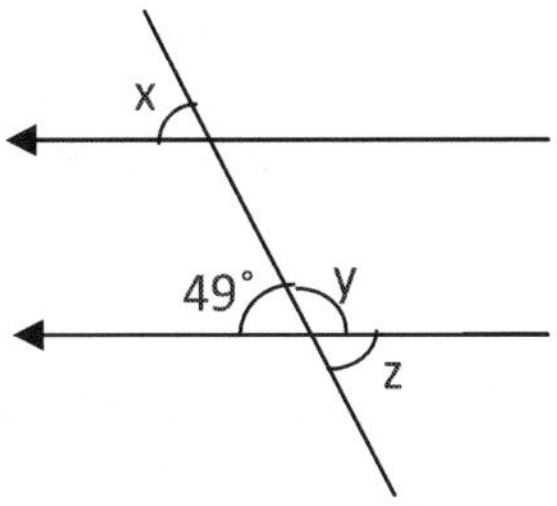

(d)

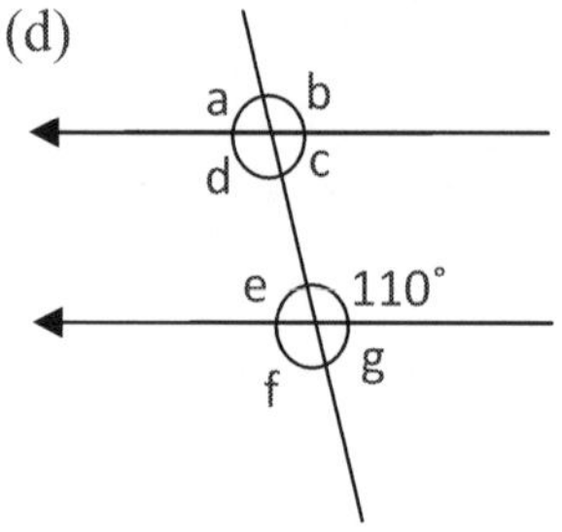

(e)

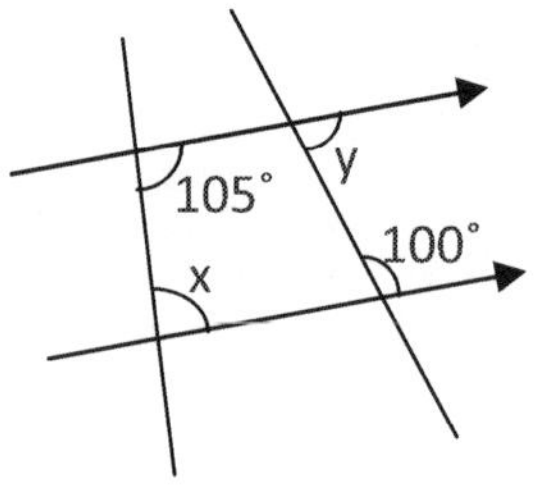

(f)

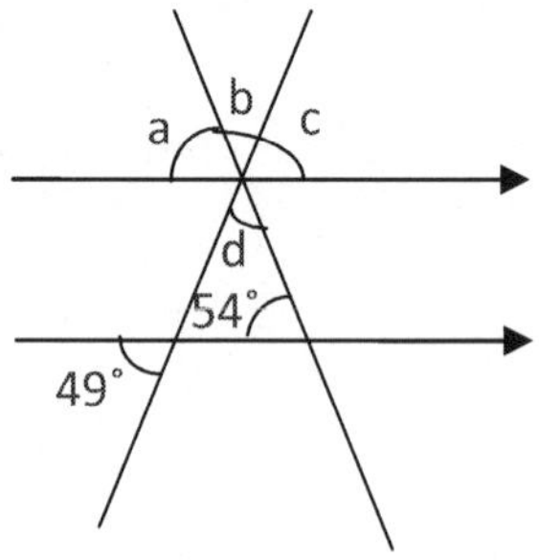

2 (a) Using the diagram write down a pair of simultaneous equations.

(b) Find the values of x and y.

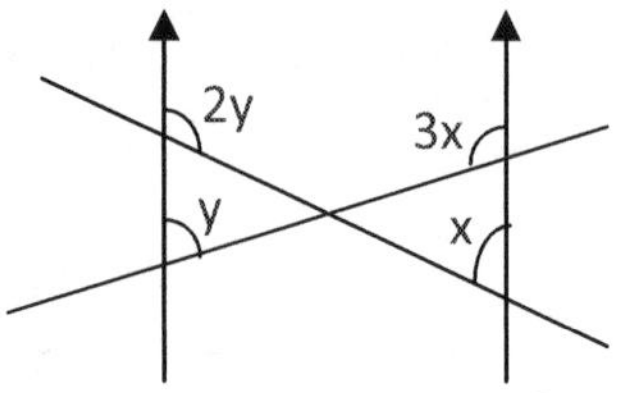

3 ABCD is a parallelogram.

Angle ACD=40 and angle BCE=55.

Work out the size of

(a) the angle BAC, (b) the angle ADC,

(c) the angle ACB, stating reason for each answer.

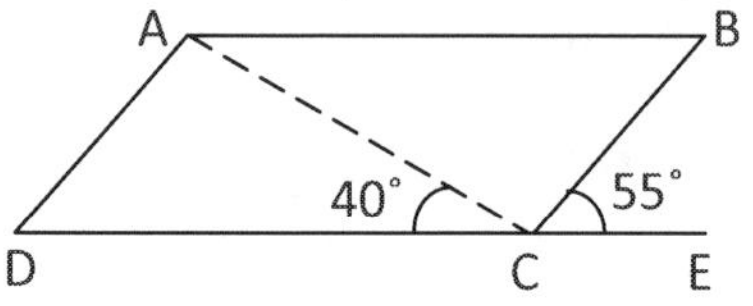

4 MNPQ is a rhombus. Angle PQM=51°.

Write down the size of

(a) angle PNM ,(b)angle QMN , giving a reason for your answer.

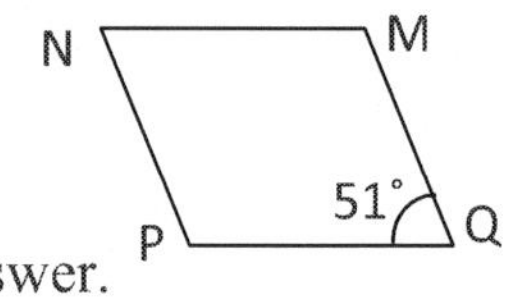

5 The diagram shows a square OABC.

Find the sizes of the angles a and b.

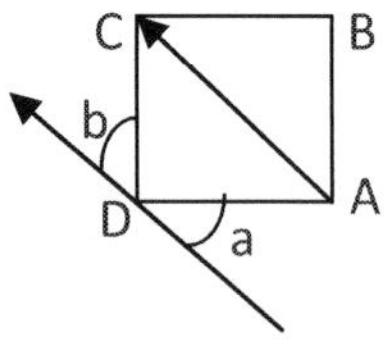

6 In the diagram AB is parallel to DE.

Angle ABC= 69° and angle ACB= 49°.

Workout the size of

(a) angle CDE, (b) angle DEC.

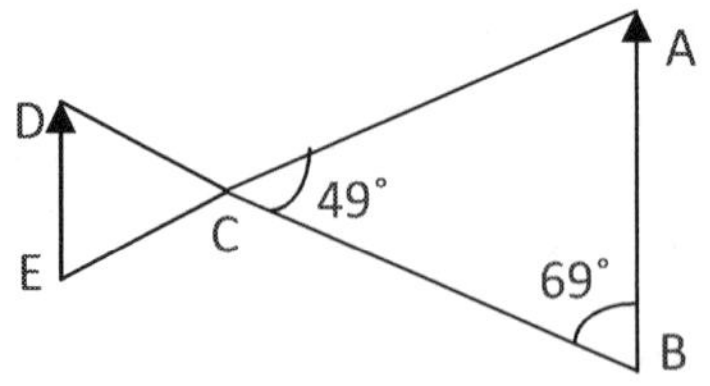

7 In the diagram BC is parallel to DE.

Angle ACB=53° and angle BAD=65°.

Work out the sizes of

(a) angle ABC, (b) angle CAE, (c) angle BAC.

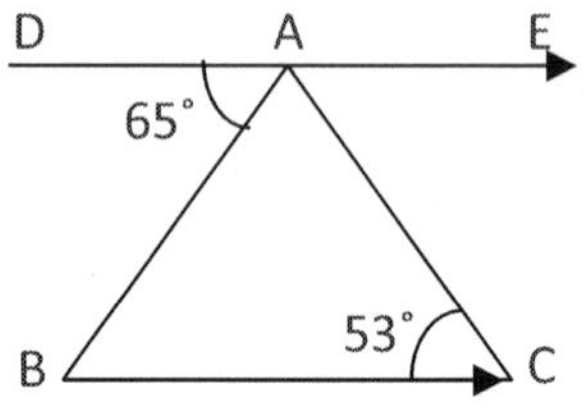

Angle properties of circle

- In the diagram, AB is the diameter of the circle.
  In the top semicircle, angle ACB=90°.
  In the bottom semicircle, angle ADB =90°.

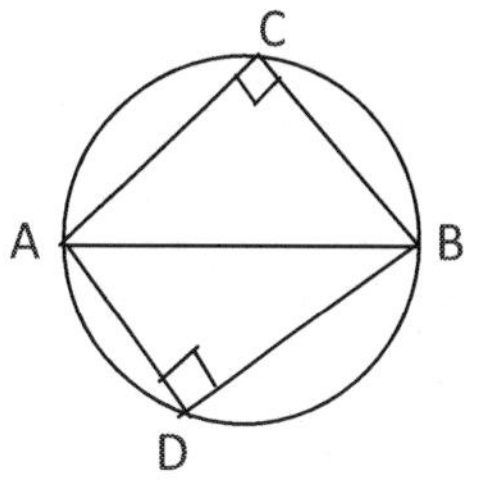

- In the diagram, OA is a radius of the circle. Line BC is a tangent to the circle.
  Then angle OAB = angle OAC = 90°.

Remember

1. The angle between a tangent to a circle and a radius through the point of contact is 90° or a right angle.
2. The angle in a semicircle is always 90° or a right angle.

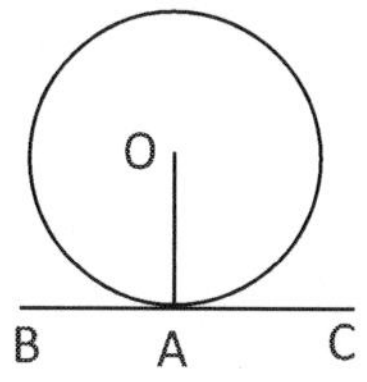

Example 1

AB is a diameter of a circle. Find the size of each angle marked x, y and z.

Working: $x = 90^\circ$ — The angle in a semicircle is 90°.

$y + 90^\circ + 55^\circ = 180^\circ$ — The angles in triangle ABC add up to 180°.

$y = 35^\circ$

$z + z + 90^\circ = 180^\circ$

$2z = 90^\circ$ or $z = 45^\circ$ — The angles in isosceles triangle ABD is 180°.

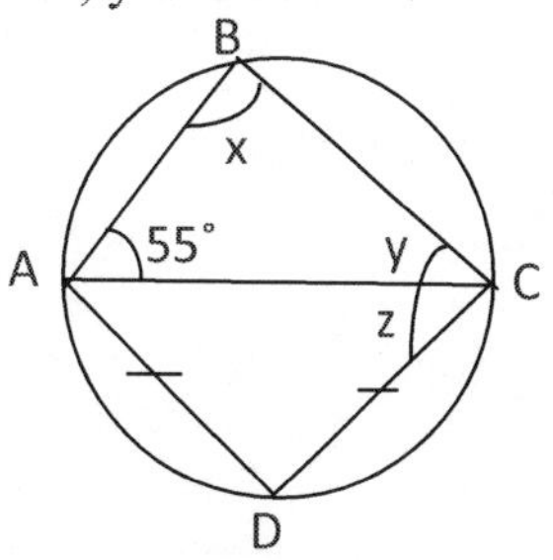

Example 2

PT is a tangent to the circle and OP is a radius.

Angle POT=65°. Find the size of angle x.

Working: Angle OPT=90°

$x+90^\circ+65^\circ=180^\circ$

$x=25^\circ$

The angle between a tangent and the radius is 90°. The angles in triangle POT add up to 180°.

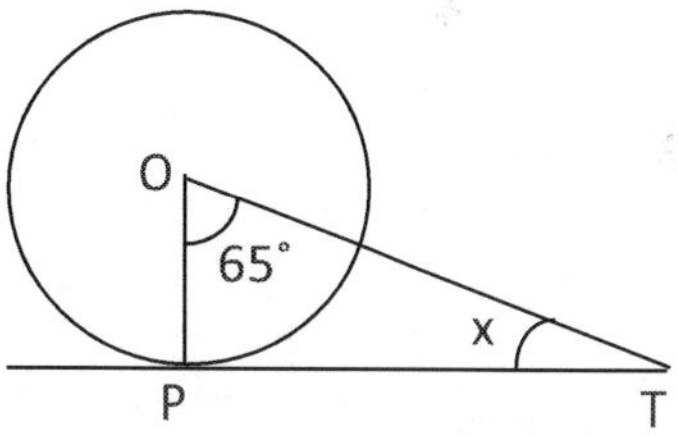

Practise exercise 13

1 Find the size of the angles marked with letters in each diagram.

(a)

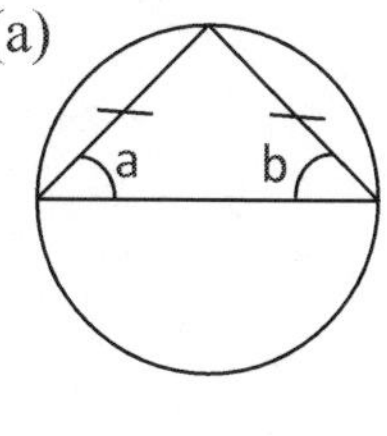

(b)

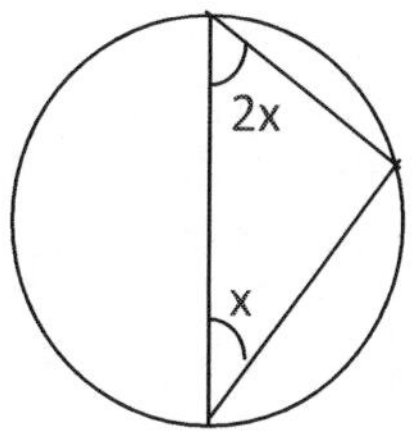

(c)

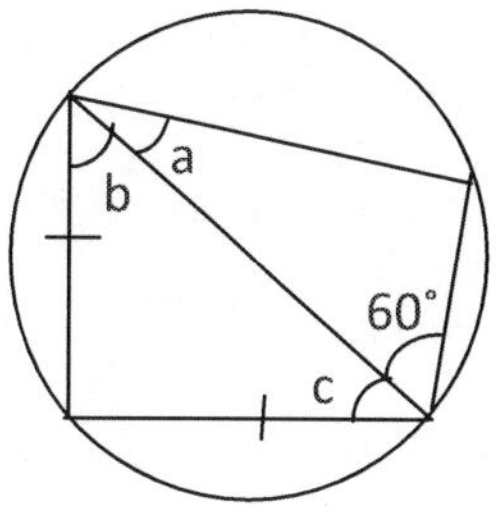

2 Find the size of the angles marked with letters in each diagram.

(a)

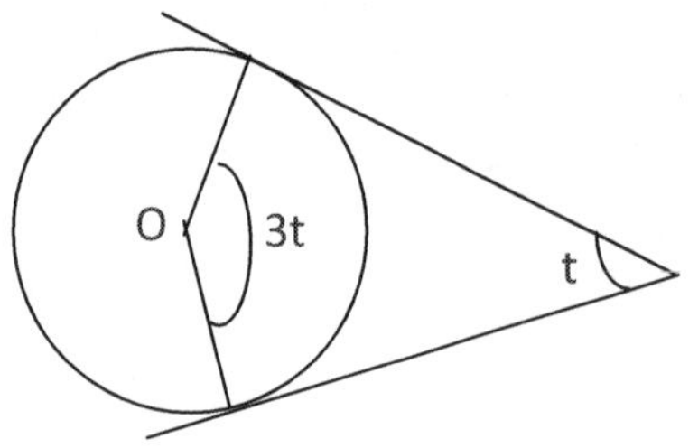

(b)

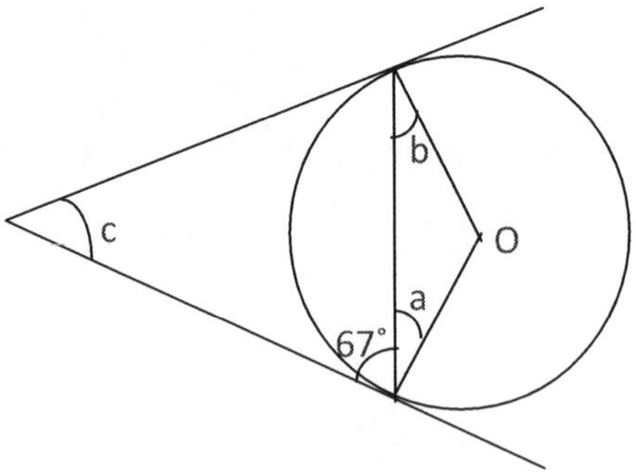

(c)

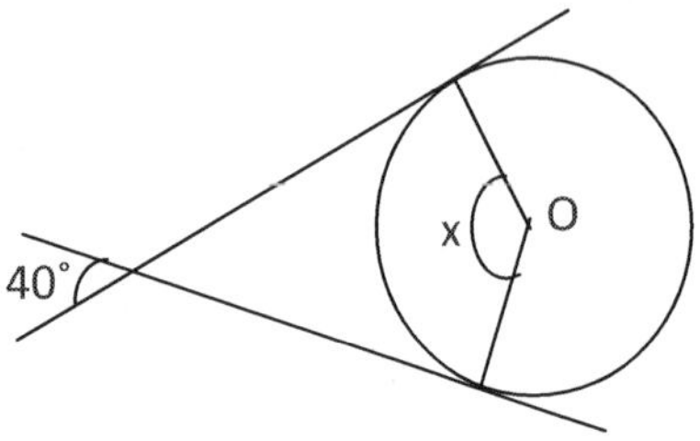

(d)

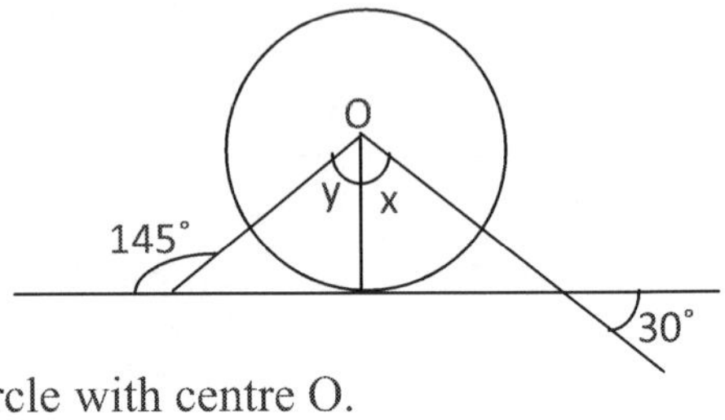

3 In the diagram AB and CD are two diameters of the circle with centre O.

Angle BOF =126° and angle CFE = 95°.

Workout the sizes of the angles x, y, z and t.

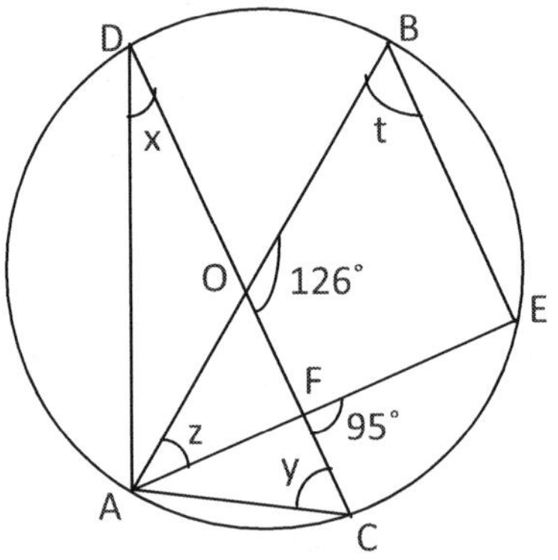

4 In the diagram BA and BC are two tangents OA is perpendicular OC.

(a) What is the special name of the angle ABC?

(b) What is the Special name of the quadrilateral OABC?

Give two reasons for your answer.

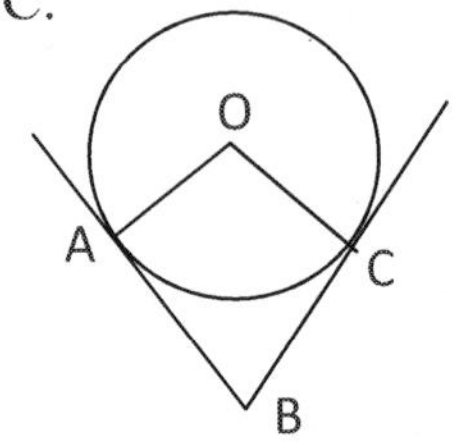

5 In the diagram, AC and BD are two diameters.

AB is parallel to CD and BC is parallel to DA.

(a) Write down the special name of the quadrilateral ABCD.

Justify your answer.

(b) Angle CAD =70°. Workout (i) angle BAO, (ii) angle COD.

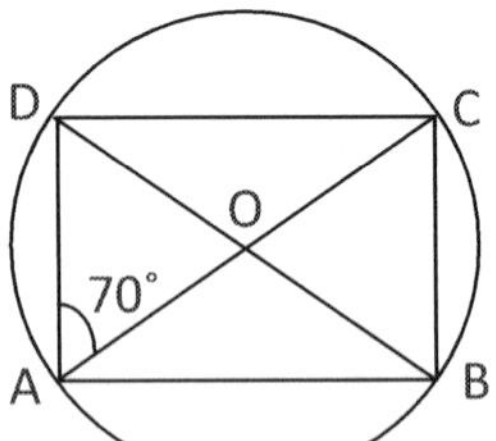

6 In the diagram PQ and PR are two tangents to the circle.

Radius OR = 9 cm. The length of OP is 15cm.

(a) Calculate the lengths of the tangents (i) PQ, (ii) PR.

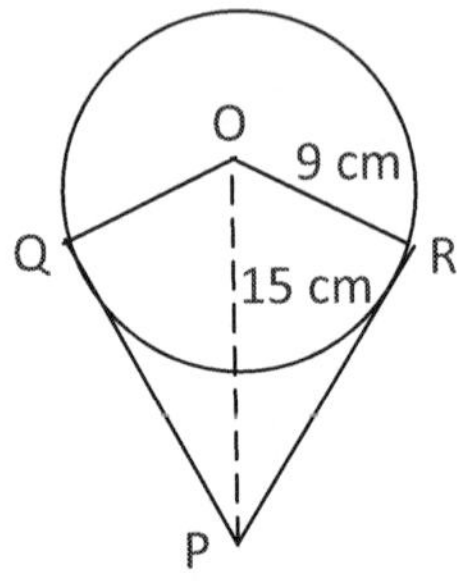

(b) What do you conclude from your answer to the parts a(i) and (ii)?

(c) What can you say about the tangents drawn from an external point to a circle?

7 In the diagram, MN is a tangent to the circle with centre O.

QN is a diameter of the circle.

(a) Write down two simultaneous equations in terms of x and y.

(b) Solve the simultaneous equations to obtain the values of x and y.

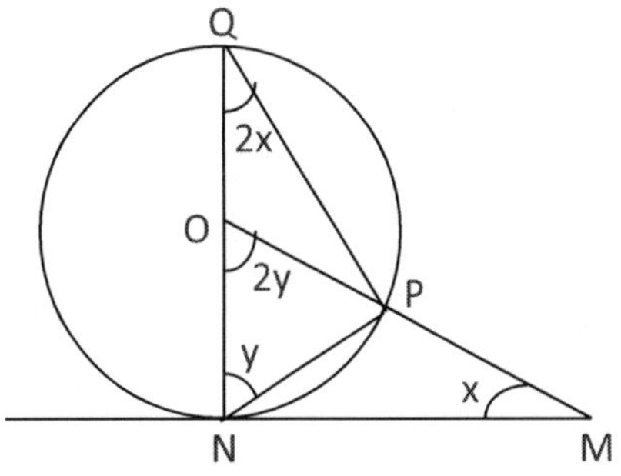

Angle properties of regular polygons

- A regular polygon has all its sides of equal length and all its angles are of equal measure.

Working rule

1) The sum of the interior angles of a n-sided polygon is $(n - 2)180^\circ$.
2) Each interior angle of a regular n-sided polygon is $(n - 2)180^\circ \div n$.
3) Each exterior angle of a regular n-sided polygon is $360^\circ \div n$.
4) If each exterior angle of any polygon is x°, then its corresponding interior angle is $(180^\circ - x^\circ)$.
5) The sum of the exterior angles of any polygon is 360°.

Example

For a regular octagon, find (a) the sum of its interior angles, (b) each of the interior angles, (c) each of the exterior angles.

Working: For an octagon $n = 8$.

Substitute n = 8 in each formula.

(a) Sum of interior angles $= (n - 2)180^\circ$
$= (8 - 2)\ 180^\circ = 1080^\circ$

(b) Each interior angle $= (n - 2)180^\circ \div n = (8 - 2)\ 180^\circ \div 8 = 135^\circ$

(c) Each exterior angle $= 360^\circ \div n = 360^\circ \div 8 = 45^\circ$

Practise exercise 14

1 For a regular pentagon, find (a) the sum of its interior angles, (b) each of the interior angles, (c) each of the exterior angles.

2 For the following regular polygons, find (i) the sum of its interior angles, (ii) each of the interior angles, (iii) each of the exterior angles.
(a) Hexagon (b) Heptagon (c) Decagon (d) Centagon

3 The sum of the interior angles of a regular n-sided polygon is 1800°.
(a) Using this statement, write down an equation in n.
(b) (i) Solve the equation for n.
(ii) Calculate each exterior angle of the polygon.

4 Each of the interior angles of a regular polygon is 140°.
(a) Find the number of sides of the polygon?

(b) Calculate each exterior angle of the polygon.

5 Each exterior angle of a regular polygon is 6°.
(a) Find the number of sides of the polygon. (b) What is the special name of the polygon?

6 (a) Find each of the interior angles of a regular quadrilateral.
(b) What is the special name of a regular quadrilateral.

7 Each interior angle of a regular polygon is 132° more than its each exterior angle.
(a) Write down an equation in n using this information.
(b) Find the value of n.

8 The diagram shows a regular pentagon.

Calculate the size of the angles a, b and c.

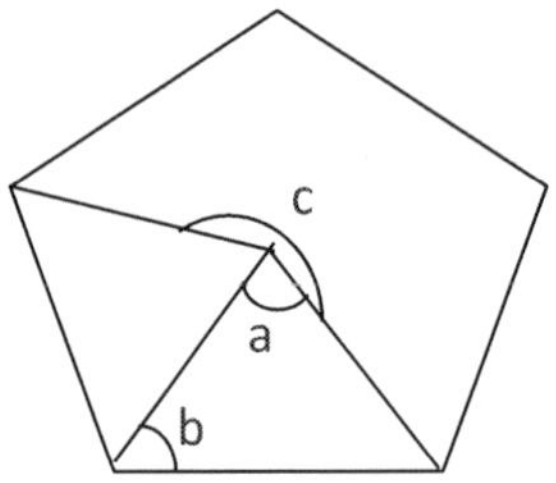

Extended curriculum

Angle properties of irregular polygons

► A polygon with at least one unequal side is called an irregular polygon.

Remember

1. In regular polygons all sides are equal and all angles are equal.
2. In an irregular polygon all sides are unequal and all angles are unequal.

Working rule

1) The sum of the interior angles of a n-sided polygon is $(n-2)180^\circ$.
2) The sum of the exterior angles of any polygon is 360°.

Example
The interior angles of a heptagon are x, 2x, 4x, 5x, 6x and 7x.
(a) Write down an equation in x and then solve it.
(b) Find (i) the smallest angle, (ii) the largest angle.

Working: (a) For a heptagon $n = 7$.
Sum of the interior angles $= (n-2)180^\circ$
$= (7-2)180^\circ = 900^\circ$
$x + 2x + 4x + 5x + 6x + 7x = 900^\circ$
$25x = 900^\circ$ or $x = 36^\circ$
(b)(i) Smallest angle $= 36^\circ$
(ii) Largest angle $= 7 \times 36^\circ = 252^\circ$

Replace n by 7 in the formula.

Use the sum of the interior angles of a heptagon as 900°.

The largest angle is 7x and replace x by 36°.

Practise exercise 15

1 The interior angles of a pentagon are x, 2x, 4x, 5x and 6x.
(a) Write down an equation in x and solve it.
(b) Find (i) the smallest angle, (ii) the largest angle.

2 A nonagon has interior angles of (x + 4), 2x, 3x, 5x, 6x, 7x, 8x and 9x.

(a) Write down an equation in x. (b) Find all the interior angles of the nonagon.
(c) Find the exterior angle at the smallest interior angle of this nonagon.

3 The exterior angles of a hexagon are 2x, 3x, 4x, 5x, 6x and 8x.
Find (i) the smallest angle, (ii) the largest angle.

4 Two of the interior angles of an octagon are 122° and 100°. The remaining interior angles are equal. Find the size of each of the remaining interior angles.

Angle properties of circle

- The angle that an arc of a circle subtends at the centre is twice the angle at the circumference.

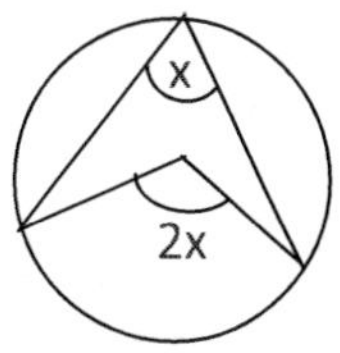

- The angles in the same segment of a circle are equal. Here a = b.

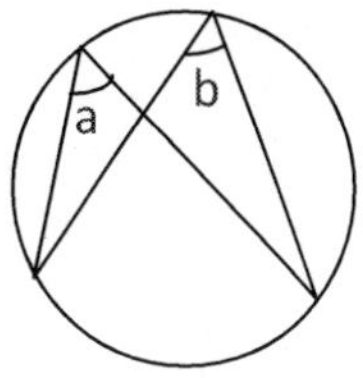

- The sum of the angles on the opposite segments of a circle is 180°.
Here $a + c = 180^\circ$ and $b + d = 180^\circ$.

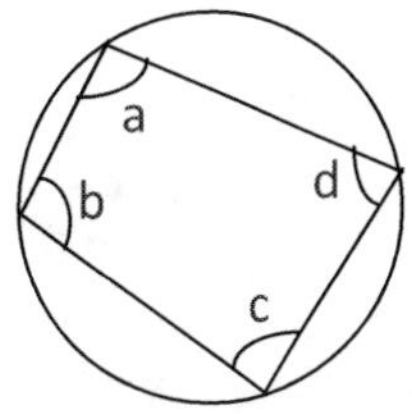

Remember

1. When all the vertices of a quadrilateral lie on the circumference of a circle, then it is called a cyclic quadrilateral.
2. The opposite angles of a cyclic quadrilateral add up to 180°.
3. Rectangle, square, kite and trapezium can be cyclic quadrilaterals.
4. Parallelogram and rhombus cannot be cyclic.
5. The angles on the same side of a chord of a circle are equal.

Example 1

In the diagram, angle ACB = 35°.

Find (i) angle AOB, (ii) angle ADB.

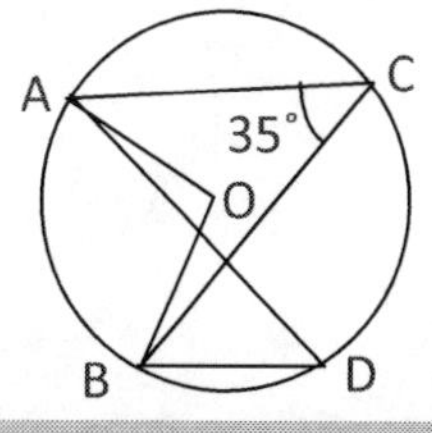

Working: (i) Angle AOB = 2 x 35° = 70°

(ii) Angle ADB = 35°

Angle at the centre = twice the angle at the circumference.
Angle ADB and angle ACB are on the same segment.

Example 2

ABCD is a cyclic quadrilateral.
Angle ABC = x, angle CDA = $2x + 9^\circ$ and angle DAB = 97°.
Calculate (a) angle BCD,
(b) angle BCE, (c) the value of x.

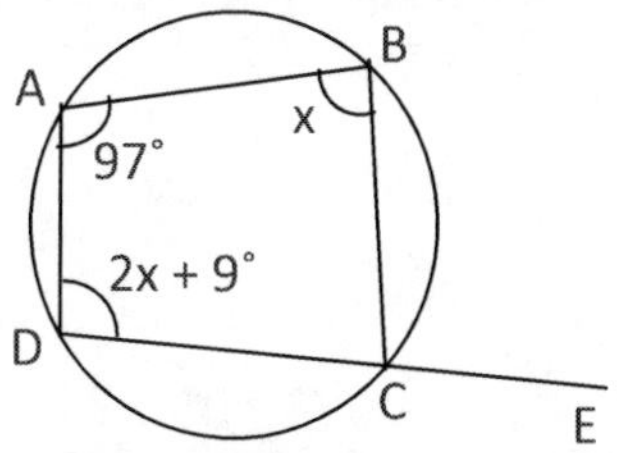

Working: (a) Angle BCD + 97° = 180°

Oposite angles of a cyclic quadrilateral add up to 180°.

Angle BCD = 83°

(b) Angle BCE + $83^\circ = 180^\circ$ — Angles on a straight line add up to 180°.

Angle BCE = 97°

(c) $x + 2x + 9^\circ = 180^\circ$ — Sum of the opposite angles of a cyclic quadrilateral is 180°. Solve the equation for x.

$3x = 180^\circ - 9^\circ$

$x = 57^\circ$

Practise exercise 16

1 Calculate the size of each angle marked in the following diagrams, where O is the centre of the circle.

(a)
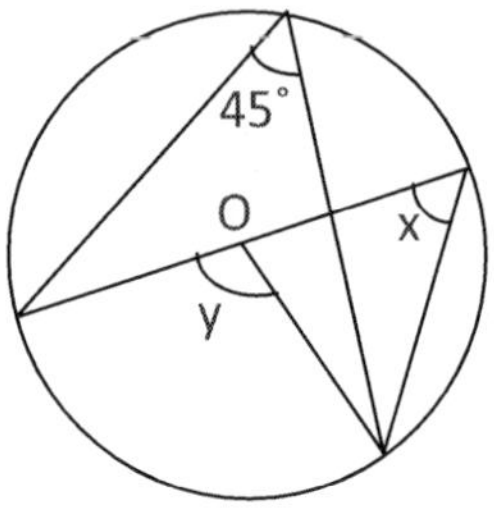

(b)
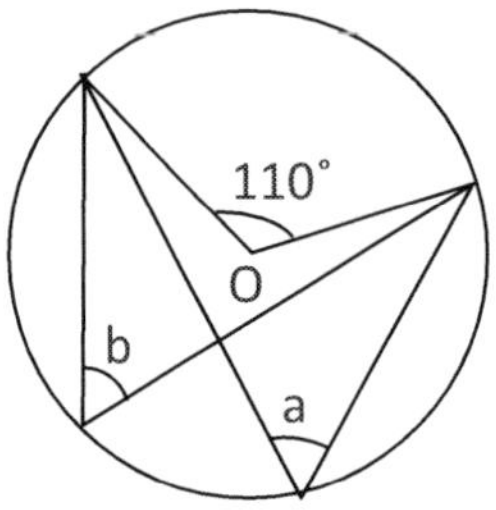

(c)
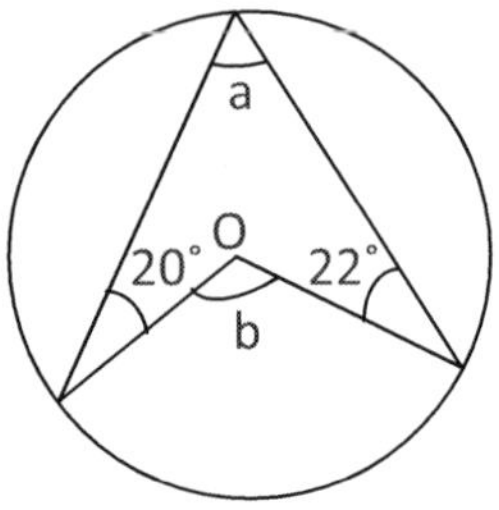

(d)
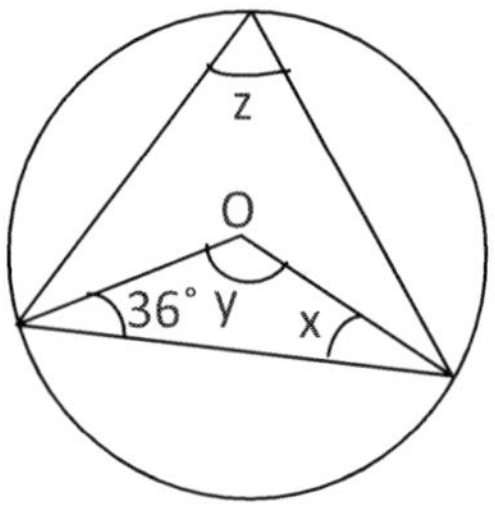

(e)
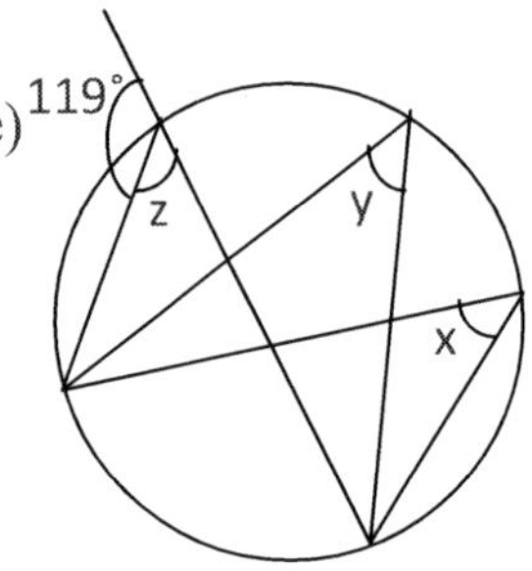

(f)
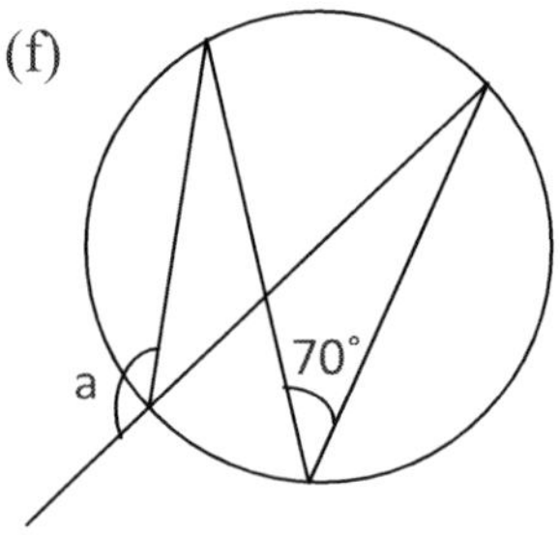

2 Calculate the size of each angle marked in the following diagrams.

(a)
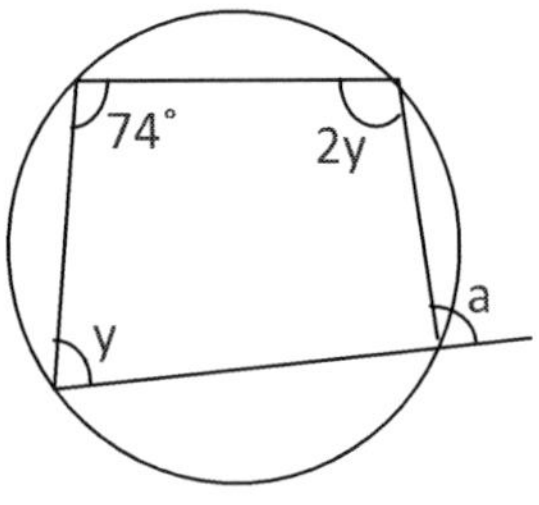

(b)
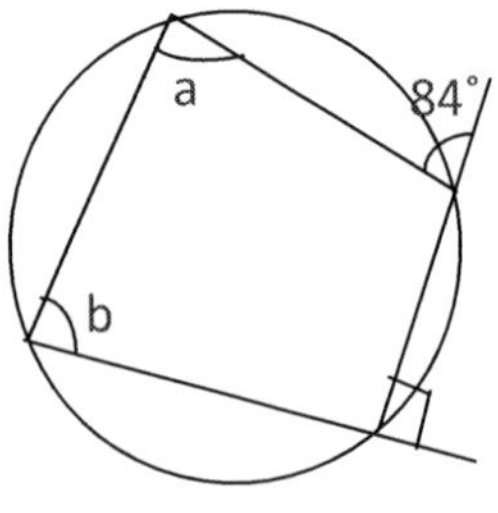

(c)
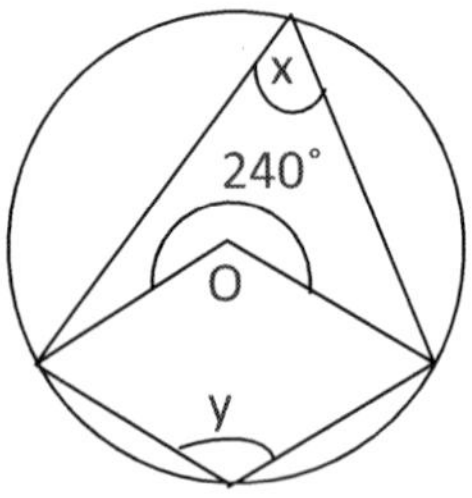

(d)
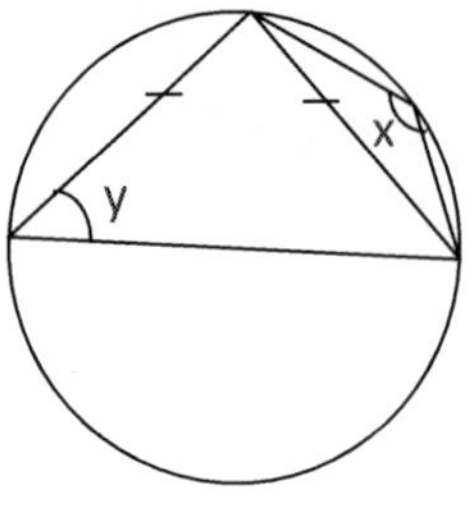

(e)
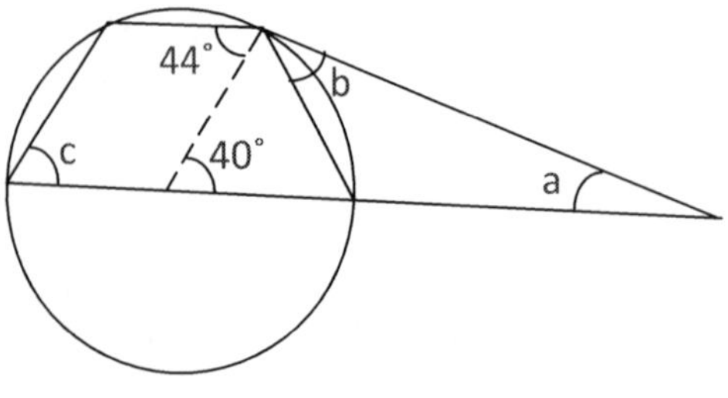

3 In the diagram, the points A, B, C, D, E and F lie on the circle.

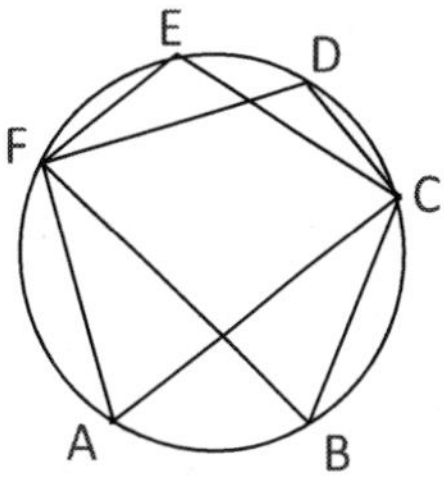

(a) Write down the any four pairs of angles which are equal.

(b) Write down the two pairs of angles whose measure add up to 180°.

4 In the diagram, MN is parallel to QP.

Angle MNR = 130°.

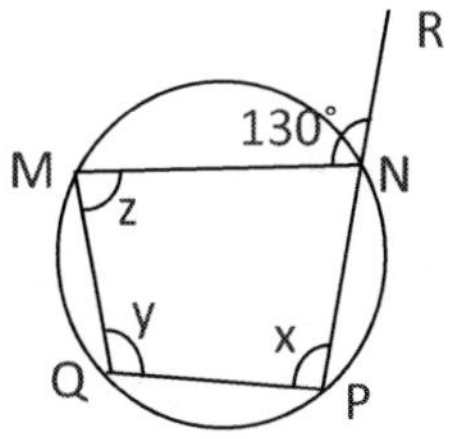

(a) What special name can you assign for the cyclic quadrilateral MNPQ?

(b) Find the sizes of the angles x, y and z.

5 PQRS is a cyclic quadrilateral and QS is a diameter.

PS is parallel to QR.

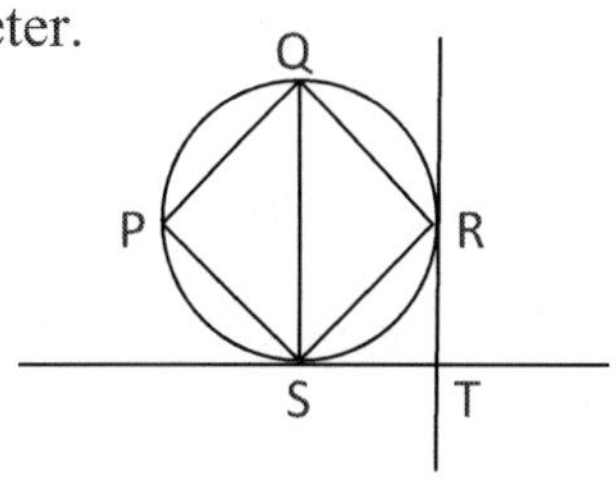

(a) What is the special geometrical name of the quadrilateral PQRS? Justify your answer.

(b) Find the angle RTS.

6 (a) Show that triangle MNR and triangle QPR are similar.

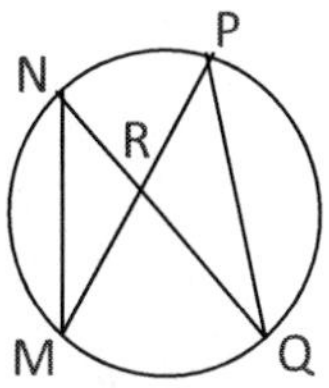

(b) If MN = 10 cm, NR = 4 cm and PR = 5 cm, find the length of QP.

3.04 Symmetry

Core curriculum

Line and rotational symmetry

- Line symmetry is the exact division of a geometric object on opposite sides of a line.

Remember

1. Any triangle cannot have exactly two lines of symmetry.
2. Any quadrilateral cannot have exactly three lines of symmetry.
3. The number of lines of symmetry a n-sided regular polygon is n.

Example

(a) A rectangle has two lines of symmetry.

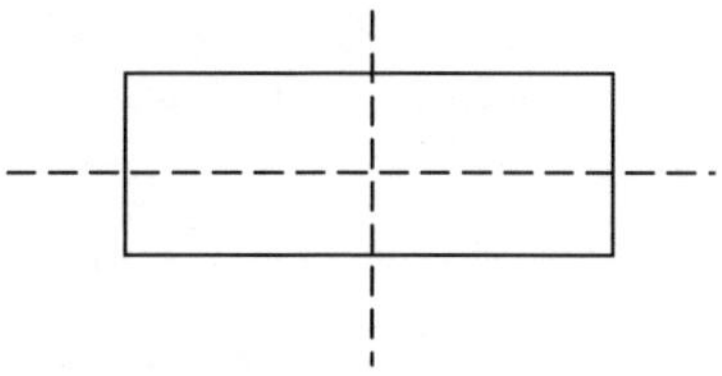

(b) The diagram below has only one line of symmetry.

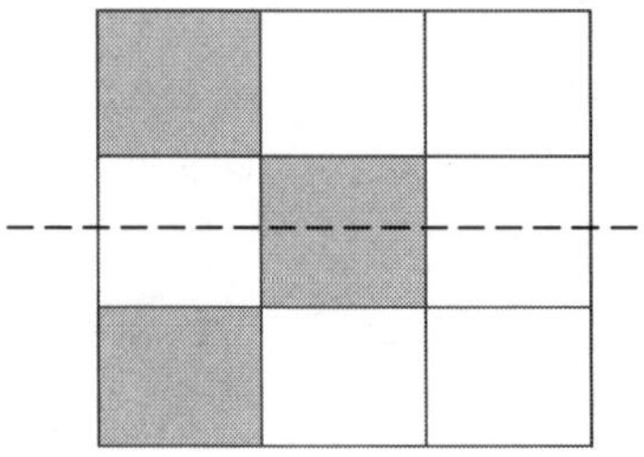

- A geometrical object has rotational symmetry, if it coincides with itself at least once before a complete rotation.

Remember

1. The order of rotational symmetry of a n-sided regular polygon is n.
2. For two-dimensional shapes the order of rotational symmetry is determined about a central point.

Example

(a) A rhombus has rotational symmetry of order 2.

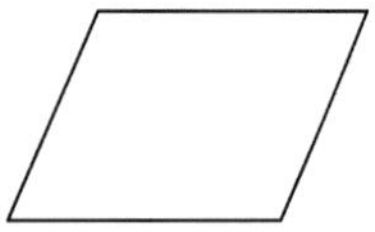

(b) There is no rotational symmetry of an isosceles trapezium.

Practise exercise 17

1 For each shape (i) Draw and write the number of line symmetry (if any),
(ii) state the order of rotational symmetry.

(a)

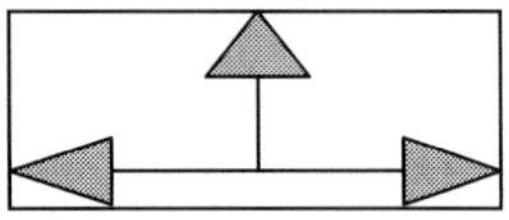

(b)

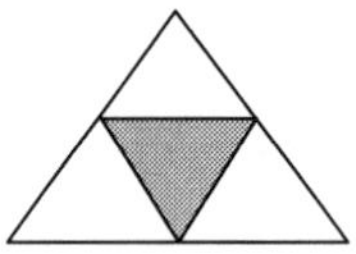

(c)

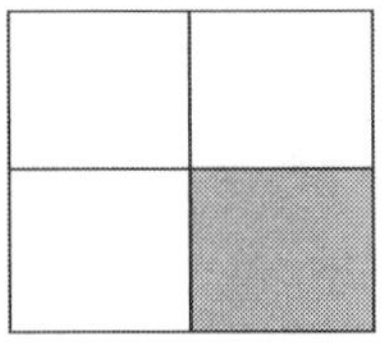

(d)

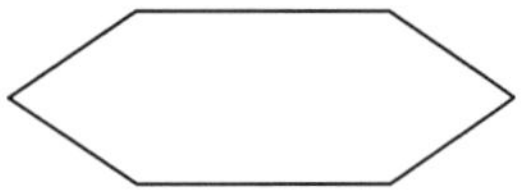

(e)

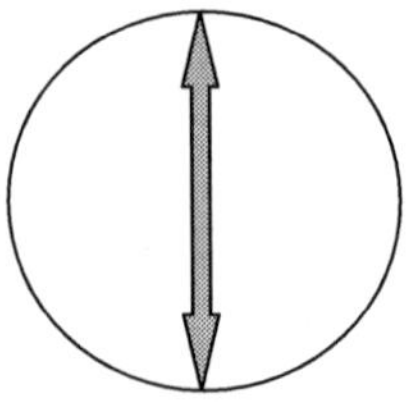

(f)

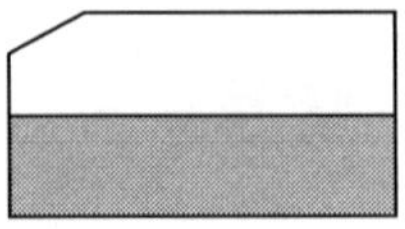

2 (a) Sketch a quadrilateral which has no line of symmetry but rotational symmetry of order 2.
(b) (i) Sketch a quadrilateral which has line of symmetry but no rotational symmetry.
(ii) What is the geometrical name of this shape?

3 Complete the following table.

Quadrilateral	Diagonals are perpendicular	No. of lines symmetry	Order of rotational symmetry
Square			
	No	2	2
Parallelogram			
Rhombus			
	Yes		0
Trapezium			
Arrow head			

4 For each shape, (i) draw the line of symmetry, (ii) state the order of rotational symmetry.

(a) Isosceles triangle

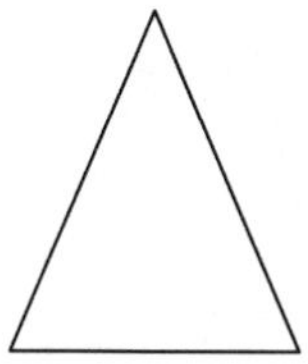

(b) Equilateral triangle

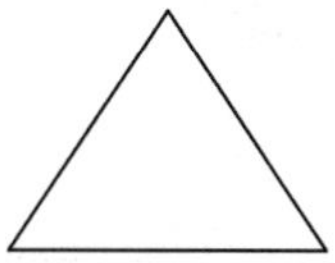

(c) Semi circle

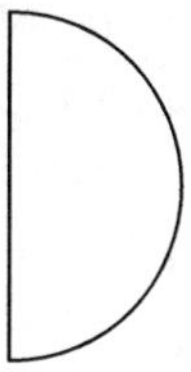

(d) Regular pentagon

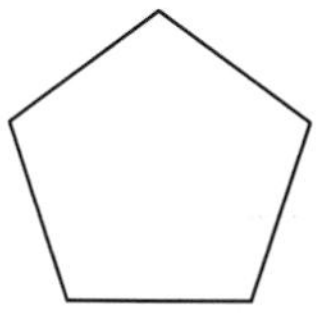

(e) Isosceles trapezium

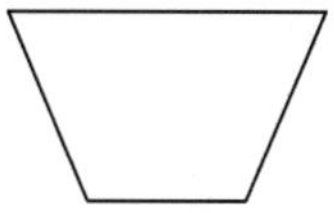

(f) Plus symbol

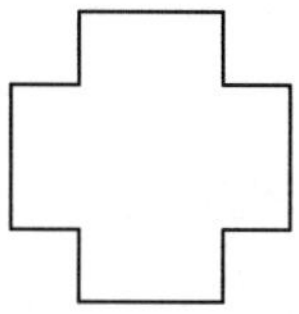

5 The diagram shows a line AB.

(i) Draw a line of symmetry for AB.
(ii) What is the special geometric terminology, you can assign for this line of symmetry?

6 The initials of International General Certificate of Secondary Education are

I G C S E

(a) Which of these letters have line of symmetry only?
(b) Which of these letters have rotational symmetry?
(c) Which of these letters have both line of symmetry and rotational symmetry?

7 Complete each shape below so that the given dotted line is a line of symmetry.

(a)

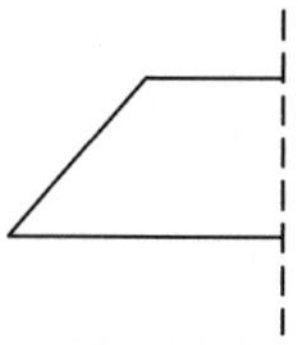

(b)

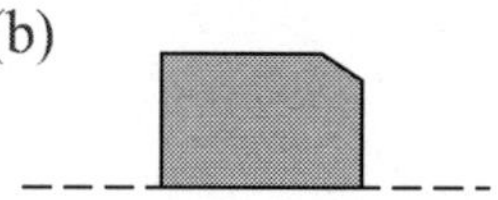

(c)

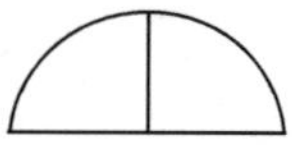

8 Complete each shape below so that they have rotational symmetry but no line of symmetry.

(a)

(b)

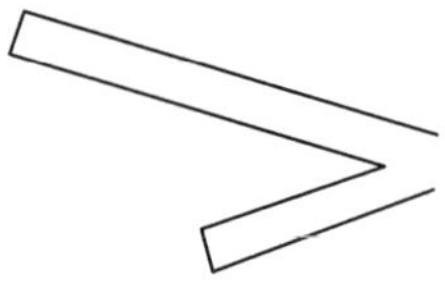

Extended Curriculum

Planes of symmetry

- Plane symmetry is the exact division of a solid shape on the opposite sides of a plane.

Remember

1. Plane of symmetry is related to three dimensional shapes.
2. For three-dimensional shapes, the order of rotational symmetry is determined about a central line called axis of symmetry.

Example
For a rectangular based pyramid,
(a) write down the order of rotational symmetry about the central axis,
(b) draw a the plane of symmetry.
Working: (a) The order of rotational symmetry is 2.
(b)

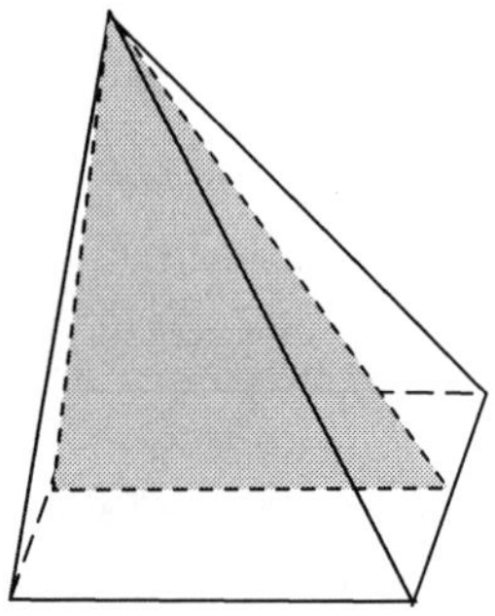

Practise exercise 18

1 For each of the shapes below
(i) write down the order of rotational symmetry about the given axis,
(ii) state the number of planes of symmetry.

(a) cube

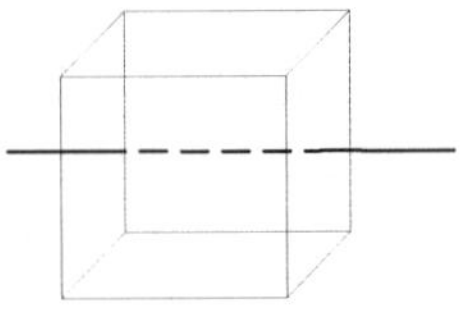

(b) Cuboid

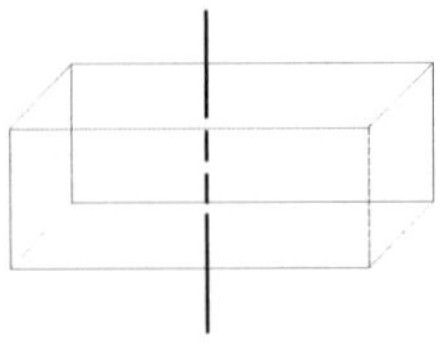

(c) Cylinder

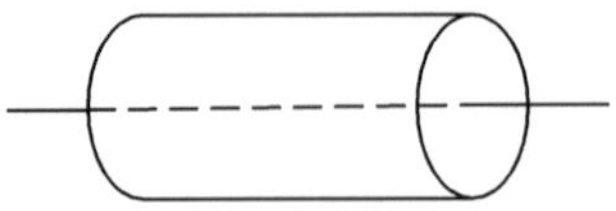

(d) Equilateral triangular prism (e) Equilateral triangular based pyramid

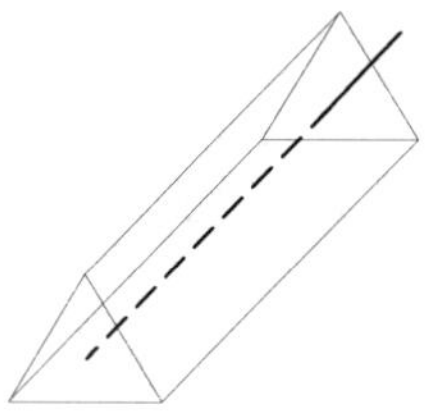

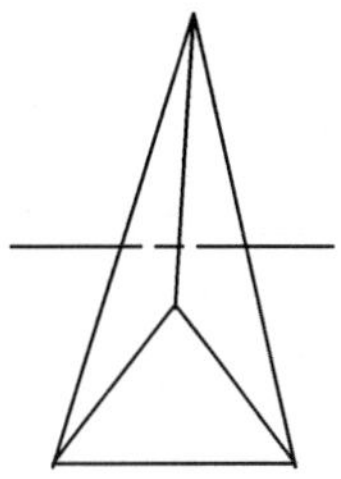

2 The diagram a compound solid consists of a cuboid and a pyramid.

(a) Find the order of rotational symmetry about the given axis.

(b) How many planes of symmetry does it have?

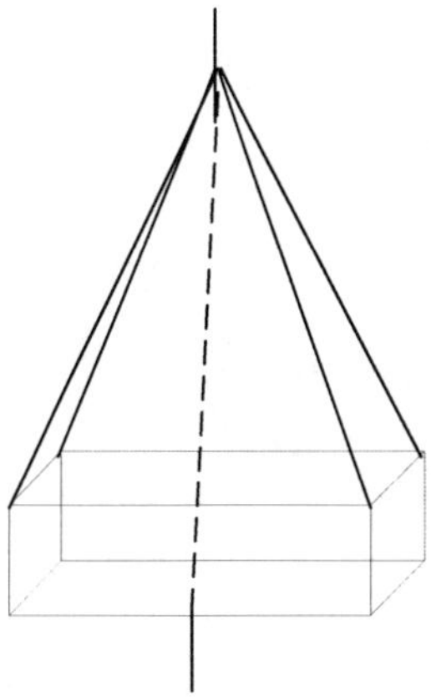

3 The diagram shows an isosceles trapezoidal prism.

How many planes of symmetry does it have?

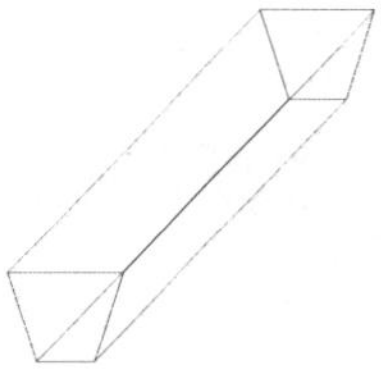

4 (a) Sketch a cuboid and show one plane of symmetry in it.
(b) Sketch a cube and show one plane of symmetry in it.
(c) Sketch a cylinder and show one plane of symmetry in it.
(d) Sketch two equilateral triangular prism and show two different planes of symmetry.

Symmetry properties of circle

- A chord is a line joins any two points on the circumference of a circle.
- In a circle, chords of same lengths are equidistant from the centre. Here chord AB = chord CD. So OE = OF.

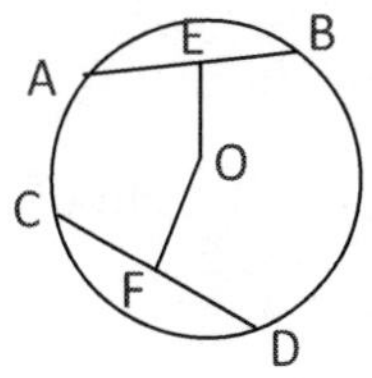

- In a circle, the perpendicular bisector of a chord passes through the centre. Here the perpendicular bisector of chord AB passes through the centre O.

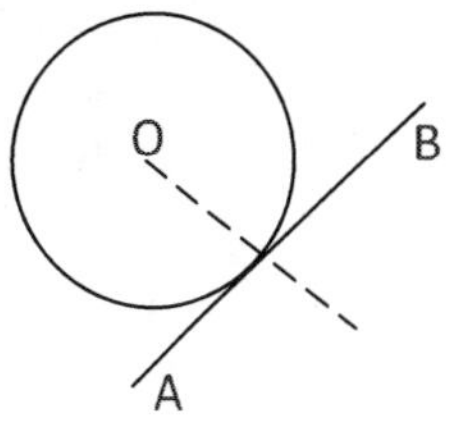

- The tangents from an external point to a circle are equal in length. Here PA = PB.

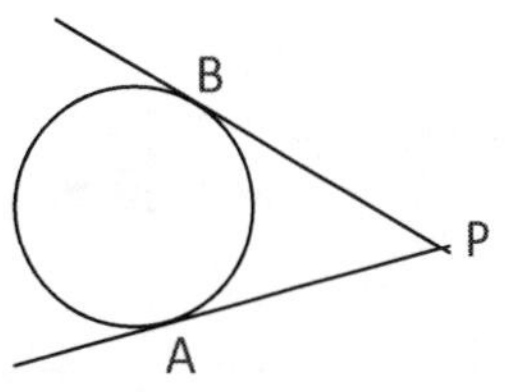

Remember

1. The longest chord of a circle is the diameter.
2. If two chords are equidistant from the centre, then they are of same length.

Example 1

The radius of a circle is 35 cm.

The perpendicular distance from the centre to the chord is 21 cm.

Calculate the length of the chord.

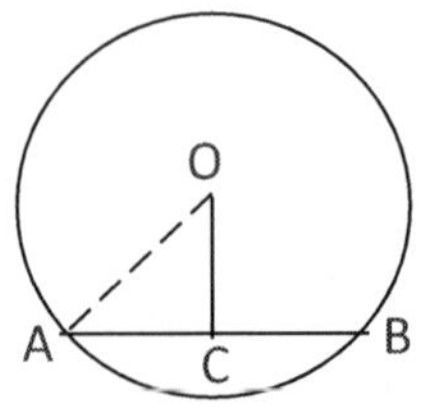

Working: OA = 35 cm and OC = 21 cm.

$AC = \sqrt{35^2 - 21^2} = 28$ cm

Chord AB = 2 x 28 cm = 56 cm

Use Pythagoras theorem.
Since OC bisects chord AB, so AB = 2 x AC.

Example 2

PA and PB are tangents to the circle. O is the centre of the circle.

OP = 20 cm and the radius of the circle is 6 cm.

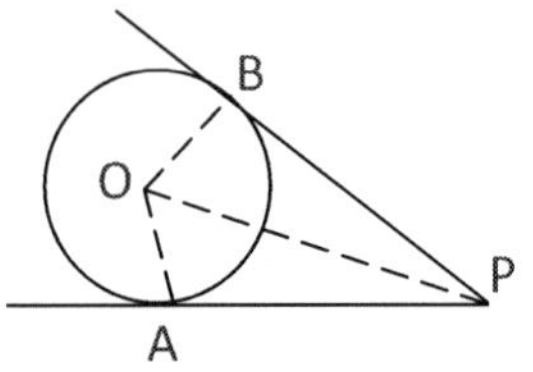

(a) Calculate the length of the tangent PA.
(b) What is the length of PB?

Working: (a) OA = 6 cm and OP = 20 cm.

$PA = \sqrt{20^2 - 6^2} = 19.1$ cm

(b) PB = 19.1 cm

Use Pythagoras theorem.
Tangents from an external point are equal.

Practise exercise 19

1 In the circle, chords MN and PQ are equidistant from the centre O. OR = 6 cm and the radius of the circle is 10 cm.
(a) Calculate the length of the chord MN.
(b) What is the perpendicular distance between PQ and the centre?
(c) What is the length of the chord PQ?

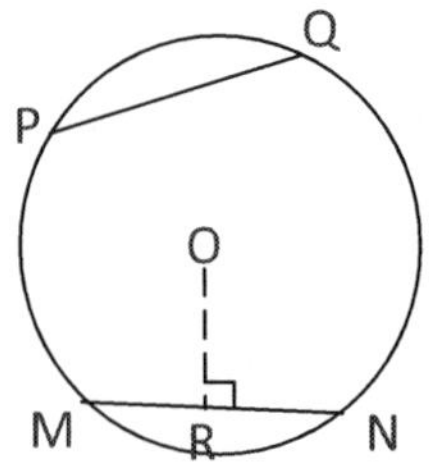

2 PQ and QR are tangents to the circle with centre O. The radius of the circle is 5 cm and the length of OP is 15 cm. Find the length of the tangent PR.

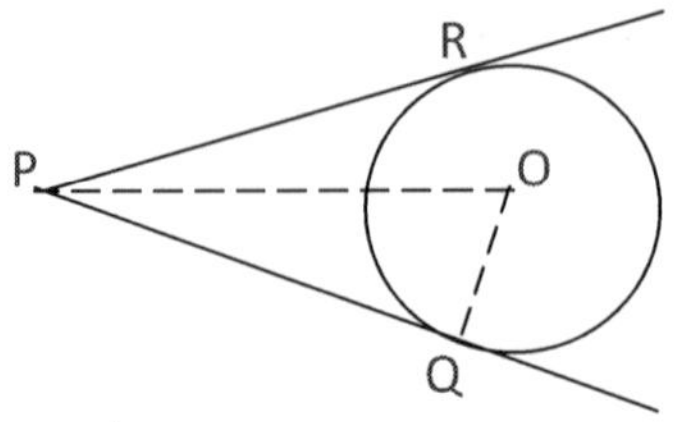

3 PL and PM are tangents to the circle with centre O.
(a) Complete the quadrilateral OLPM.
(b) What is the special name of the quadrilateral OPLM? Give reasons to your answer.

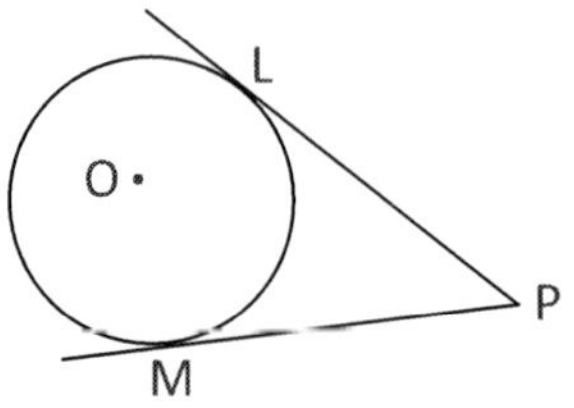

4 UT and US are the tangents to a circle centre O. Angle TUS = 37°.
Find (i) angle TOS,
(ii) angle TQS,
(iii) angle TRS,
(iv) angle OTS

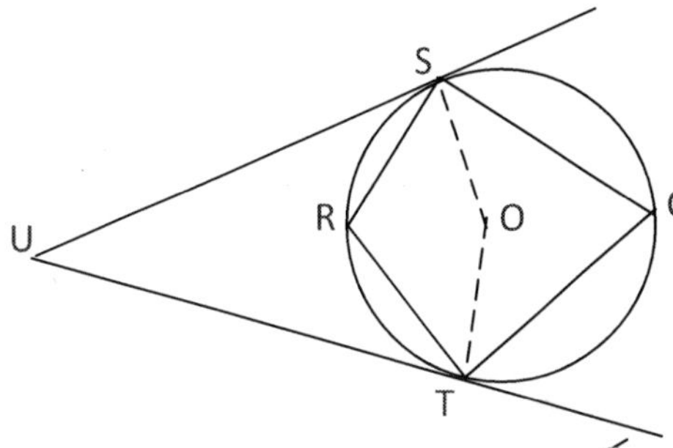

5 AX and AY are tangents to a circle centre O.
Z is a point on the circumference.
AX is parallel to YZ.
Angle AXY is 70°.
Work out the values of a, b and c.

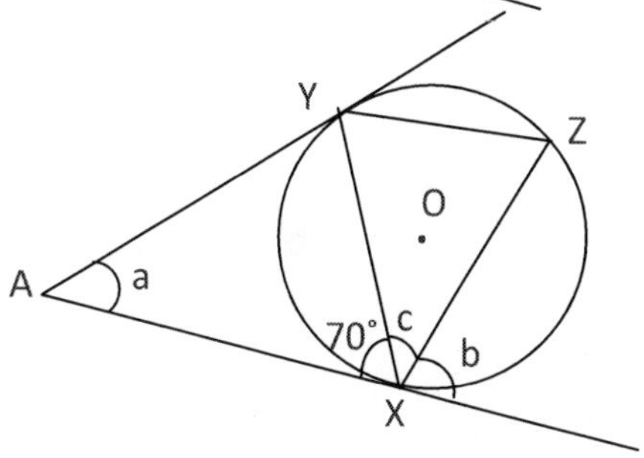

6 PQR is a right angled triangle.
A circle with centre O touches the sides of the triangle PQR.
QR = 18 cm and PQ = 24 cm.
(a) Work out the length of PR.
(b) Find the radius of the circle.
(c) What is the length of the tangent PU?

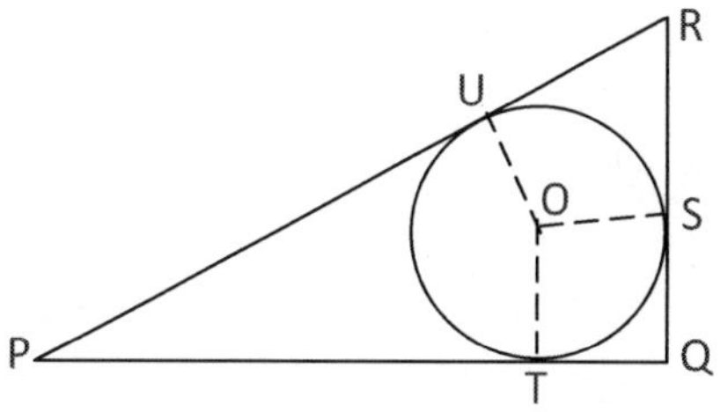

3.05 Trigonometry

Core curriculum

Pythagoras theorem

- A constant multiple of 3, 4, and 5 are called Pythagorean numbers.

  E.g. 6, 8 and 10 Pythagorean.

- Pythagoras theorem states that, in a right angled triangle, the square of the hypotenuse equals the sum of the squares of the other two sides. Here $h^2 = a^2 + b^2$.

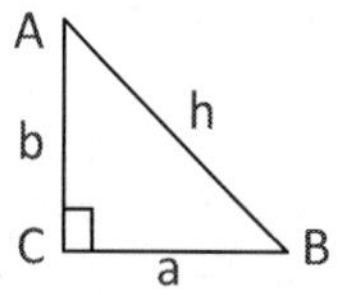

Remember

1. In a right angled triangle, hypotenuse is the longest side.
2. Pythagoras theorem is based on right angled triangle only.
3. Pythagoras theorem is used to find a side of a right triangle, when other two sides are given.
4. Don't use Pythagoras theorem in a triangle without a right angle.

Example

Calculate the length of the side BC.

Working:

$h^2 = a^2 + b^2$

$25^2 = 15^2 + b^2$

$b^2 = 625 - 225$

$b = \sqrt{400} = 20$ cm

Use Pythagoras theorem.
Substitute the values of h and a.
Rearrange for b.
Solve for b.

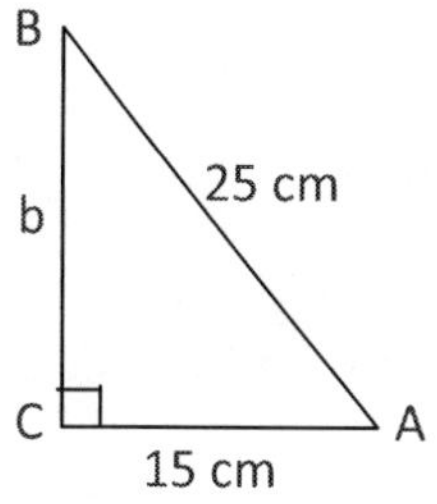

Practise exercise 20

1 Calculate the length of the unknown side in each triangle. All lengths are given in cm.

(a)
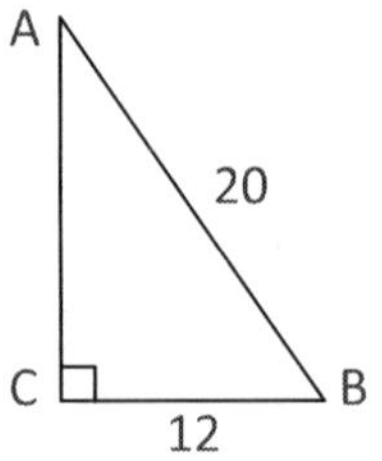

(b)
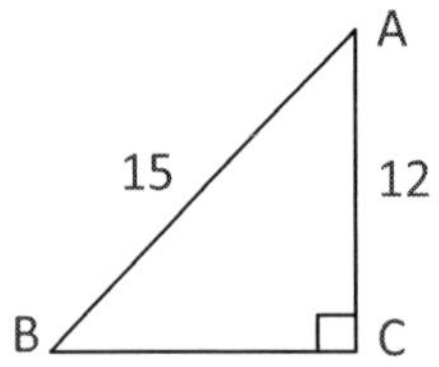

(c)
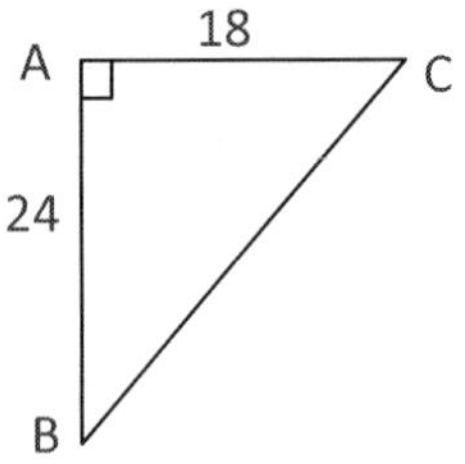

(d)
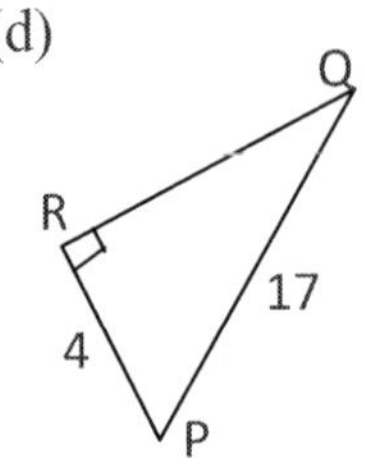

(e)
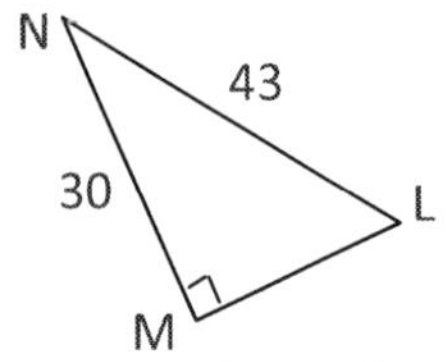

(f)
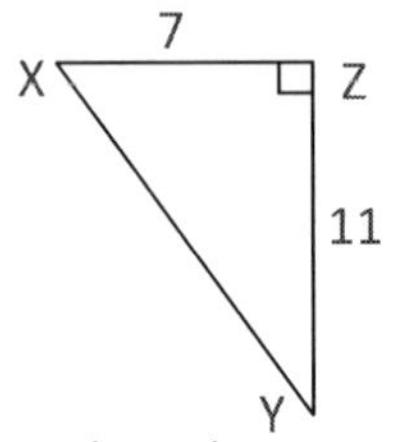

2 Workout the values of the letters marked x and y. All lengths are given in cm.

(a)
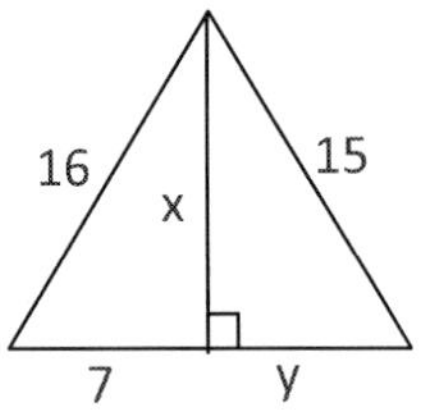

(b)
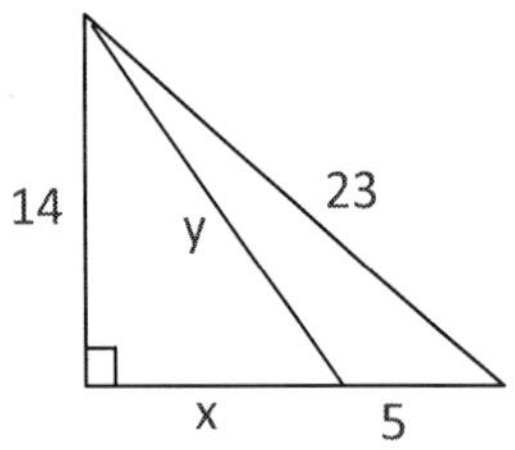

(c)
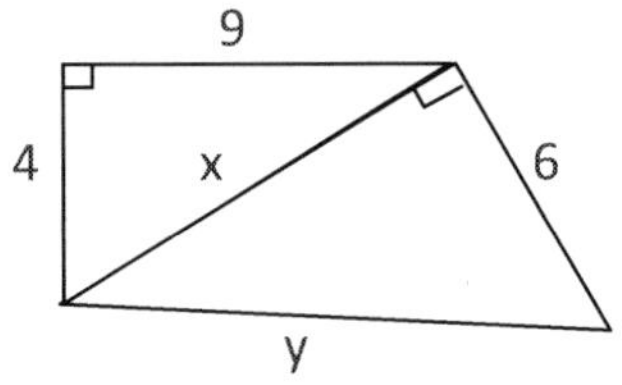

(d)
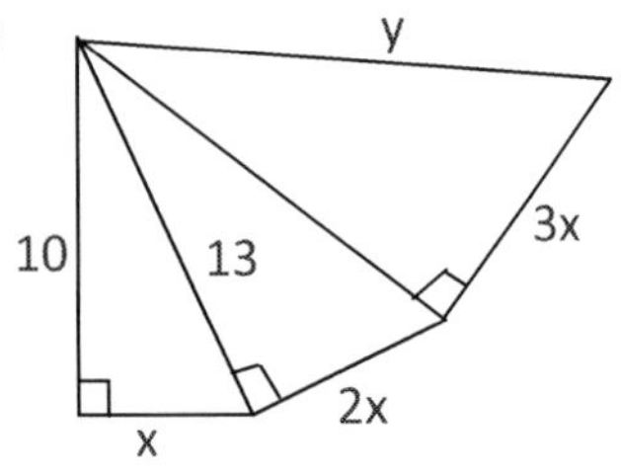

(e)
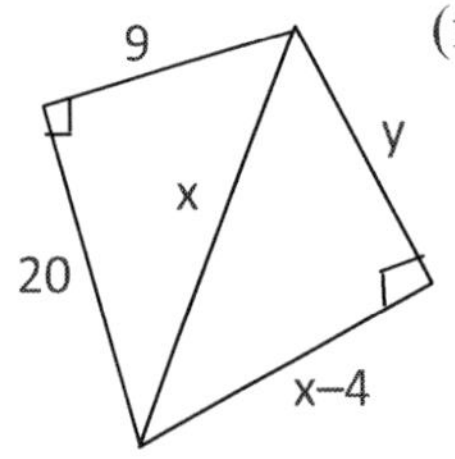

(f)
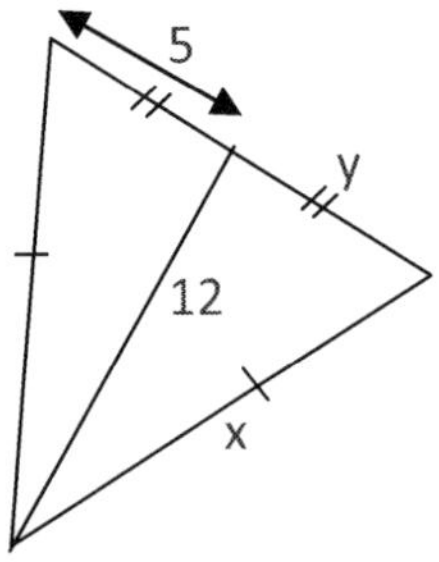

3 (a) The length of the sides of a square is 28 cm. Work out the length of its diagonal.
(b) The dimensions of a rectangular garden are 60 m by 40 m. Calculate the total lengths of its diagonals.
(c) The diagonals of a rhombus are 36 cm and 22 cm. Calculate the length of its side.
(d) The height of a palm tree is 12 m. A ladder of 15 m touches the top of the tree. Calculate the distance between the foot of the ladder and the tree.

4 ABCD is a kite. Diagonals AC and BD intersect at point E.

AD = 10 cm, BC = 26 cm and AE = 4 cm.

Work out the length of (a) BD, (b) CE.

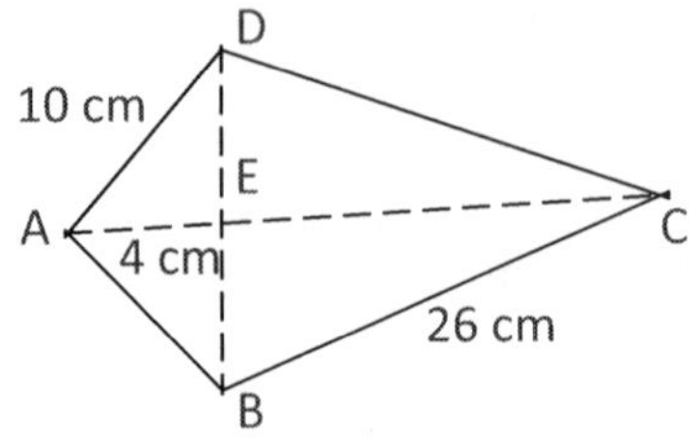

5 A triangle has sides of 45 cm, 27 cm and 36 cm. What is the special name that can be assigned to this triangle? Justify your answer.

Trigonometric ratios

- The three trigonometric ratios are sin (sine), cos (cosine) and tan (tangent).
- Trigonometric ratios are based on right angled triangles.
- In right angled triangle ABC,

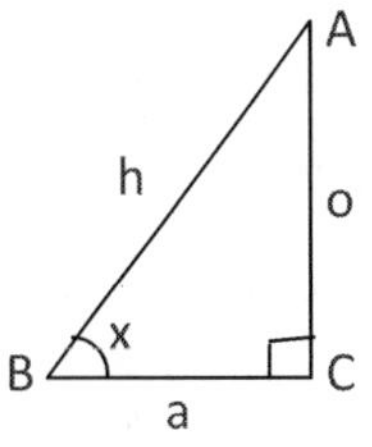

h = hypotenuse,
o = opposite (side to angle x),
a = adjacent (side to angle x).

Then $\sin x = \frac{o}{h}$, $\cos x = \frac{a}{h}$ and $\tan x = \frac{o}{a}$.

Working rule

In a right triangle, to calculate
1) an unknown side, use direct trigonometric ratios such as sin, cos and tan.
2) an unknown angle, use inverse trigonometric ratios such as $\sin^{-1}$, $\cos^{-1}$ and $\tan^{-1}$.

Remember

1. To work out the trigonometric ratios, first label the sides by h, o and a.
   h is always straight to the right angle, but o and a changes depending on the acute angle you consider.
2. For trigonometry questions, you must use degree mode in your calculator instead of radian or grads mode.
3. When the required answer is an angle, it should be given correct to 1 decimal place.

Example 1

In right triangle ABC, AC = 20 cm and angle ABC = 58°.

Work out the lengths of (i) AB, (ii) BC.

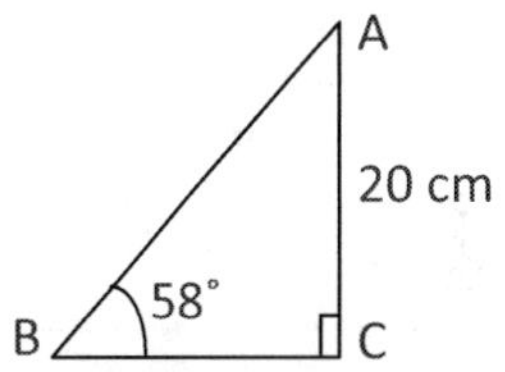

Working: (i) $\sin x = \frac{o}{h}$ or $\sin 58^\circ = \frac{20}{h}$

$h = \frac{20}{\sin 58^\circ} = 25.31 = 25.3$ cm.

So AB = 25.3 cm.

Use sin ratio for o and h.
Solve the equation for h.
Give answer up to 3 s.f.

(ii) $\cos x = \frac{a}{h}$ or $\cos 58^\circ = \frac{a}{25.31}$

$a = 25.31 \cos 58^\circ = 15.5$ cm

So BC = 15.5 cm

Use cos ratio for a and h.
Solve the equation for a.
Give answer up to 3 s.f.

Example 2

In triangle MNP, MN = 22 cm, NP = 17 cm

and PMN = x. Find the size of angle PMN.

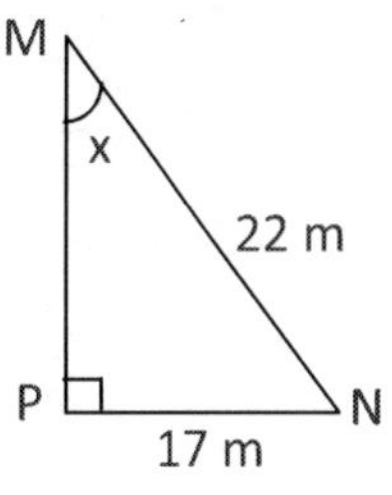

Working: $\cos x = \frac{a}{h}$

$x = \cos^{-1}\left(\frac{17}{22}\right) = 43.8^\circ$

So angle PMN = 43.8°.

> Use cos for a and h.
> Solve the equation for x.
> For angle, give your answer correct to 1 d.p.

Practise exercise 21

1 For each triangle, calculate both the unknown sides. All sides are given in cm.

(a)
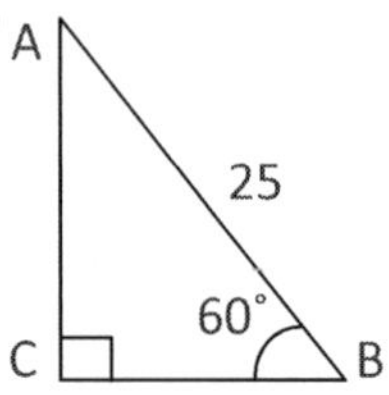

(b)
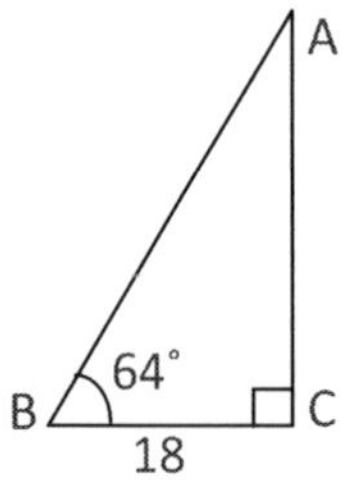

(c)
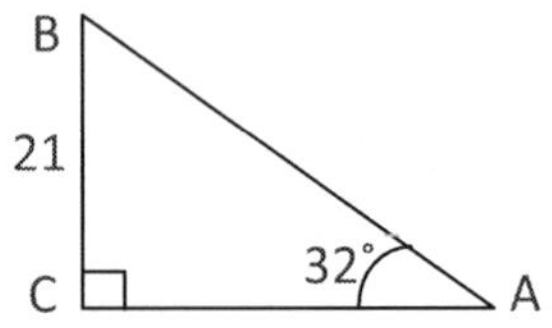

(d)
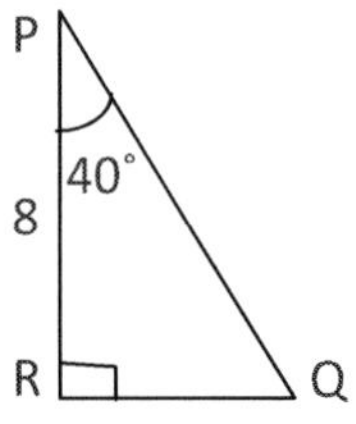

(e)
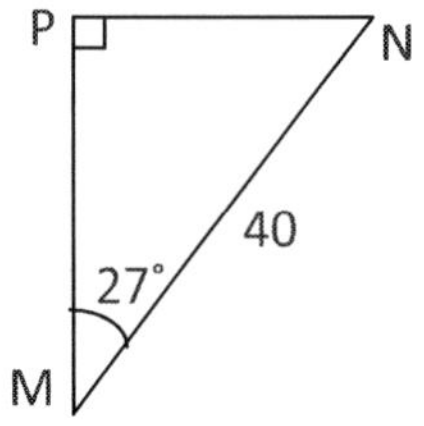

(f)
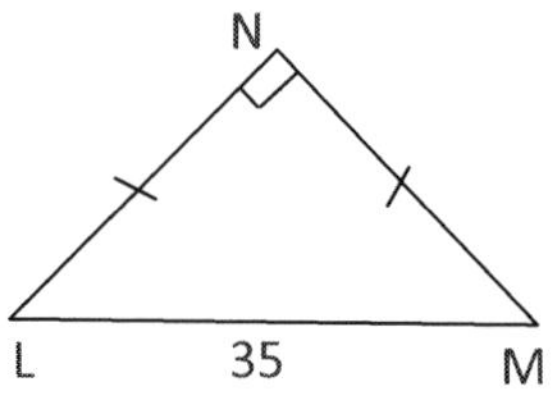

2 Calculate the size of the angle marked x in each triangle. All sides are given in cm.

(a)
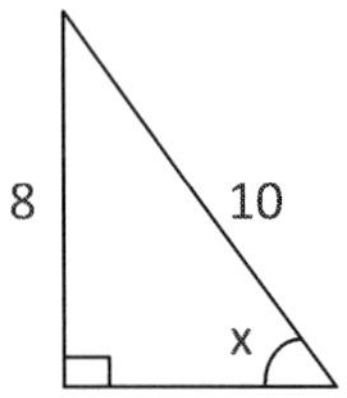

(b)
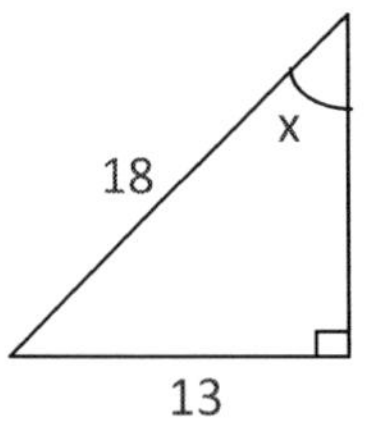

(c)
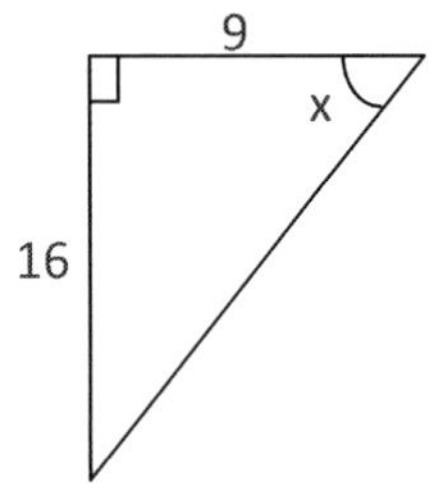

3 Find the value of each of the following.

(a) $\sin 49^\circ$ (b) $\sin 62.4^\circ$ (c) $\cos 72^\circ$ (d) $\cos 37.6^\circ$

(e) $\tan 70^\circ$ (f) $\cos^{-1}(0.84)$ (g) $\sin^{-1}\left(\frac{24}{31}\right)$ (h) $\tan^{-1}\left(\frac{19}{12}\right)$

(i) $\sin^2 26^\circ + \cos^2 26^\circ$ (j) $\tan^2 42^\circ - \frac{1}{\cos^2 42^0}$ (k) $\frac{1-\cos^2 20^0}{\sin^2 20^0}$ (l) $\frac{2\tan 37^0}{1-\tan^2 37^0}$

4 Calculate the size of the angle x in each case.

(a) $\sin x = 0.47$ (b) $\cos x = 0.652$ (c) $\tan x = 4.25$

(d) $\sin x = \frac{11}{15}$ (e) $\cos x = \frac{22}{27}$ (f) $\tan x = \frac{10}{19}$

5 In the diagram, AB = 34 cm, CD = 29 cm and angle ABC = 41°.

(a) Work out the length of AD.

(b) Work out the size of angle CAD.

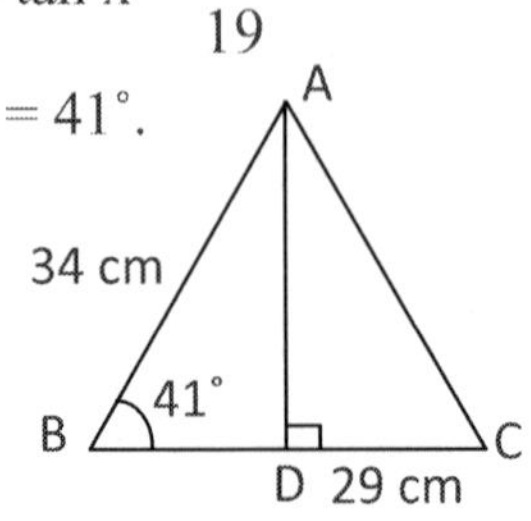

6 AB and AC are two tangents to the circle with centre O. OA = 13 cm and AOC = 69°.

(a) Calculate the length of AB.

(b) Find the radius of the circle.

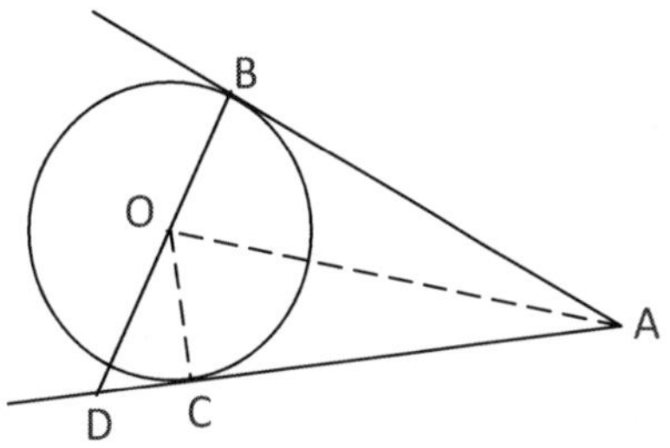

(c) (i) Calculate the length of OD. (ii) Work out the size of angle ODC.

7 In the diagram, QR = 24 m, angle QPR = 24° and angle PST = 31°.

(a) Calculate the lengths of

(i) PQ, (ii) RS.

(b) What is the size of angle RPS?

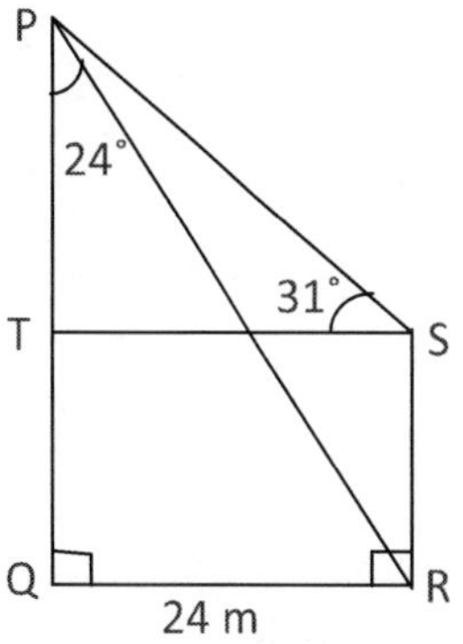

Bearing

- A bearing is a three-digit angle measured clockwise from the north.
- The bearing of North-east is 045°.
  The bearing of East is 090°.
  The bearing of South-east is 135°.
  The bearing of West is 270°.
  The bearing of North-west is 315°.

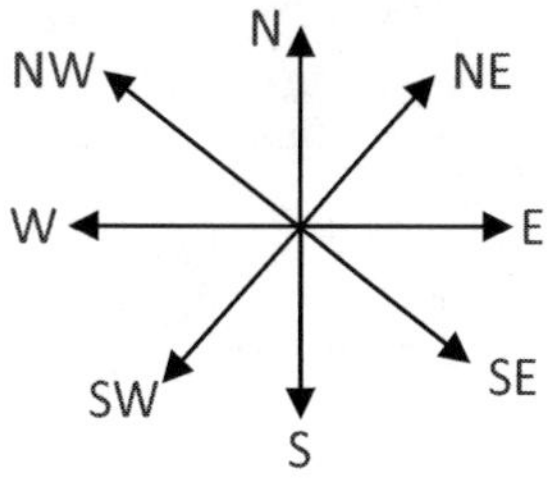

Remember

1. The bearing must be a three-digit angle.
2. The bearing of A form B means the angle is at B.
3. Bearings are used to calculate the position or direction of movement of some body or something using a map or a compass.

Example

A plane flies 400 km due east to reach point A and then another 550 km at a bearing of 225° to reach at point B. It then directly returns to its starting destination.
(a) Using a scale of 1 cm to represent 50 km, draw an accurate drawing of this travel.
(b) Find the total distance travels by the plane. (c) Calculate the bearing of A from B.
Working: (a)

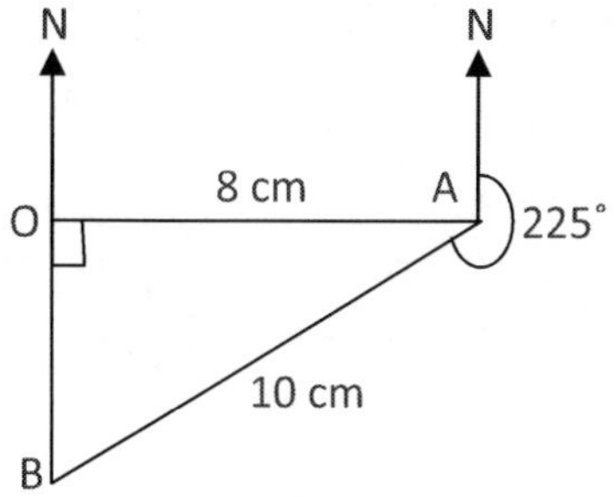

The scale length of 400 km 8 cm and that 500 km is 10 cm.

Draw OA = 8 cm.

The bearing of B form A is 225°.

Draw AB = 10 cm.

Join OB to get a right triangle.

(b) In right triangle OAB,

$OB = \sqrt{500^2 - 400^2} = 300$ km

Total distance = 400 + 500 + 300 = 1200 km

(c) The bearing of A from B = $225^\circ - 180^\circ = 045^\circ$

> Use Pythagoras theorem.
> Since the bearing of B from A is more than 180°, just subtract 180° from 225°.

Practise exercise 22

1 A and B are two points.

The bearing of B from A is 215°.

Calculate the bearing of A from B.

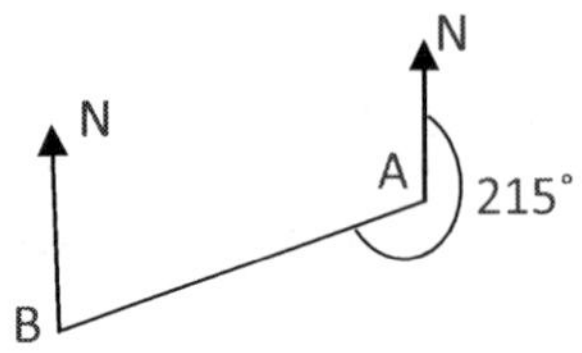

2 P and Q are two Islands.

The bearing of P from Q is 124°.

Calculate the bearing of Q from P.

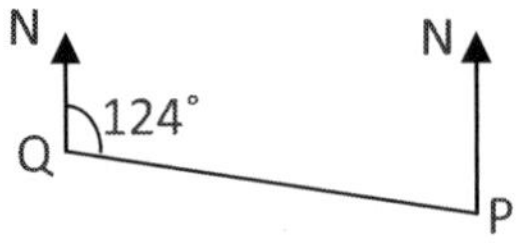

3 A point L is 200 km due east of point P. Another point M is 350 km from L. The bearing of M from L is 210°.
(a) Using a scale of 1 cm to represent 40 km draw an accurate drawing to show this information.
(b) Measure and write down the actual length of MP.
(c) Calculate the bearing of L from M.

4 An aeroplane flies 600 km due west to reach a point X and then a further 800 km at a bearing of 135° to reach the point Y.
(a) Use a scale of 1 cm = 100 km, draw an accurate sketch of this travel.
(b) Find the total distance travelled by the aeroplane in your sketch.
(c) Calculate the bearing of X from Y.

5 A ship starts from Island A and travels 300 km on a bearing of 050° to reach at Island B. Then it travels another 250 km on a bearing of 210° to reach at Island C.
(a) Using a scale of 1 cm to represent 20 km draw an accurate drawing of this travel.
(b) Find the straight distance of the ship from the starting point.

6 The diagram shows the position of three Villages P, Q and R. Angle QPR = 45°, PQ = QR and PR = 30 km.

(a) Which village is due east of village P?

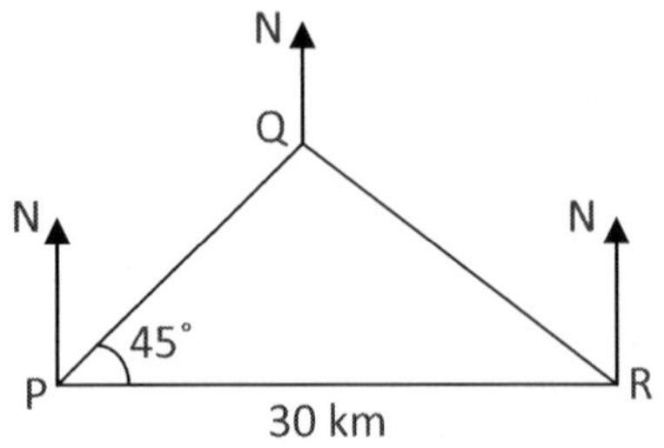

(b) Show that PQR is a right angle.
(c) Calculate the length of PQ.
(d) Calculate the bearing of (i) Q from P, (ii) P from Q, (iii) Q from R, (iv) P from R.

Extended curriculum

Angles of elevation and depression

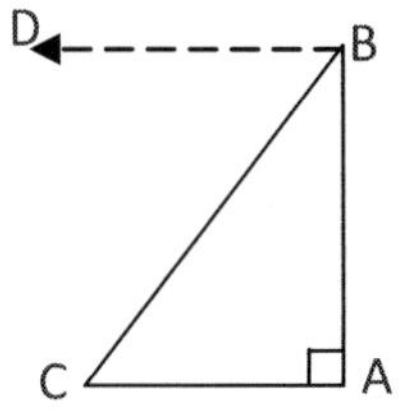

- Let AB is a tree and C is a point on the ground. B is the top of the tree.
  The angle of elevation of B from the point C is the angle ACB.
  The angle of depression of the point C from B is the angle CBD.

Remember

The angle of elevation of B from C = the angle of depression of C from B.

Example

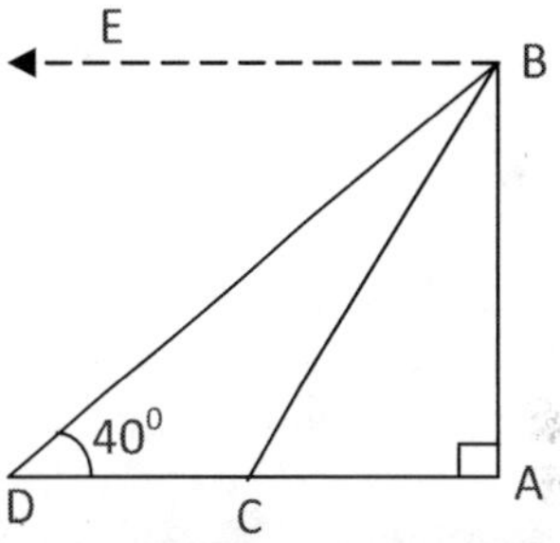

In the diagram, AB is a tower. AD = 220 m and CD = 70 m.

The angle of elevation of B from D is 40^0.

Calculate (a) the height of the tower,

(b) the angle of depression of C from B.

Working: In the right angled triangle ABD,

$$\tan 40^0 = \frac{AB}{AD} = \frac{AB}{220}$$

Use tan ratio for opposite and adjacent.

Substitute AD = 220

AB = 220 x tan 40^0 = 185 m

In the right angled triangle ABC,

$$\tan x = \frac{AB}{AC} = \frac{185}{220 - 70}$$

Let angle CBE = x. So angle ACB = x.

Substitute AB = 185 and AC = AD – CD = 220 – 70.

$$x = \tan^{-1}\left(\frac{184.6}{150}\right) = 50.9^0$$

Practise exercise 23

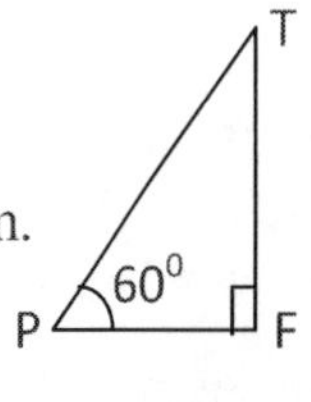

1 The angle of elevation of the top of a tree T from a point P is 60^0.

The distance between the bottom of the tree F and the point P is 24 m.

Calculate the height of the tree.

2 The height of a vertical tower is 120 m. A point P is 32 m away from the foot of the tower. Calculate the angle of elevation of the top of the tower from P.

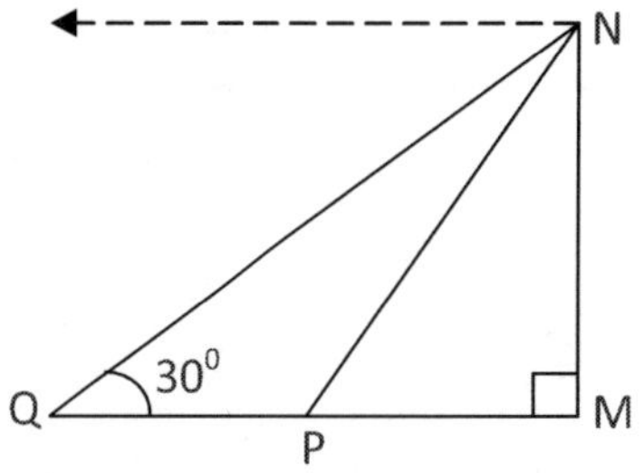

3 In the diagram, MN is a pole.

The angle of elevation of N from Q is 30^0.

MN = 27 m and PQ = $\frac{1}{3}$ MN. Calculate

(a) the length of PM, (b) the angle of depression of P from N.

4 The length of the shadow of a vertical flag post FP on the level ground is increased by 6 m when the altitude of the sun changes from 40^0 to 60^0.

Calculate the height of the flag post.

Give your answer correct to 2 decimal places.

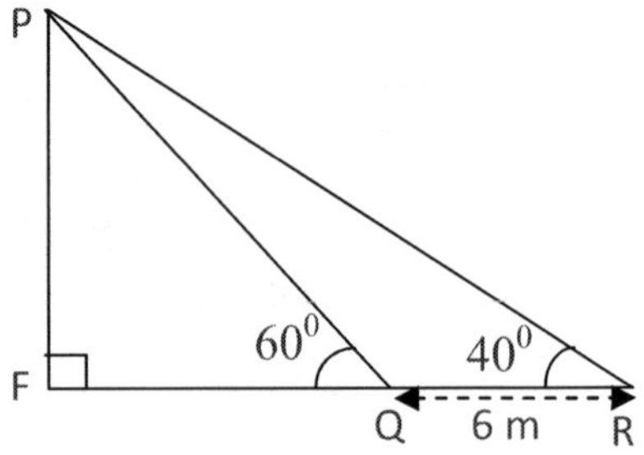

5 A kite, K, flying at an altitude of 90 m from the level ground is attached to a straight string LK.

The angle of depression of K from L is 67^0.

Calculate the length of the string KL.

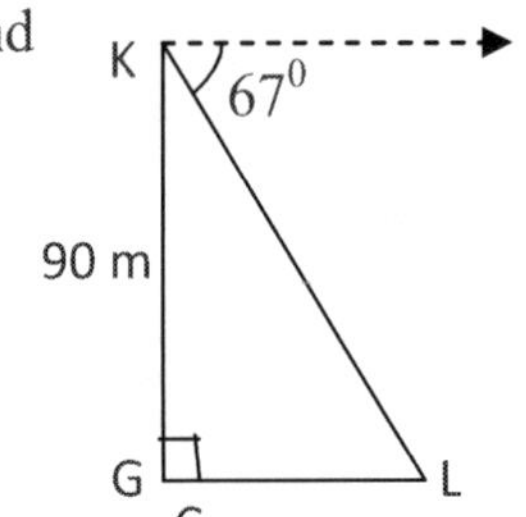

6 In the diagram, CD is a light house of 150 m high.

A and B are two ships on opposite sides of the tower.

The angle of depression of A from C is 37^0.

The angle of elevation of C from B is 69^0.

Find the distance between A and B.

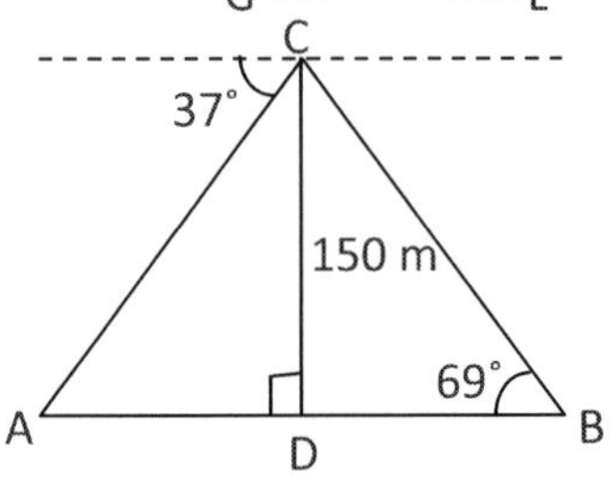

7 In the diagram, MN is a building of 80 m high and PQ is another building. The angles of depressions of P and Q from N are 77^0 and 42^0 respectively.

(a) Find the horizontal distance between both the buildings.

(b) Calculate the height of the building PQ.

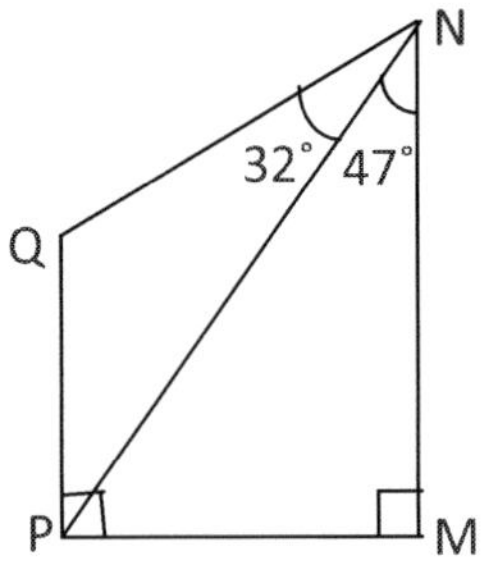

The sine and cosine rules

- The sine rule states that $\frac{a}{\sin A} = \frac{b}{\sin B} = \frac{c}{\sin C}$.
- The cosine rule states that
  $a^2 = b^2 + c^2 - 2bc \cos A$ (Use when two sides and the included angle are given.)

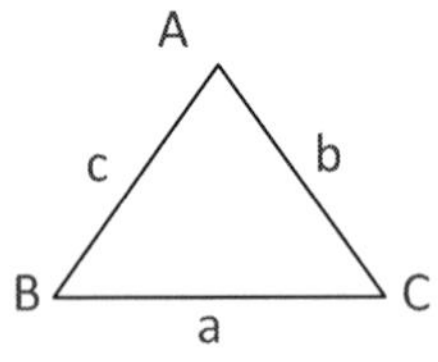

or $\cos A = \frac{b^2 + c^2 - a^2}{2bc}$ (Use when all the three sides are given.)

- When two sides and the included angles are given, the area of any triangle is $A = \frac{1}{2}ab \sin C$.

Remember

1. The sine and the cosine rules are used with non-right angled triangles.
2. The sine rule can be used to find any side or any angle of a triangle.
3. The cosine rule can be used to
   (i) find a side, when other two sides and their included angle are given,
   (ii) find an angle, when all the three sides are given.

Example 1

In triangle ABC, angle ABC = 50^0, angle CAB = 74^0 and BC = 14 cm.

Find the length of (i) AC, (ii) AB, correct to 1 decimal place.

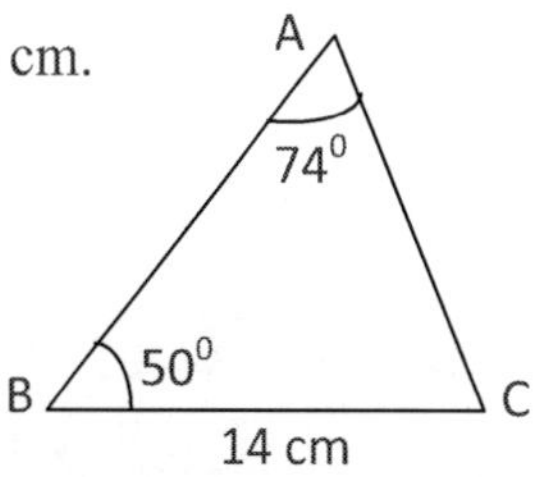

Working: Angle ACB = $180^0 - (50^0 + 74^0) = 56^0$.

$$\frac{a}{\sin A} = \frac{b}{\sin B} = \frac{c}{\sin C}.$$

$$\frac{14}{\sin 74^0} = \frac{AC}{\sin 50^0} = \frac{AB}{\sin 56^0}.$$

Taking first and second ratios, we have $\frac{14}{\sin 74^0} = \frac{AC}{\sin 50^0}$

AC = 11.2 m

Taking first and third ratios, we have $\frac{14}{\sin 74^0} = \frac{AB}{\sin 56^0}$

AB = 12.1 m

Use the angles in a triangle add up to 180^0 to find the third angle.

Use sine rule, since all angles are given.

Substitute the known sides and the angles.

Example 2

In triangle ABC, angle BAC = 62^0, AB = 7 cm and AC = 10cm.

(i) Find the length of BC.

(ii) Calculate the area of triangle ABC.

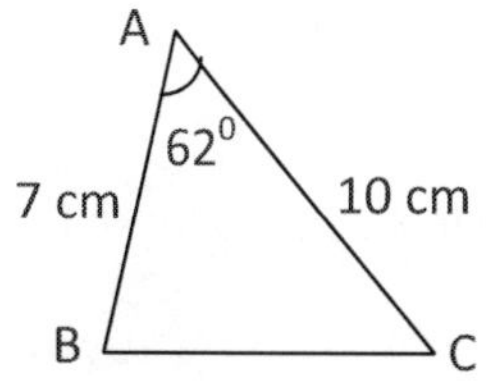

Working: (i) $a^2 = b^2 + c^2 - 2bc \cos A$

$= 10^2 + 7^2 - 2 \times 10 \times 7 \times \cos 62^0 = 83.27$

$a = \sqrt{83.27} = 9.13$cm

Area of triangle ABC = $\frac{1}{2}$ bc sin A

$= \frac{1}{2} \times 10 \times 7 \sin 62^0 = 30.9 \text{ cm}^2$

Since two sides and their included angles are given, use cosine rule.

Use sine rule to find the side.

Use area of a triangle formula.

Practise exercise 24

1 Find the value of x in each triangle.

(a)

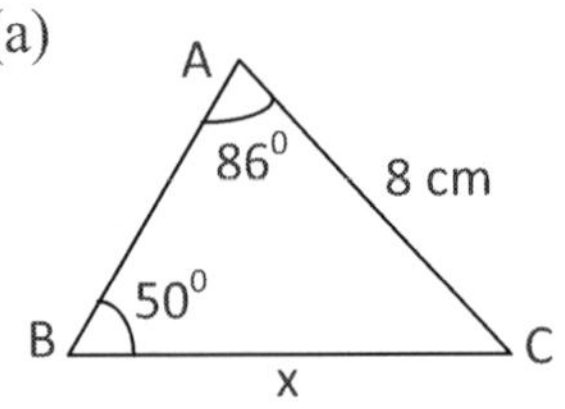

(b)

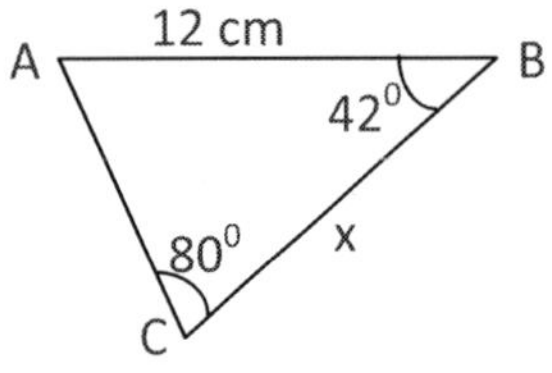

(c)

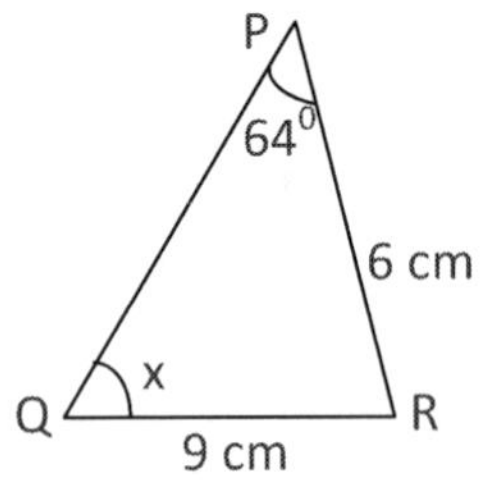

(d)

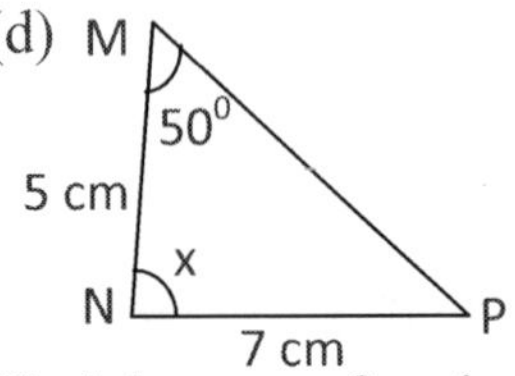

`(e)

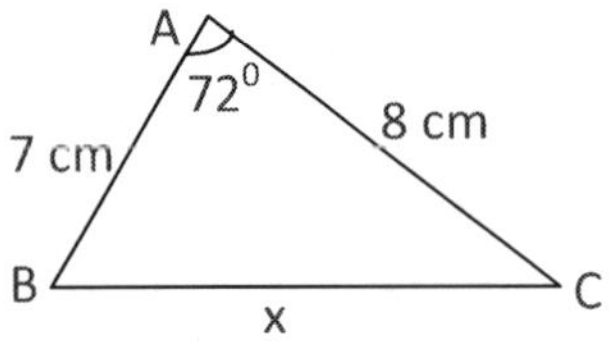

(f)

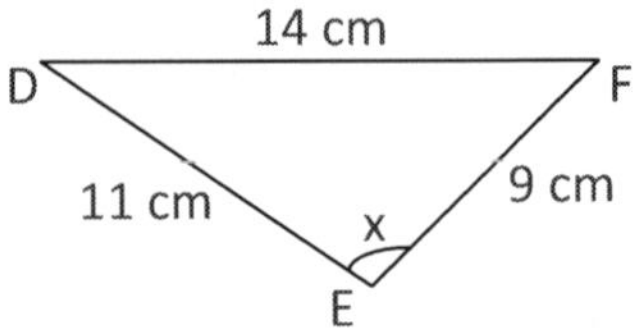

2 Find the area of each triangle.

(a)

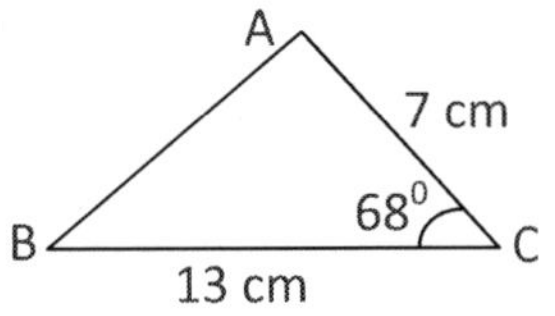

(b)

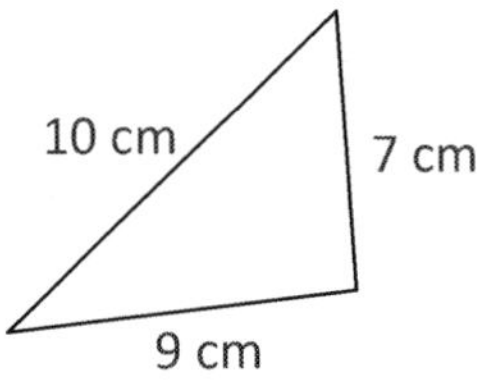

3 In the triangle GHI, angle GHI= 98^0, angle HIG = 60^0 and IG = 16 cm.

(a) Find the lengths of

(i) GH, (ii) HI, correct to 1 decimal place.

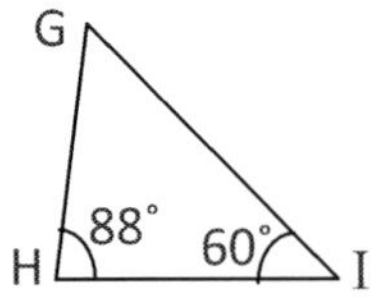

(b) Find the area of triangle GHI.

4 In triangle PQR, PQ = 11 cm, QR =6 cm and angle RPQ = 40^0.

(a) (i) Calculate the size of angle QRP.

(ii) Find the length of PR.

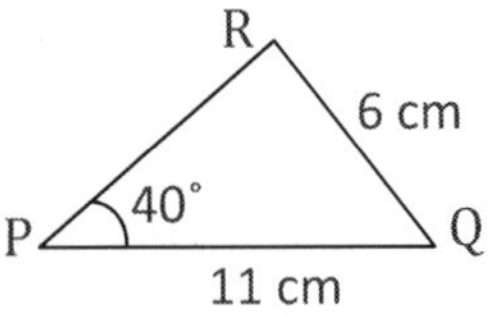

(b) Find the area of triangle PQR.

5 In triangle MNP, MN = 10 cm, NP = 13 cm and angle MNP = 100^0.

(a) Find the length of PM.

(b) Find the size of angle NPM.

(c) Find the area of triangle MNP.

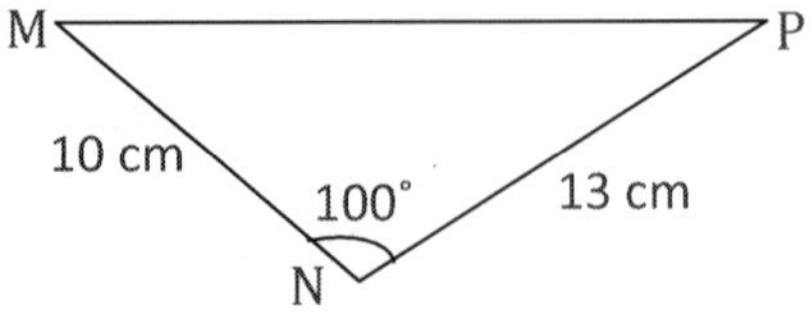

6 The diagram shows a rhombus ABCD.

AB = 10 cm and angle DAB = 50^0.

(a) Find the lengths of the diagonals (i) BD, (ii) AC.

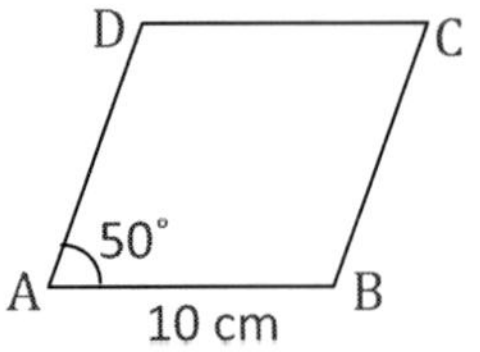

(b) Calculate the area of the rhombus.

7 In parallelogram QRST, ST = 14 cm, TQ = 11 cm and angle QST = 44^0.

(a) Find the sizes of

(i) angle SQT, (ii) angle TQR, (iii) angle QRS.

(b) Find the lengths of the diagonals (i) QS, (ii) TR.

(c) Find the areas of (i) triangle QST,

(ii) parallelogram QRST.

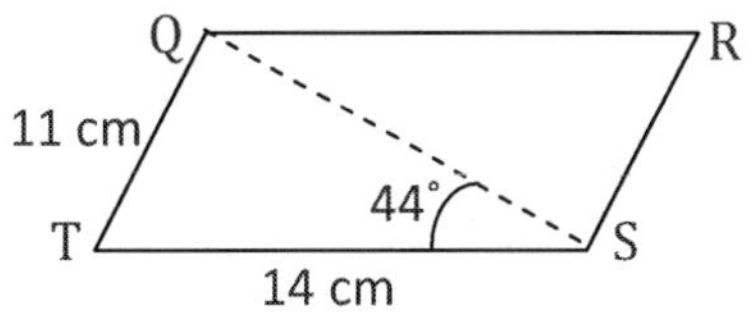

8 Three ships A, B and C leave a port, P, at the same time.

Ship A travels at 20 km/h due east.

Ship B travels at 30 km/h on a bearing of 130^0.

Ship C travels at 22 km/h on a bearing of 200^0.

The diagram shows the travel of the ships for one hour.

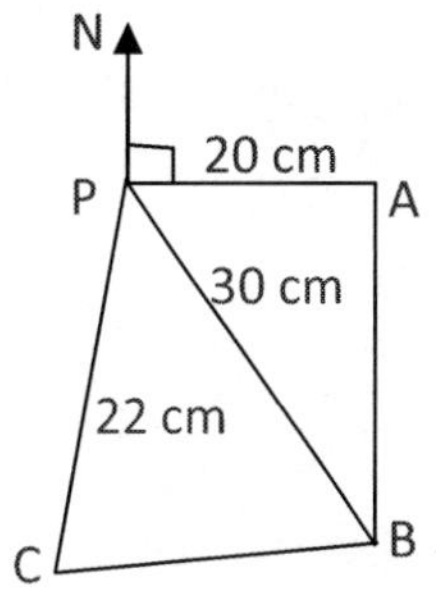

(a) Calculate the distances between
(i) A and B, (ii) B and C, (iii) A and C.

(b) Calculate the area of the region bounded by PABC.

The sine and the cosine curves

▶ The sketches of the sine curve and the cosine curve are shown below.

y = sin x

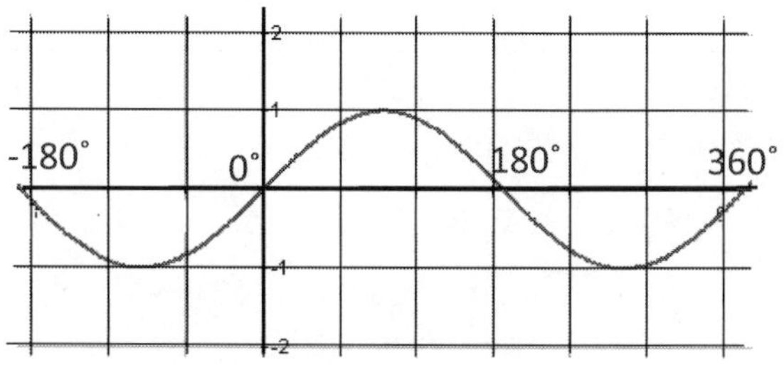

y = cos x

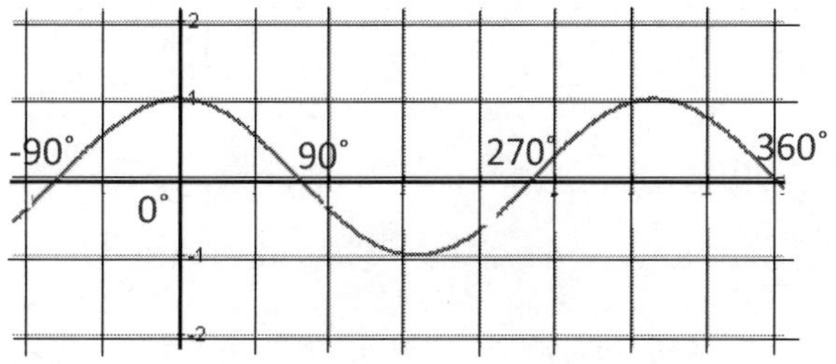

Remember

1. Both the sine curve and the cosine curve are wavy in nature and lie between +1 and −1.
2. Both the curves repeat with themselves in every 360^0 intervals.
3. The value of sine is 0 at multiple of 180^0.
4. The value of cosine is 0 at odd multiple of 90^0.
5. The sine and cosine curves are used to calculate two different values for an angle.
6. $\sin x = \sin(180^0 - x)$ and $\cos x = -\cos(180^0 - x)$.

Example

Using the sketch of $y = \sin x$, find the two values of x for the equation $\sin x = 0.5$ and $0^0 \leq x \leq 360^0$.

Working: $x = 30^0$ or 150^0

Draw a horizontal line at $y = 0.5$. The points of intersection of the line and the curve of $y = \sin x$ are 30^0 and 150^0.

Alternative method

$\sin x = 0.5$ or $x = \sin^{-1}(0.5) = 30^0$

$\sin x = \sin(180^0 - x) = \sin(180^0 - 30^0) = \sin 150^0$ or $x = 150^0$

Therefore $x = 30^0$ or 150^0.

Use $\sin x = \sin(180^0 - x)$.

Practise exercise 25

1 Using the sketch of $y = \sin x$, find the two values of x (if possible) for which each of the following equations is true and $0^0 \leq x \leq 180^0$.

(i) $\sin x = 0.4$ (ii) $\sin x = -\frac{1}{2}$ (iii) $\sin x = 1$ (iv) $\sin x = 0.88$ (v) $\sin x = \frac{1}{\sqrt{2}}$

2 Using the sketch of $y = \cos x$, find the two values of x (if possible) for which each of the following equations is true and $0^0 \leq x \leq 180^0$.

(i) $\cos x = 0.8$ (ii) $\cos x = 1$ (iii) $\cos x = 0.98$ (iv) $\cos x = -\frac{3}{4}$ (v) $\cos x = \frac{\sqrt{3}}{2}$

3 In triangle ABC, AB = 13 cm, CA = 8 cm and angle ABC = 35^0.

There are two possible positions for vertex C.

Calculate the two possible values of angle BCA.

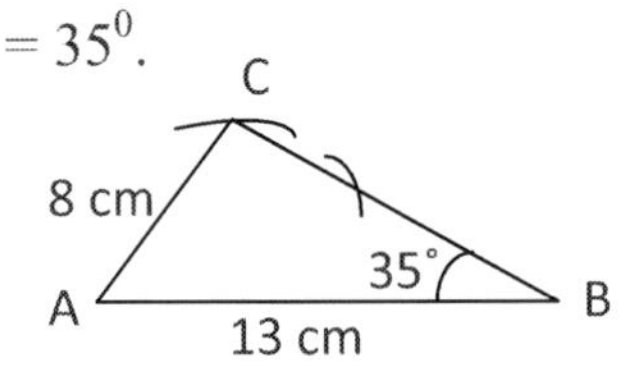

Solid Trigonometry

- To solve trigonometrical problems in three dimensions, you need to extract a triangle from it and then use either Pythagoras theorem or trigonometry or both.

Example

The diagram shows a square based pyramid.

AB = 20 cm and the perpendicular height of the pyramid EF = 15 cm.

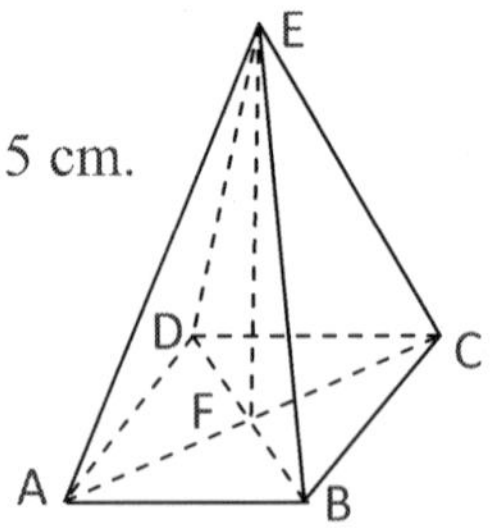

(a) Calculate the length of BD.

(b) Calculate the angle between CE and the base ABCD.

(c) If G is the midpoint BC, calculate the length of EG.

Working: (a) In right triangle ABD,

$BD^2 = AB^2 + DA^2$

$= 20^2 + 20^2 = 800$

$BD = \sqrt{800} = 28.3$ cm

Extract a square from the square based pyramid.

Use Pythagoras theorem to calculate BD.

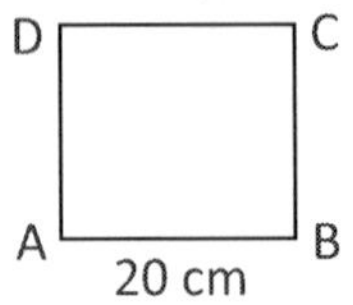

(b) In right triangle EFC,

Extract right triangle EFC from the pyramid.

Use tan ratio for opposite and adjacent.

$$\tan x = \frac{15}{14.15}$$

$$x = \tan^{-1}\left(\frac{15}{14.15}\right) = 46.7^0$$

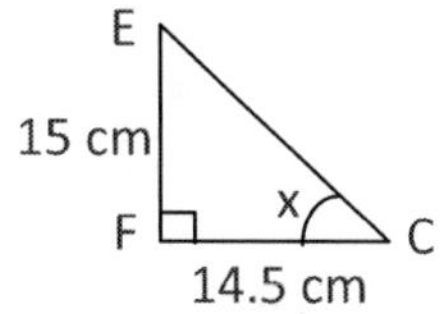

So angle ECF is 46.7^0.

(c) In right triangle EFG,

Extract right triangle EFG from the square based pyramid.

Use Pythagoras theorem to calculate EG.

$$EG^2 = EF^2 + FG^2 = 15^2 + 10^2$$

$$= 325$$

$$EG = \sqrt{325} = 18 \text{ cm}$$

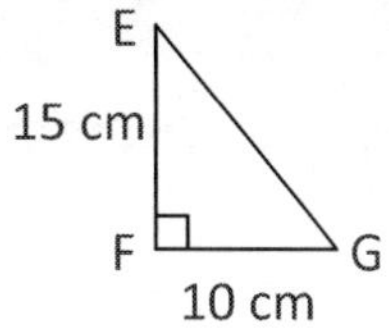

Practise exercise 26

1 The diagram shows a cuboid. AB = 16 cm, BC = 9 cm and BF = 7 cm.

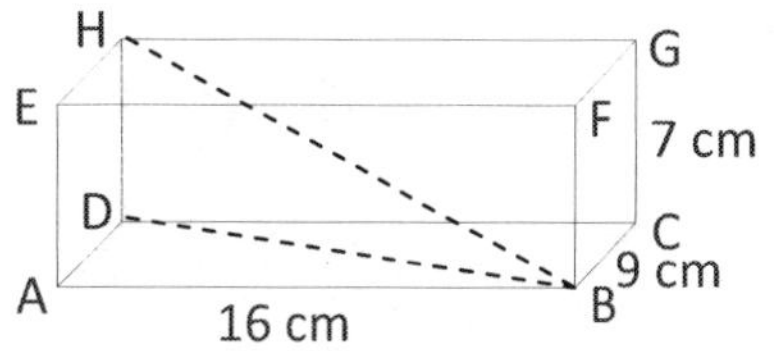

(a) Calculate the length of BD.

(b) Calculate the length of the diagonal BH.

(c) Calculate the angle between the line BH and the plane ABCD.

2 The diagram shows a cube in which MN = 12 cm.

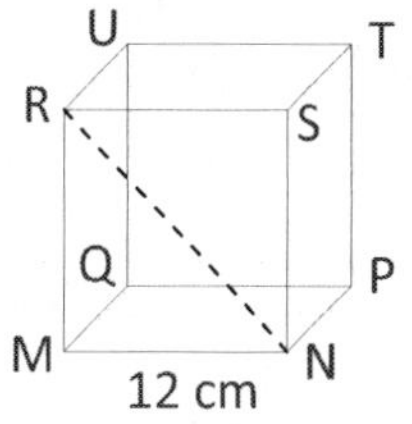

(a) Calculate the lengths of (i) MP, (ii) MT.

(b) Calculate the angle between NU and the base MNPQ.

(c) Calculate the area of the face RSTU.

3 The diagram shows a pyramid whose base is a square. AB = 8 cm and the vertical height PQ = 11 cm.

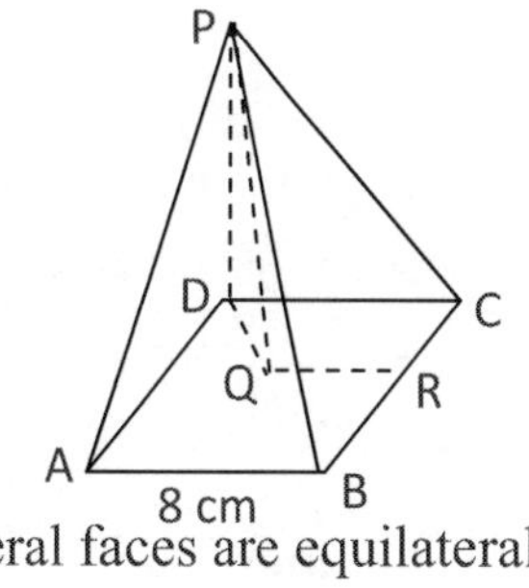

R is the midpoint BC.

(a) Calculate the length of (i) AC, (ii) PD.

(b) Calculate the angle that PA makes with the base.

(c) Calculate the angle between PR and the plane ABCD.

4 The diagram shows a square based pyramid whose all the lateral faces are equilateral triangles.

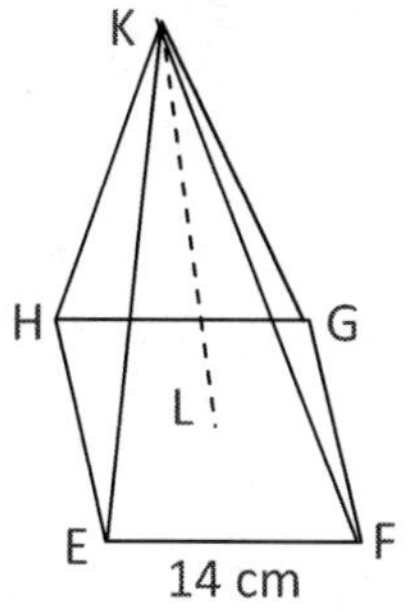

(a) Calculate the lengths of the vertical height KL.

(b) Calculate the angle between KF and the base EFGH.

(c) Calculate the angle between KL and the face KFG.

(d) Find the area of the triangle KFG.

(e) What is the total surface area of the pyramid?

5 The diagram shows a pyramid whose base is rectangle MNPQ.

MN = 18 cm, NP = 12 cm and PR = 24 cm.

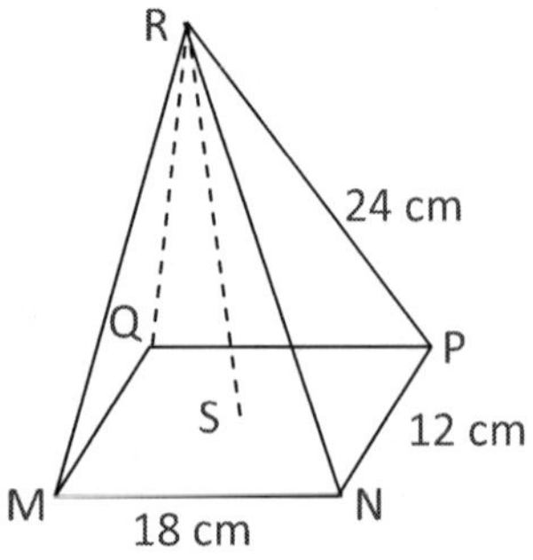

(a) Calculate the lengths of

(i) MP, (ii) SP, (iii) RS.

(b) Calculate the angles that

(i) RP makes with the base MNPQ,

(ii) RS makes with the face MNR.

(c) If T is the midpoint of MN, calculate
(i) the length of RT, (ii) the angle between RT and the plane MNPQ.

6

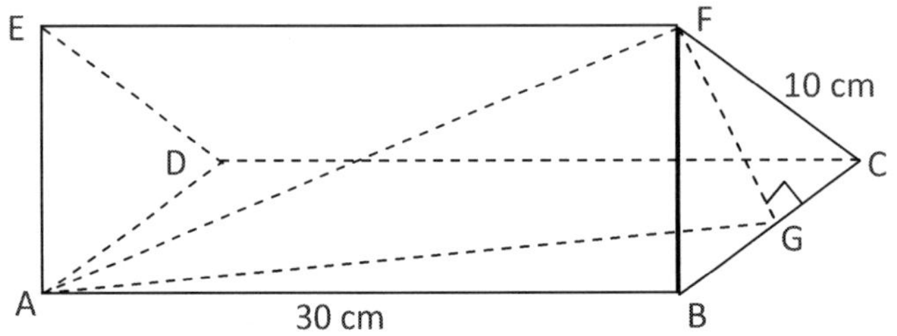

The diagram shows an equilateral triangular prism. AB = 30 cm and FC = 10 cm.
(a) Find the height of the triangular face labelled FG.
(b) Calculate the lengths of (i) AG, (ii) AF.
(c) Find the total surface area of this triangular prism.

7 The diagram shows a pyramid whose base is a

regular pentagon with each side 9 cm.

The vertical height XY = 16 cm and

the slant height XR = 23 cm.

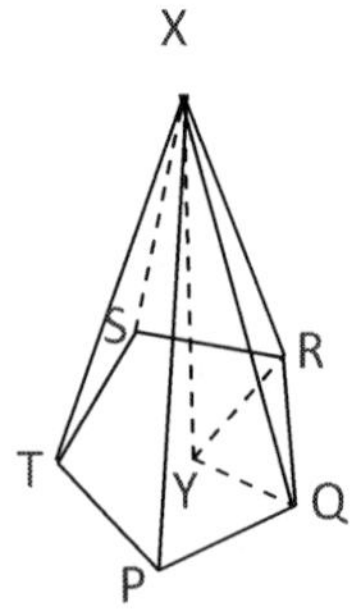

(a) Calculate the size of the angles (i) QYR, (ii) YRQ.
(b) Calculate the angle between (i) XR and the base, (ii) XY and the face XPT.

3.06 Mensuration

Core curriculum

Perimeter and area

- Perimeter is the boundary length of a plane surface. It is expressed in linear units such as mm, cm, m km etc.
- Area is the bounded region of a plane surface. It is expressed in square units such as mm^2, cm^2, m^2, km^2 etc.

- For a square, Area = l^2, Perimeter = $4l$ and diagonal = $\sqrt{2l^2}$.

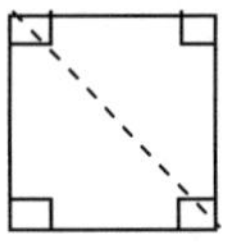

- For a rectangle, Area = lb, Perimeter = $2(l + b)$ and diagonal = $\sqrt{l^2 + b^2}$.

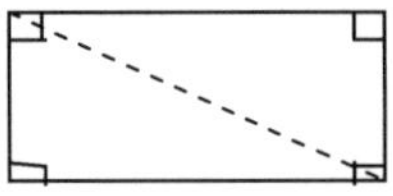

- For a triangle, Area = $\frac{1}{2}$ bh and Perimeter = a + b + c.

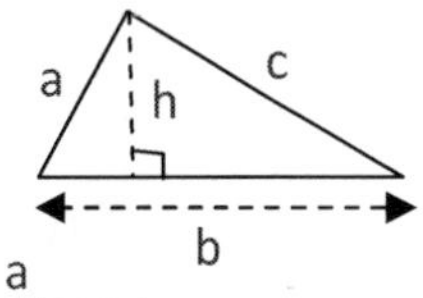

- For a trapezium, Area = $\frac{1}{2}$ h(a+b) and Perimeter = a + b + c + d.

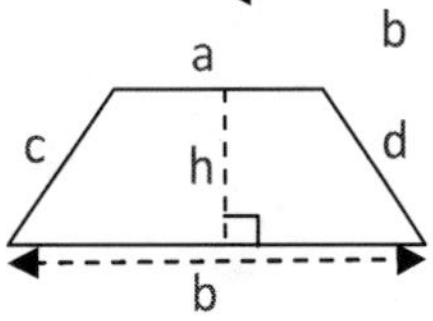

- For a parallelogram, Area = bh or ab sin x and Perimeter = 2a + 2b.

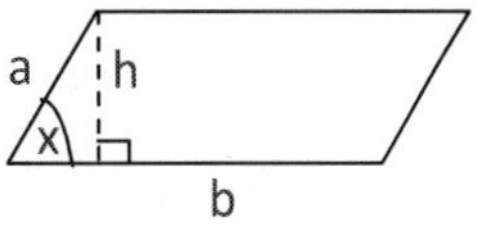

- For a kite, Area = $\frac{1}{2}$ d_1d_2 and Perimeter = 2a + 2b.

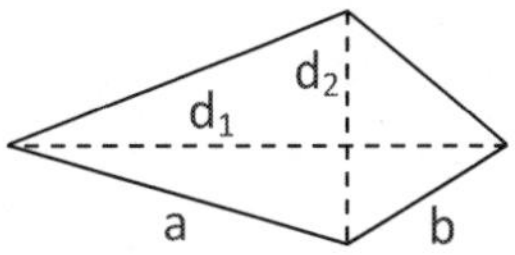

- For a rhombus, Area = bh or b^2 sin x or $\frac{1}{2}$ d_1d_2 and Perimeter = 4b.

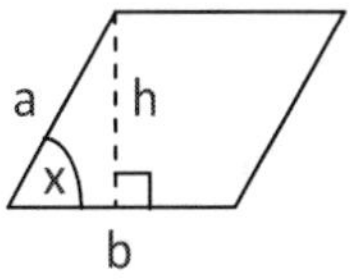

Example

The diagram is a compound figure made up of a square of side 10 cm, a parallelogram and a trapezium.

(a) Calculate (i) the perimeter of the square, (ii) the area of the square, (iii) the area of the trapezium.

(b) What is the height of the parallelogram?

(c) Find the area of the whole shape.

(d) What is the height of the parallelogram?

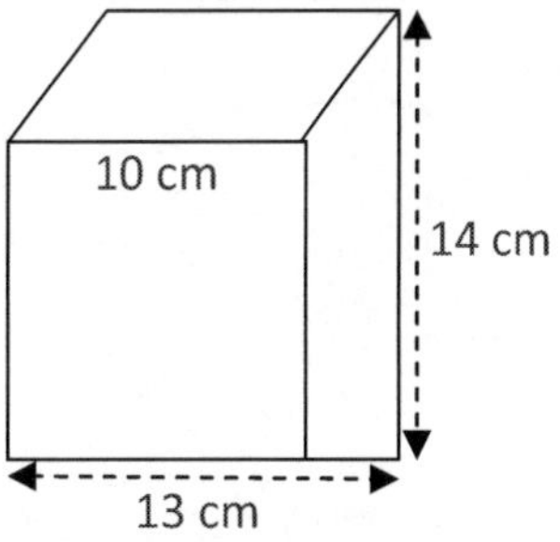

Working: (a) (i) Perimeter = 4 x 10 = 40
(ii) Area = 10^2 = 100 cm^2
(iii) Area = $\frac{1}{2}$ x 3(14+10) = 36 cm^2
(b) Height of the parallelogram = 14 — 10 = 4 cm.
(c) Area = 10 x 4 = 40 cm^2
(d) Area of the whole shape = 100 + 36 + 40 = 176 cm^2

Use the formulae for perimeter and area of a square.
For trapezium, h = 13 — 10 =3 cm.
Use the formula for area of a trapezium.
Use the formula for area of a parallelogram.
Add the areas of all parts.

Practise exercise 27

1 For each shape, calculate (i) the perimeter, (ii) the area.
All the dimensions are given in centimetres.

(a) Square

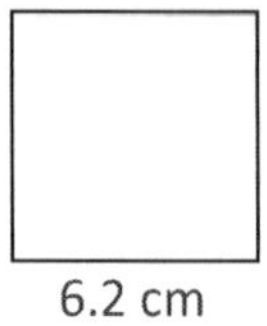

(b) Rectangle

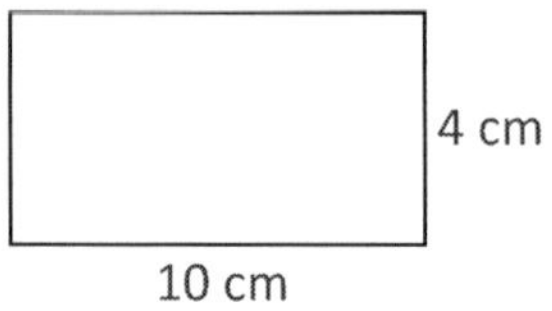

(c) Triangle

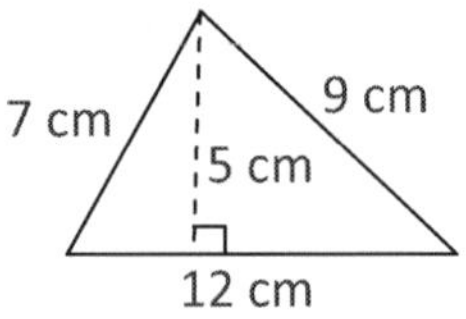

(d) Parallelogram

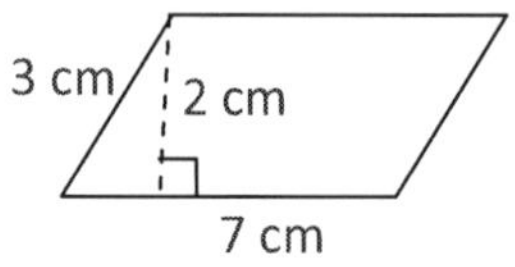

(e) Parallelogram

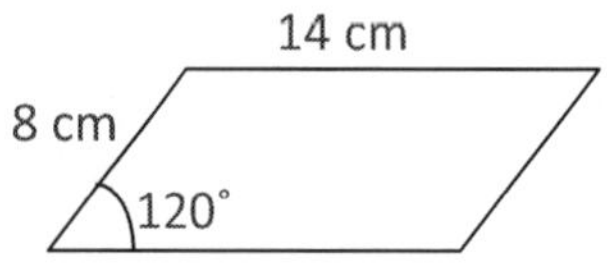

(f) Trapezium

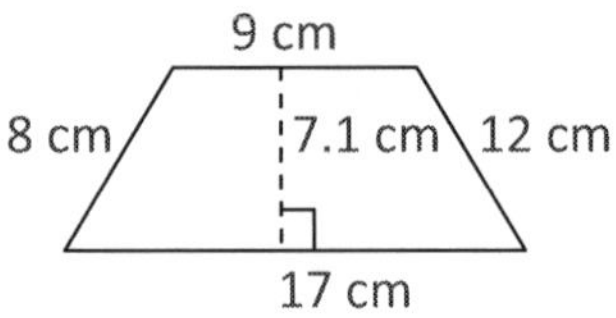

(g) Rhombus

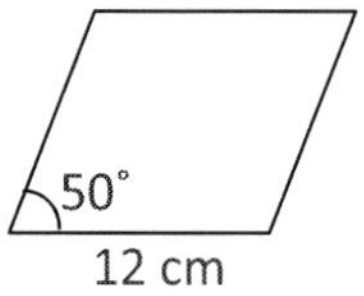

(h) Rhombus

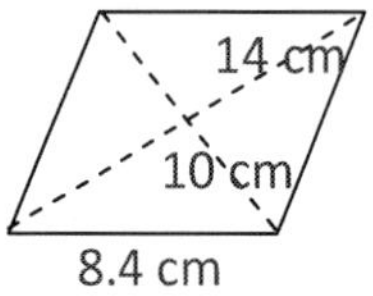

(i) Kite

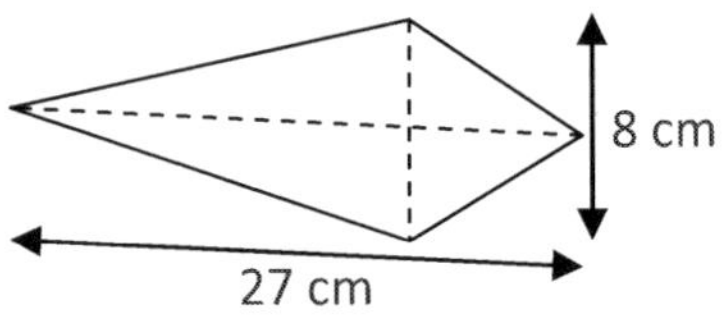

2 Calculate the area of each compound figure.

(a)

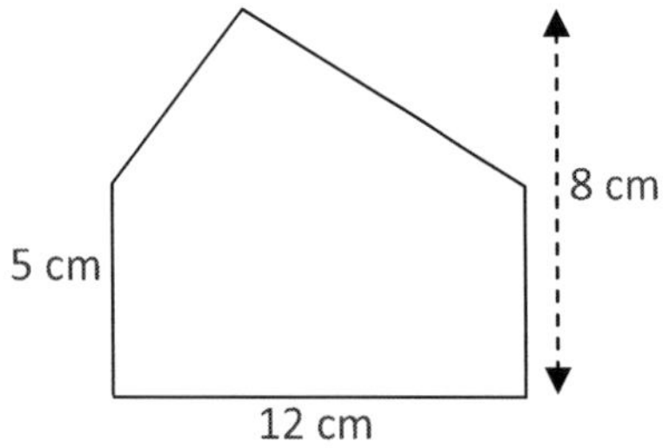

(b)

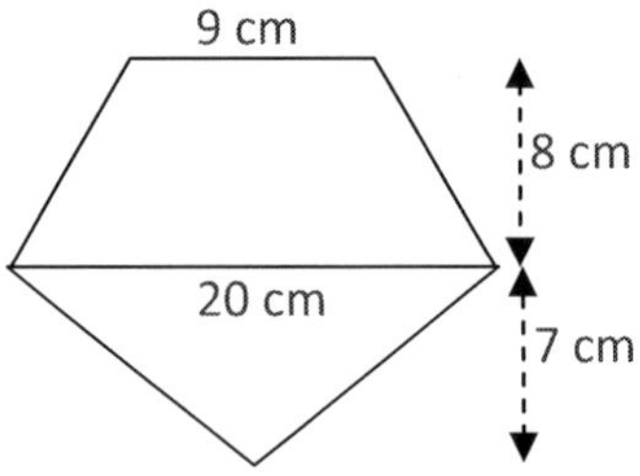

(c)

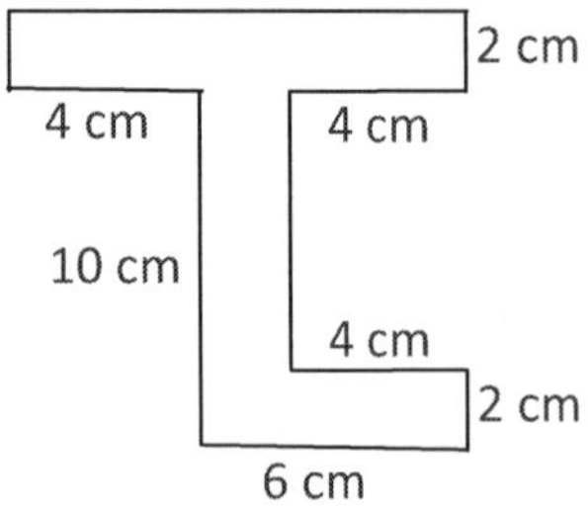

(d)

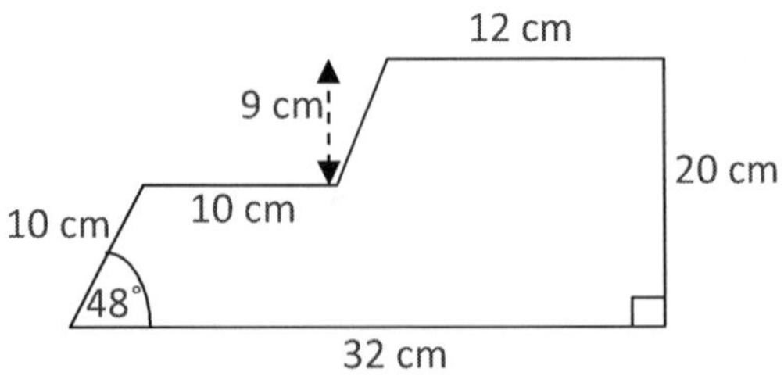

3 Calculate the shaded area of each figure.

(a)

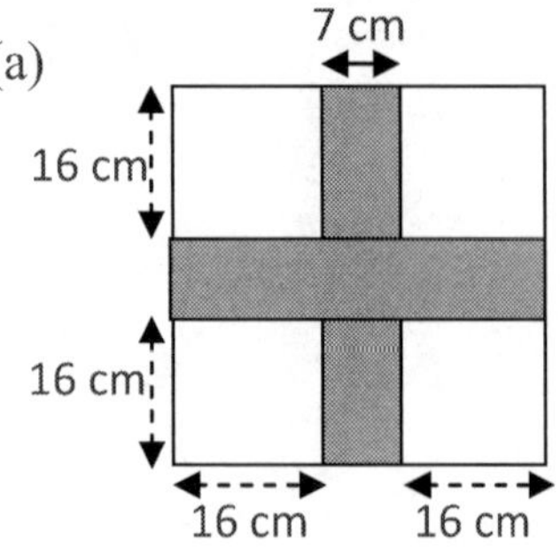

(b)

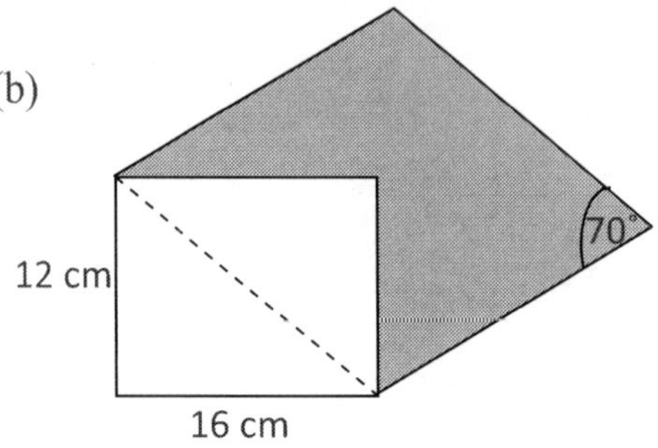

(c)

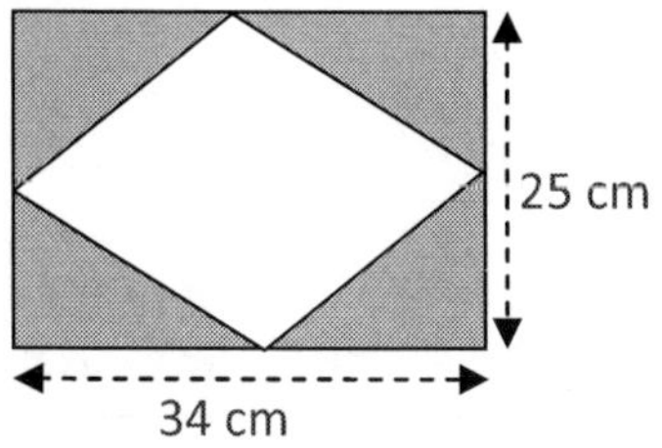

(d)

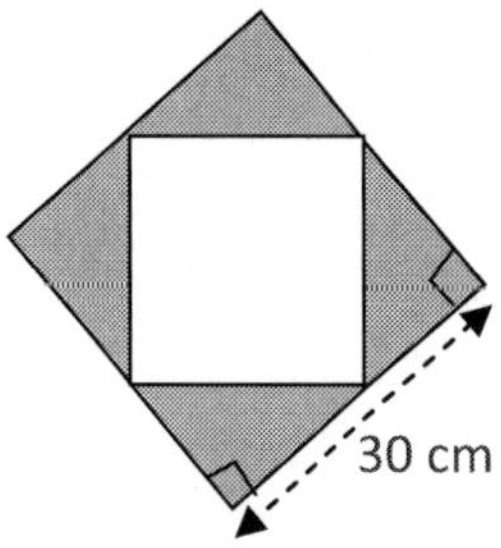

4 (a) The area of a square is 179.56 cm^2. Calculate its (i) side, (ii) perimeter.
(b) The perimeter of a rectangle of 30 m long is 100 m. calculate its (i) width, (ii) area.
(c) The diagonals of a kite are 45 cm and 22 cm. Calculate its area.
(d) The area of a rhombus is 1344 m^2 and its one diagonal is 42 m. Calculate the other diagonal.
(e) The parallel sides of a trapezium are 29 cm and 43 cm. The area of the trapezium is 648 cm^2. Calculate the altitude of the trapezium.

5 The diagram shows a compound figure made up of a rectangle, a parallelogram and a trapezium. All the dimensions are given in centimeters.

(a) Calculate

(i) the perimeter of the rectangle,

(ii) the area of the rectangle.

(b) Find the heights of

(i) the parallelogram, (ii) the trapezium.

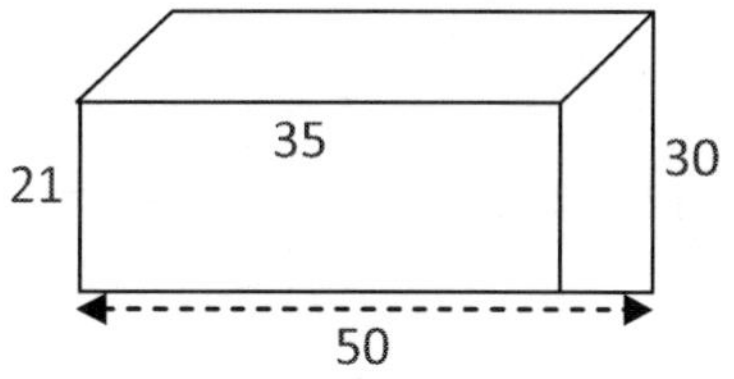

(c) Calculate the areas of (i) the parallelogram, (ii) the trapezium.

6 The diagram shows a square ABCD and AB = $3x^2$.

(a) The perimeter of the square is 432 m. Write down an equation in x.

(b) Solve the equation for x. (c) Calculate the area of the square.

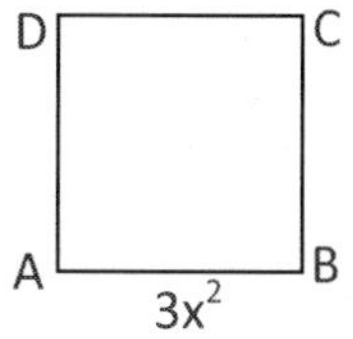

7 The diagram shows a parallelogram MNPQ.

MN = 3y + 4, NP = y, PQ = 8x + 1,

QM = 2x + 2 and angle PQM = 132^0.

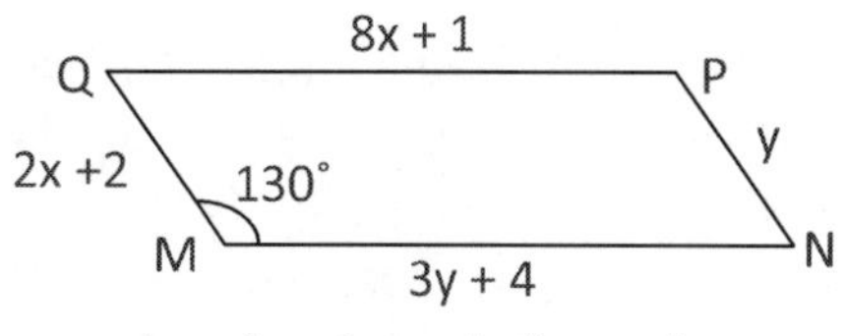

(a) Write down a pair of simultaneous equations using the above information.
(b) Solve the simultaneous equations from your answer to part(a).
(c) Write down the dimensions of the parallelogram.
(d) Calculate the area of the parallelogram MNPQ.

The circumference and area of a circle

- For a circle, Circumference = $2\pi r$ and Area = πr^2.

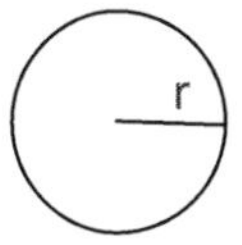

- For a semi circle, Perimeter = $\pi r + 2r$ and Area = $\frac{1}{2}\pi r^2$. (Where $r = d/2$)

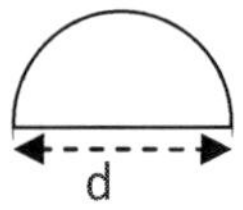

Remember

1. A semi circle is half of a circle.	2. A quarter circle is one-fourth of a circle.

Example

The diagram shows a rectangle of 12 m long and 5 m wide and is inscribed in a circle.

(a) Calculate (i) the area of the rectangle,

(ii) the length of AC,

(iii) the area of the circle, correct to 1 decimal place.

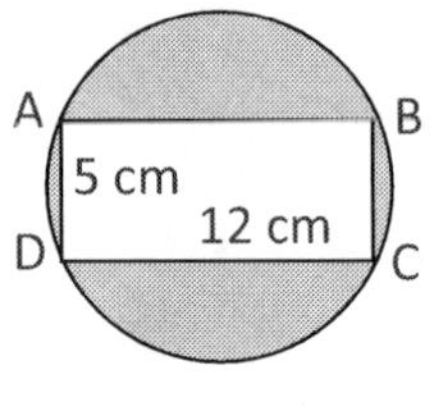

(b) What is the area of the shaded region?

(c) Calculate the circumference of the circle giving your answer up to 3 significant figures.

Working: (a) (i) Area = 12 x 5 = 60 m^2

(ii) AC = $\sqrt{12^2 + 5^2}$ = 13 m

(ii) Area = π x 6.5^2 = 132.7 m^2

(b) Shaded area = 132.7 — 60 = 72.7 m^2.

(c) Circumference = π x 13 = 40.8 m

Use the formula for area of a rectangle.
Use the formula for diagonal of a rectangle.
The diameter of circle is 13 m. So radius is 6.5 m.
Use the formula for area of a circle.
Shaded area = Area of circle — Area of rectangle.
Use the formula for circumference of a circle.

Practise exercise 28

1 For each circle, calculate (i) the area, (ii) the circumference.

(a)

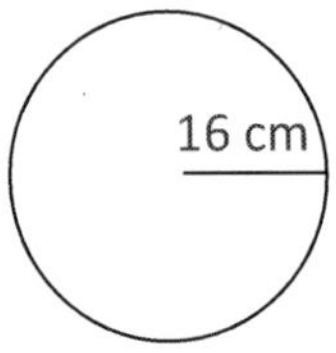

(b)

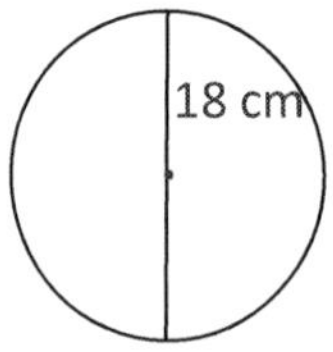

(c)

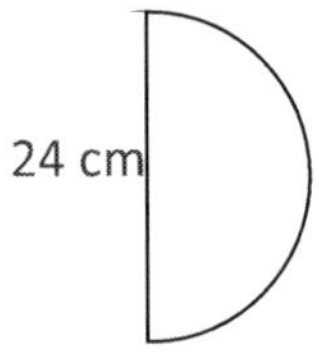

2 (a) The diameter of a circle is 21 cm. calculate its (i) area, (ii) circumference.

(b) The circumference of a circular ground is 377 m. Calculate its (i) radius, (ii) area.

(c) A garden is in the shape of a circle. The area of the garden is 3217 m^2.
Calculate its (i) diameter, (ii) circumference.

3 The diagram shows two concentric circles of radii 8 cm and 11 cm.
Calculate the area of the shaded region.

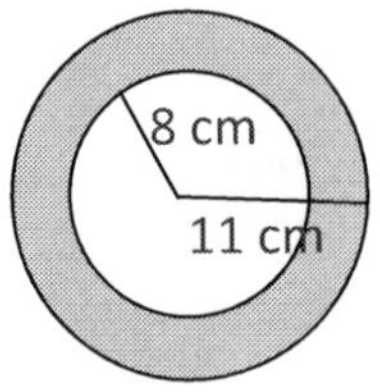

4 The diagram shows a rectangle ABCD and two congruent circles in it. AB = 80 cm.

(a) What is the length of BC?

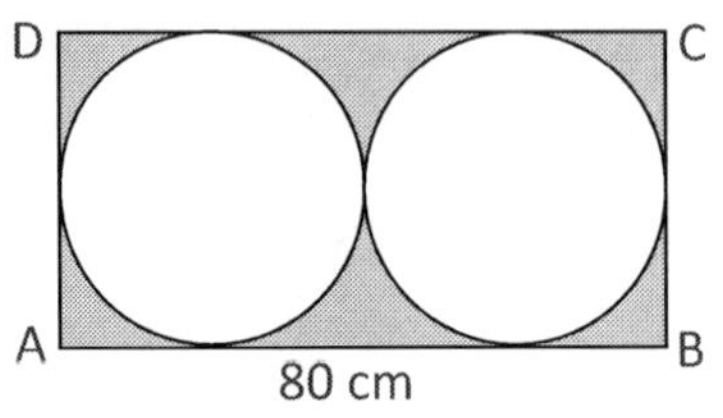

(b) Calculate the area of the shaded region.

5 The diagram is made up of two semi circles and two equilateral triangles.

(a) Calculate the shaded area.

(b) Find the outer boundary of the diagram.

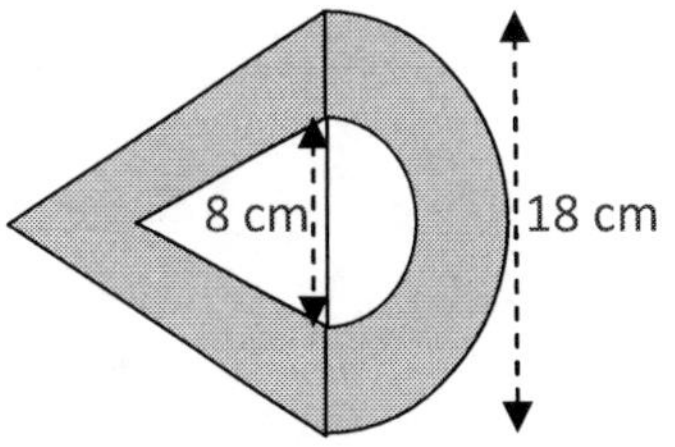

6 The diagram shows a moving track consists of two rectangular paths and two semi circular paths.

The width of the track is 1 m.

(a) Calculate (i) the outer length of the track,

(ii) the inner length of the track.

(b) Calculate the total area of the track.

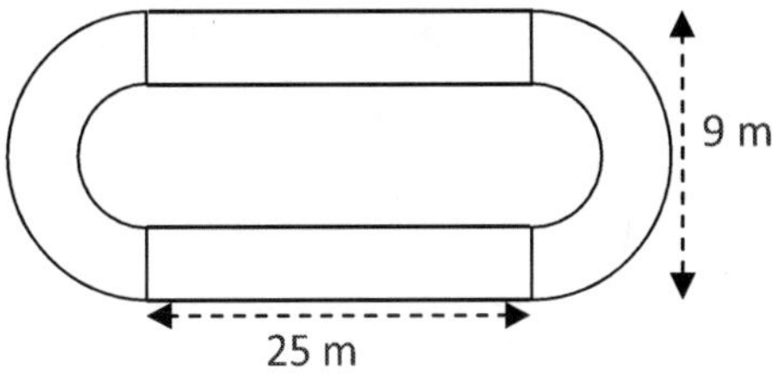

7 The radius of the wheels of a truck is 50 cm. The truck covers a distance of 68.4 km in one hour. Find the average number of revolutions made by a wheel in one minute.

Surface area and volume

- Volume is the bounded space of a solid. It is expressed in mm^3, cm^3, m^3, km^3 etc.
- For a prism, Volume = Area of cross-section x Length or V = AL
- For a cuboid, Volume = lbh, Lateral surface area = 2(bh+lh),
  Total surface area = 2(lb + bh + lh), Diagonal = $\sqrt{l^2 + b^2 + h^2}$.

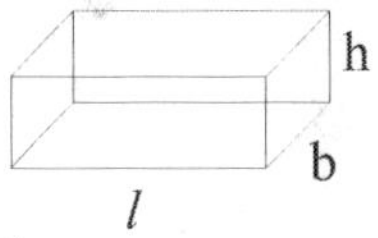

- For a cube, Volume = l^3, Lateral surface area = $4l^2$,
  Total surface area = $6l^2$, Diagonal = $\sqrt{3}\, l$.

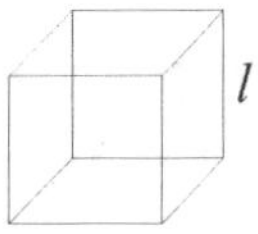

- For a cylinder, Volume = $\pi r^2 h$,
  Curved surface area = $2\pi rh$,
  Total surface area = $2\pi rh + 2\pi r^2$.

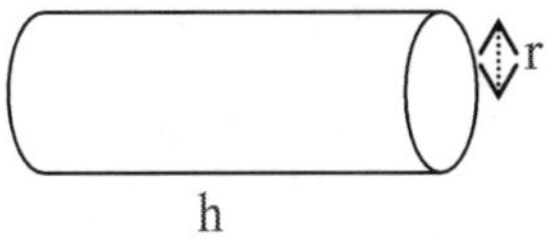

Example

The diagram shows a prism of triangular cross-section ABC. AB = 11 cm and CD = 8 cm.

The prism is of 16 cm long.

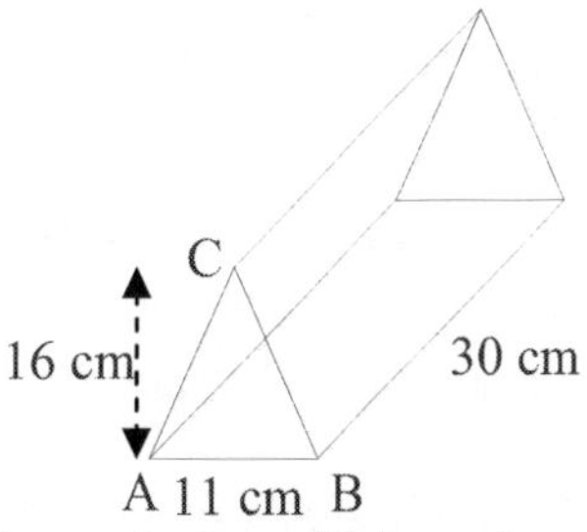

(a) (i) Calculate the area of the cross-section ABC,

(ii) Calculate the volume of the prism.

(b) The volume of this prism is equal to the volume of a cylinder of 14 cm long.

(i) Work out the radius of the cylinder,

(ii) Calculate the total surface area of the cylinder.

Working: (a) (i) Area of cross-section $= \frac{1}{2} \times 11 \times 8$

$= 44 \text{ cm}^2$

(ii) Volume $= 44 \times 16 = 704 \text{ cm}^3$

(b) (i) $\pi r^2 h = 704$ or $r^2 = \frac{704}{14\pi} = 4$ cm

(ii) Total surface area $= 2\pi \times 4 \times 14 + 2\pi \times 4^2 = 452 \text{ cm}^2$

> Use the formula for area of a triangle.
> Use the formula V = AL for volume of a prism.
> Set an equation by equating both the volumes and then solve for r.
> Use the formula for total surface area of a cylinder.

Practise exercise 29

1 For each shape, calculate it (i) volume, (ii) lateral surface area, (iii) total surface area.

(a)

(b)

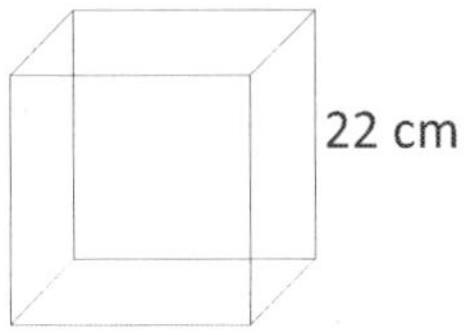

(c)

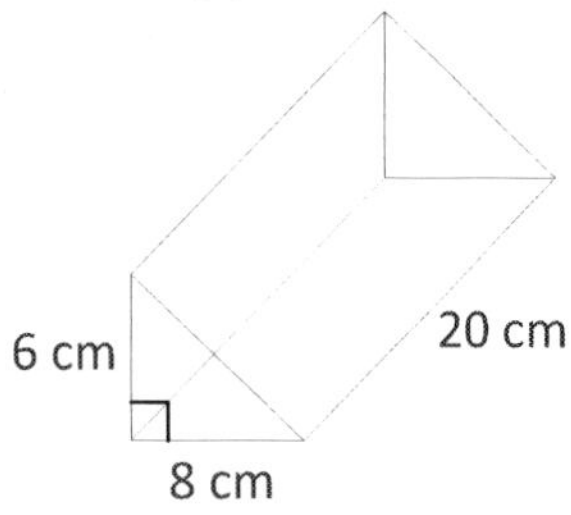

(d)

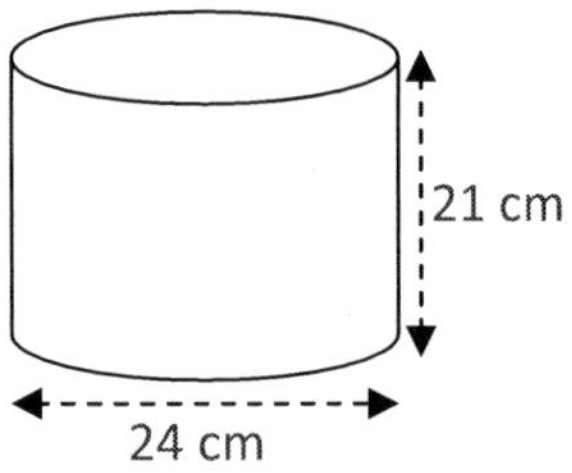

2 The length, breadth and height of a cuboid are 9 cm, 8 cm and 3 cm respectively.
(a) Calculate the volume of the cuboid.
(b) The lateral surface area of a cube is equal to the volume of this cuboid.
(i) Find the edge of the cube. (ii) Calculate the volume of the cube.
(iii) Volume which shape is greater and by how much?

3 The diagram shows a rectangular box of dimensions 20 cm by 12 cm by 8 cm.

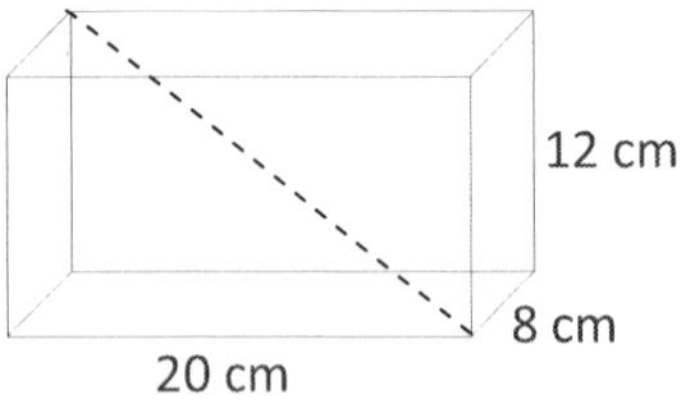

(a) Calculate (i) the lateral surface area of the box,

(ii) the total surface area of the box,

(iii) the diagonal of the box.

(b) The curved surface area of a cylindrical pipe of length 16 cm is same as the lateral surface area of the box. Find (i) the base radius of the pipe,
(ii) the base area of the pipe, (iii) the total surface area of the pipe.

4 The diagram shows a trapezoidal prism.

AB = 11 cm, CD = 7 cm and the perpendicular distance between AB and AC is 8 cm.

The length of the prism is 17 cm.

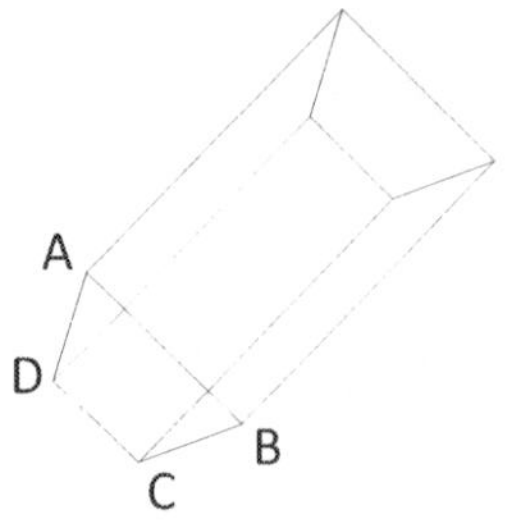

(a) Calculate the area of the cross-section ABCD.

(b) Calculate the volume of the prism.

5 The diagram shows the cross-section of a prism.

MN = 30 cm, NP = 15 cm, QR = 24 cm

and RM = 18 cm.

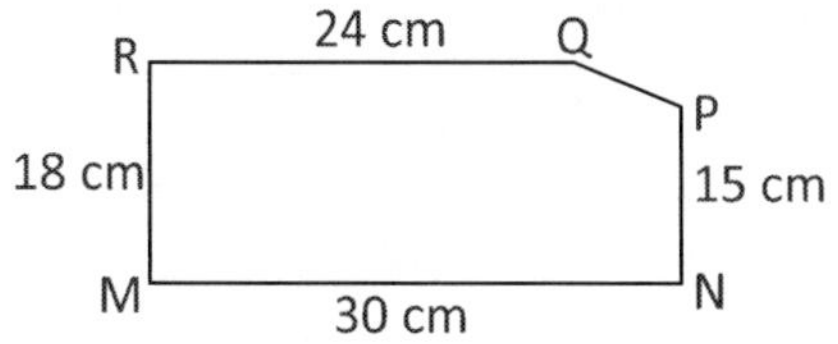

(a) Calculate the area of this cross-section.
(b) The length of the prism is 64 cm. Calculate the volume of the prism.

6 (a) The volume of a cuboid is 945 cm^3. The length and width of the cuboid are 15 cm and 9 cm. Calculate the height of the cuboid.
(b) The surface area of a cube is 384 m^2.
(i) Calculate the edge of the cube. (ii) Calculate the volume of the cube.
(c) The volume of a cylindrical pipe of 0.25 m long is 2827 cm^3.
Calculate (i) the radius of a circular end, (ii) the surface area of the pipe.
(d) The dimensions of a wooden box are 90 cm by 35 cm by 19 cm. The thickness of the wood used is 1.5 cm. Calculate the volume of air present in an empty box.

Nets

Remember

1. A net can be drawn by using the exact faces of a solid in which some edges are separated to make it a two dimensional shape.
2. The net of a solid can be folded along its edges to get back the original shape.
3. The net of a solid can be drawn in different ways.
4. Sketch the net of a solid means you need to draw it accurately.
5. Draw an accurate net means you need to use the given exact measurements.

Example

Sketch the net of a cuboid in three different ways.

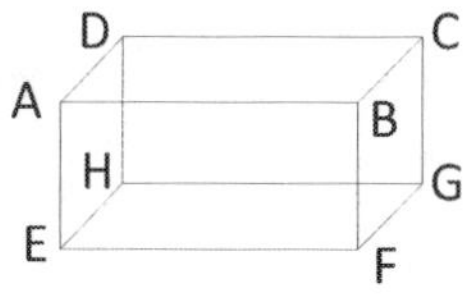

Working:

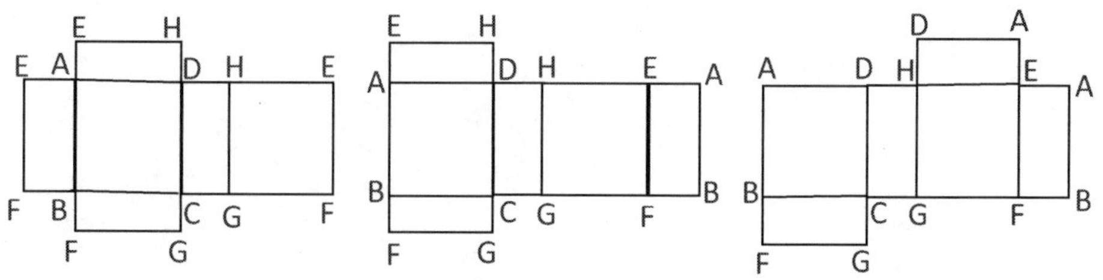

Practise exercise 30

1 The diagram shows a cube.

Sketch the net of this cube

in three different ways.

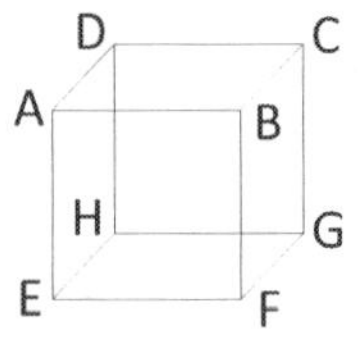

2 (a) The diagram shows a square based pyramid

and all its sloping edges are equal.

Draw an accurate net of the pyramid.

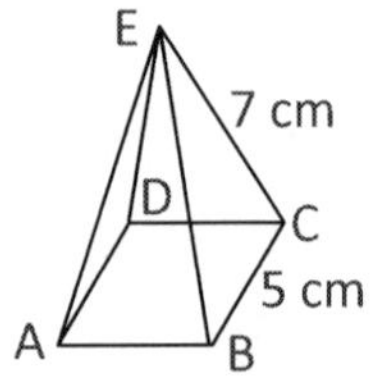

(b) The diagram shows a pyramid whose base

is a rectangle and all the other faces are congruent

triangles. Sketch a net of this pyramid.

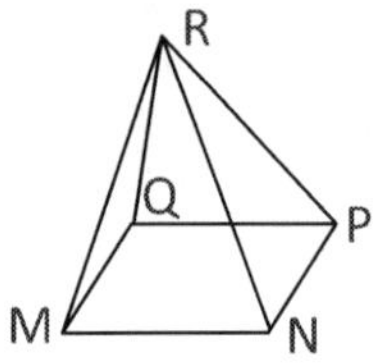

(c) (i) Sketch the net of a cylinder.

(ii) In your net, what is the special name of the geometrical figure that shows the curved surface?

3 The diagram shows a tetrahedron whose each edge is 6 cm long.

(a) Draw an accurate net of this tetrahedron.

(b) Using your net, calculate

(i) the area of triangle ABC,

(ii) the total surface area of the tetrahedron.

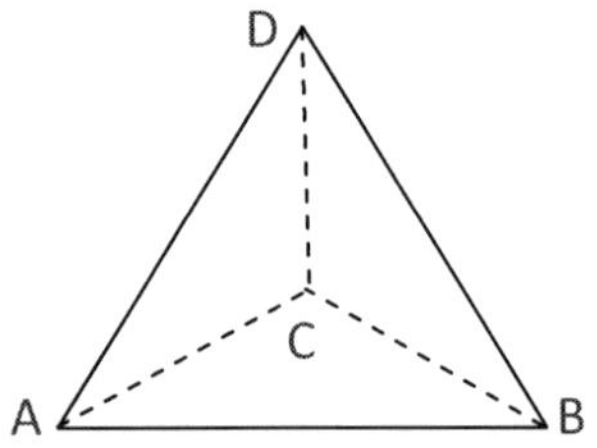

4 (a) Sketch the net of a triangular prism. (b) Sketch the net of a regular hexagonal prism.
(c) Sketch the net of a cone. (d) Sketch the net of a square based pyramid.

5 The diagram shows a regular pentagonal prism.

Sketch its net.

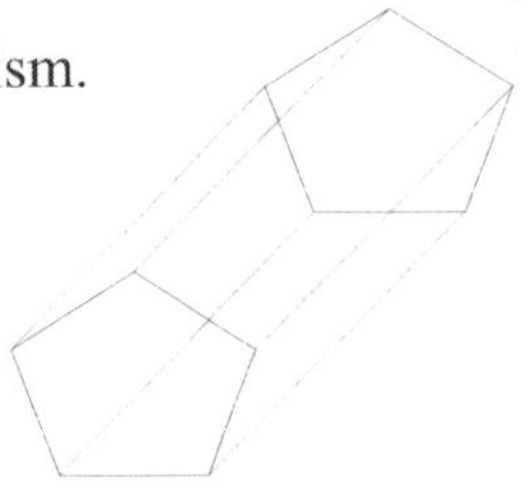

6 The diagram shows the net of a solid with the labelled dimensions.

(a) Write down the name of the solid

that the net will make.

(b) Calcualate

(i) the total surface area of the solid,

(ii) the volume of the solid.

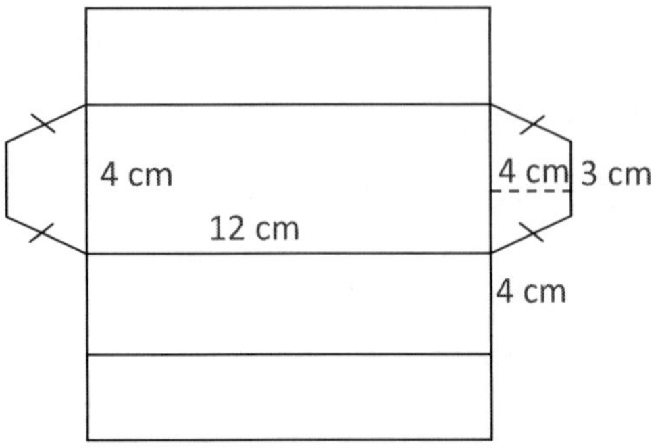

Extended curriculum

Arc length and area of a sector

- An arc length is the part of the circumference of a circle.
- A sector is a part of the area of a circle bounded by two radii.
- For a sector of radius r and central angle x,

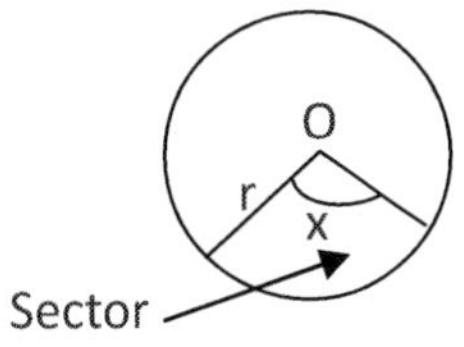

Arc-length = $\frac{x}{360^0} 2\pi r$,

Area = $\frac{x}{360^0} 2\pi r^2$,

Perimeter = Arc-length + 2r.

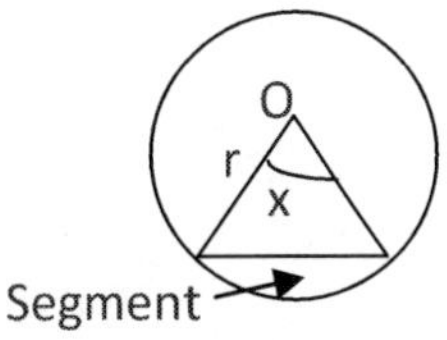

- Area of a segment = Area of sector – $\frac{1}{2} r^2 \sin x$.

Example

The diagram shows a circle of radius 14 cm.

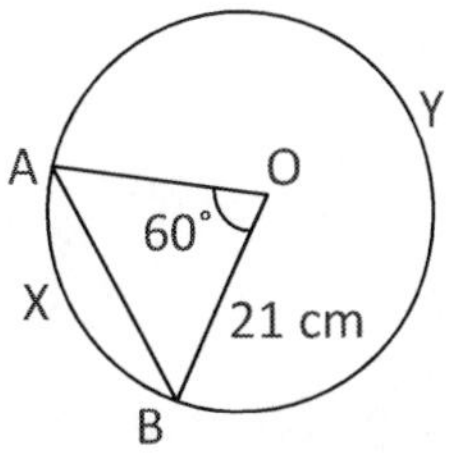

AOBX is a sector with central angle 60°.

(a) Find (i) the arc-length AXB,

(ii) the arc-length AYB.

(b) Calculate (i) the area of the sector AOBX,

(ii) the area of the segment ABX, giving your answer up to 2 decimal places.

Working: (a) (i) Arc-length AXB = $\frac{x}{360^0} 2\pi r$

$= \frac{60}{360^0} 2\pi \text{ x } 21 = 22.0$ cm

(ii) Arc-length AYB = $\frac{300}{360^0} 2\pi \text{ x } 21 = 110$ cm

(b) (i) Area of sector AOBX = $\frac{x}{360^0} 2\pi r^2$

$= \frac{60^0}{360^0} 2\pi \text{ x } 21^2 = 231 \text{ cm}^2$

(ii) Area of the segment ABX

$= 231 - \frac{1}{2} \text{ x } 21^2 \sin 60^0 = 40.04 \text{ cm}^2$

> Substitute $x = 60^0$ and r = 21 cm in the formula for arc-length.
>
> For major arc, central angle = $360^0 - 60^0 = 300^0$.
>
> Substitute $x = 60^0$ and r = 21 cm in the formula for area of a sector.
>
> Use the formula for area of a segment as Area of sector — $\frac{1}{2} r^2 \sin x$.

Practise Exercise 31

1 For each of the sectors labelled S, calculate (i) arc-length, (ii) area, (iii) perimeter.

(a)

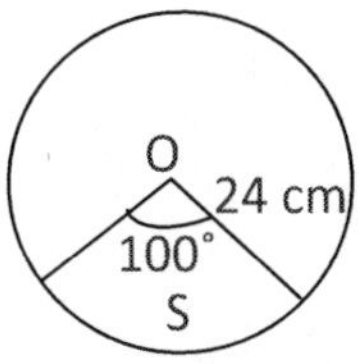

(b)

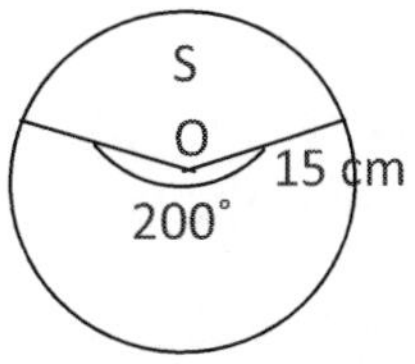

(c)

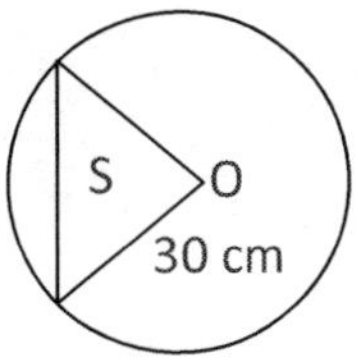

2 (a) The central angle of the sector of a circle is 85^0. The radius of the circle is 12 cm.
Calculate (i) the arc-length of the sector, (ii) the area of the sector.
(b) A sector of a circle has the radius of 20 cm and the central angle of 140^0.
Calculate (i) the length of the arc of the sector,
(ii) the area of the sector, (iii) the perimeter of the sector,
(iv) the area of the remaining part of the circle.

3 The diagram shows a circle of radius 34 cm.

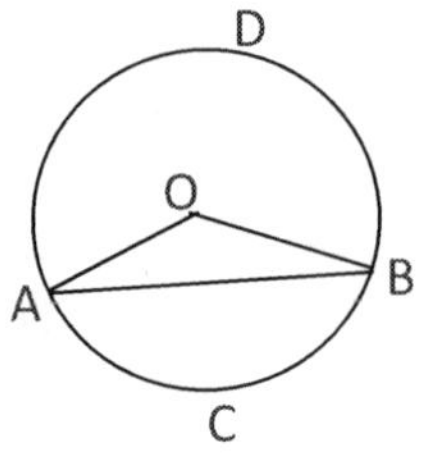

Angle AOB = 120^0.

(a) Find (i) the arc-length ACB,

(ii) the length of the arc ADB.

(b) Calculate (i) the area of the sector OACB, (ii) the area of the sector OADB,
(iii) the area of the triangle OAB, (iv) the area of the segment ABC.

4 The square ABCD is inscribed in a circle.

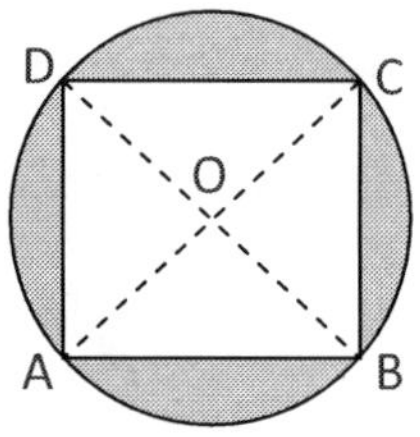

The length of AB is 15 cm.

(a) What is the special name of the triangle AOB?

Give a reason for your answer.

(b) Calculate (i) the arc-length of the sector AOB,
(ii) the length of the arc that bounds any of the three sectors,
(iii) the area of the sector AOB, (iv) the total area of the four segments.

5 The diagram shows two concentric sectors.

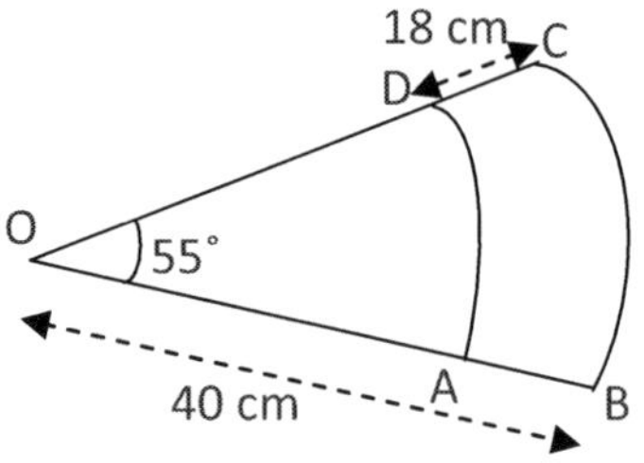

OB = 40 cm, CD = 18 cm and angle BOC = 55^0.

(a) Calculate the area of the region ABCD.

(b) Find (i) the perimeter of the sector OBC,

(ii) the difference between the lengths of arcs of both the sectors.

6 The diagram shows a regular hexagon ABCDEF inscribed in a circle. FC is a diameter of the circle and FC = 24 cm.

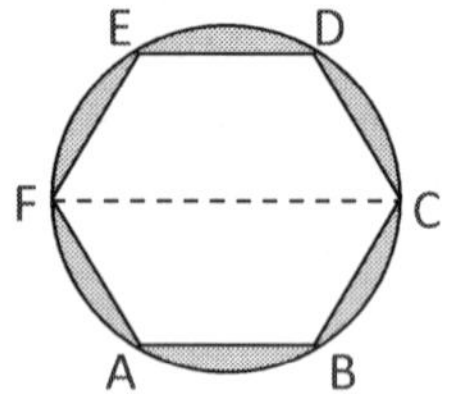

Find the area of the shaded region.

7 A cake is in the shape of a prism.

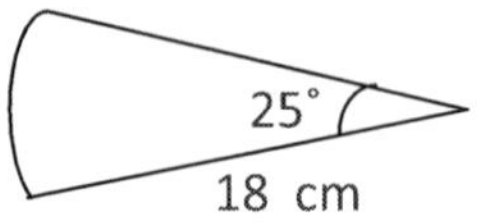

The cross-section of the cake is a sector of radius 18 cm and sectorial angle 25^0.
(a) Calculate the cross-sectional area of the cake.
(b) The length of the cake is 1 m. Calculate the volume of the cake.

(c) To make the cake each cubic centimeter costs 60 cents. Calculate the amount of money to be spent to make the whole cake.

Surface area and volume

- For a pyramid, Volume = $\frac{1}{3}$ base area x height.
- A cone is a pyramid whose base is a circle.
  For a cone, Curved surface area = πrl,
  Total surface area = $\pi rl + \pi r^2$,
  Volume = $\frac{1}{3}\pi r^2 h$.

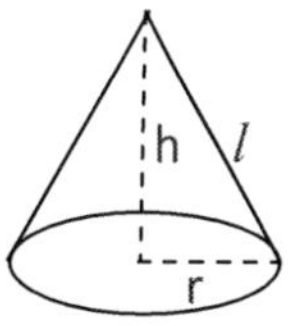

- For a sphere, Surface area = $4\pi r^2$,
  Volume = $\frac{4}{3}\pi r^3$

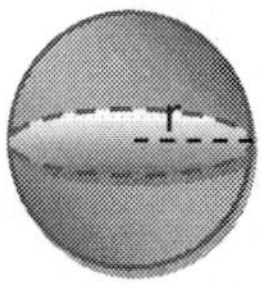

- A hemi sphere is half of a sphere.
  For a hemi sphere, Total surface area = $3\pi r^2$, Volume = $\frac{2}{3}\pi r^3$.

Remember

1. You need not by heart the formulae for pyramids and sphere. Te formula will be stated in the question and you should know how to use it.
2. Use Pythagoras theorem to find an unknown dimension related to slant height, perpendicular height and radius of a cone, i.e. $l^2 = h^2 + r^2$.

Example 1

The diagram shows a tent whose base is a circle of radius 7 cm. The slant height of the tent is 25 m.

For the tent, Calculate

(i) the perpendicular height,

(ii) the curved surface area,

(iii) the total surface area,

(iv) the volume.

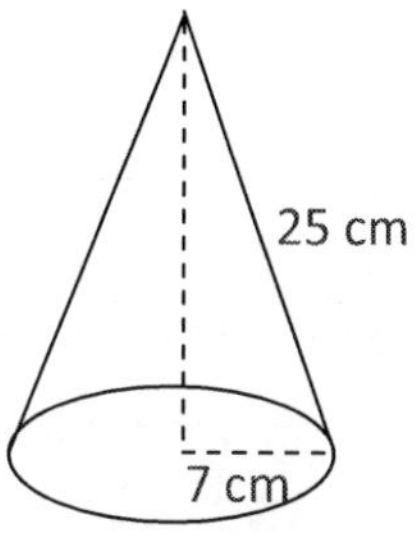

Working: $l = 25$ cm and r = 7 cm.

(i) $25^2 = h^2 + 7^2$

$h^2 = 25^2 - 7^2$ or h = 24 m

(ii) Curved surface area = $\pi \times 7 \times 25 = 528$ m^2

(iii) Total surface area = $528 + \pi \times 7^2 = 682$ m^2

(iv) Volume = $\frac{1}{3}\pi \times 7^2 \times 24 = 1232$ m^3

> Use Pythagoras theorem to find the unknown dimension.
>
> Substitute r = 7, l = 25 and h = 24 in the formulae for a cone.

Example 2

The radius of a sphere is 6.5 cm. Calculate its

(i) surface area, (ii) volume, up to 1 decimal place.

Working: r = 6.5 cm

(i) Surface area = $4\pi \times 6.5^2 = 531$ cm^2

> Use the formulae for surface area and put r = 6.5.
>
> Use the formula for volume and put r = 6.5.

(ii) Volume = $\frac{4}{3}\pi \times 6.5^3 = 1150.3\ cm^3$

Practise exercise 32

1 Calculate the volume of each pyramid.
(a) Square based pyramid (b) Rectangular based pyramid

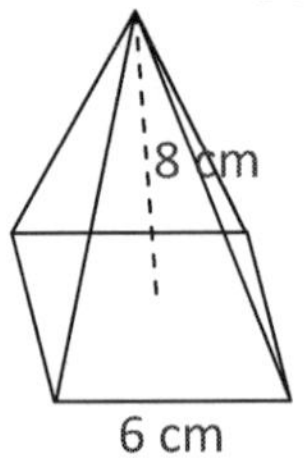

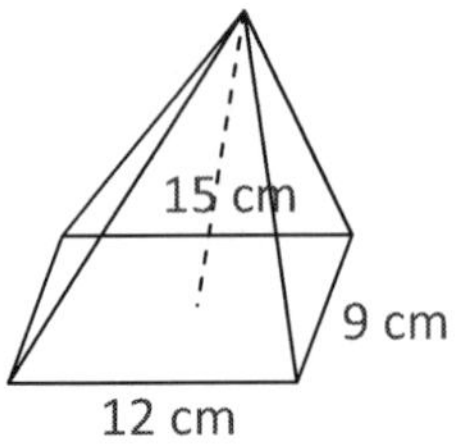

2 The base of a pyramid is a square of each side 10 cm. The other four faces of the pyramid are equilateral triangles. Calculate
(a) the perpendicular height of the pyramid, (b) the total surface area of the pyramid,
(c) the volume of the pyramid.

3 (a) The length and breadth of the base of a rectangular based pyramid are 18 m and 11 m. The height of the pyramid is 26 m. Calculate the volume of the pyramid.
(b) The volume of a square based pyramid is 3920 m^3. The height of the pyramid is 20 m. Calculate the length of each side of the base of the pyramid.
(c) The volume of a regular tetrahedron is 841.57 m^3. The perpendicular height of the tetrahedron is 11.5 m. Calculate (a) the area of its base, (b) its total surface area.

4 The diagram shows a conical glass with the labelled dimensions.

Calculate its (i) slant height, (ii) curved surface area,

(iii) base area, (iv) total surface area, (v) volume.

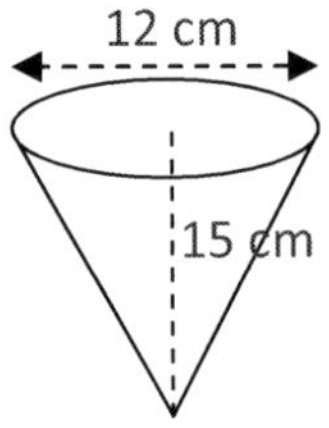

5 (a) The curved surface area of a cone is 578.05 cm^2. The diameter of its base is 16 cm.
Calculate its (i) slant height, (ii) perpendicular height,
(iii) total surface area, (iv) volume.
(b) The slant height of a cone is 13 cm and its perpendicular height is 12 cm.
Calculate its (i) base radius, (ii) curved surface area,
(iii) total surface area, (iv) volume.

6 (a) The radius of a sphere is 21 cm. Calculate its (i) surface area, (ii) volume.
(b) The volume of a sphere is 2144.7 cm^3. Calculate its (i) radius, (ii) surface area.
(c) The diameter of a hemi sphere is 17 cm.
Calculate its (i) radius, (ii) curved surface area, (iii) total surface area, (iv) volume.
(d) The outer and inner radii of a hemi spherical metal vessel are 24 cm and 22 cm respectively.
(i) Calculate the capacity of the vessel.
(ii) Calculate the amount of metal used to make the vessel.
(iii) Calculate the mass of the vessel if 1 cm^3 of metal has a mass of 2 grams.
(iv) How many spheres of radius 3 cm can be made by the metal used in this vessel?

7 The diagram shows a sphere covered by a hollow cone.
The diameter of the cone is 8 cm and its height is 11 cm.
Calculate (a) the curved surface area of the cone,
(b) the radius of the sphere, (c) the volume of the sphere.

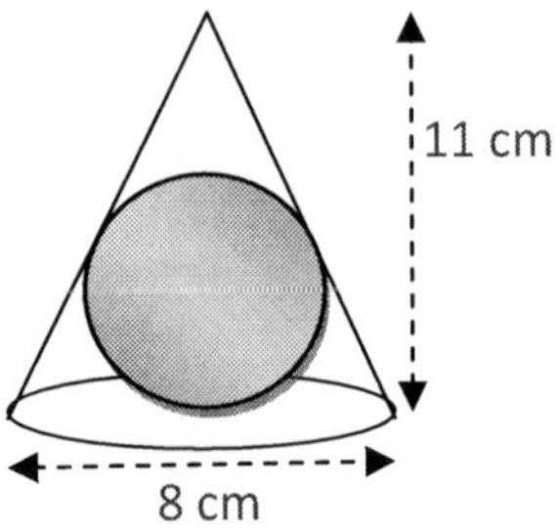

3.07 Matrices ⟵ Extended curriculum ⟶

Matrix formation

Remember

1. The order of a matrix with m rows and n columns is m x n (read as m by n).
2. Numerical information can be stored in a matrix.
3. The numbers that form a matrix are called its elements. A matrix of the order m x has mn elements.
4. You must not use commas to separate the elements in a matrix.

Example

A shopkeeper sells 9 books and 15 pens on the first day, 5 books and 17 pens on the second day, 20 books and 12 pens on the third day of a week.
Show this information in the form of a matrix of (i) order 3 x 2, (ii) 2 x 3.

Working: (a) $\begin{pmatrix} 9 & 15 \\ 5 & 17 \\ 20 & 12 \end{pmatrix}$

Write down the books in the first column and pens in the second column.

(b) $\begin{pmatrix} 9 & 5 & 20 \\ 15 & 20 & 12 \end{pmatrix}$

Write down the books in the first row and pens in the second row.

Prastice exercise 33

1 Write down the order of each matrix.

(a) $\begin{pmatrix} 2 & 3 \\ 4 & 5 \end{pmatrix}$ (b) $\begin{pmatrix} 1 & 7 & 3 \end{pmatrix}$ (c) $\begin{pmatrix} 3 \\ x \\ 8 \end{pmatrix}$ (d) $\begin{pmatrix} 3 & 7 & 11 \\ 9 & 2 & 1 \\ 0 & 4 & 2 \end{pmatrix}$

2 A shopkeeper sells 24 pencils and 10 erasers, 18 pencils and 9 erasers, 7 pencils and 11 erasers on three different days. Represent this information in the form of a matrix of (i) order 3 x 2, (ii) order 2 by 3.

3 Alisha and Alka wrote two tests. In the first test, Alisha scored 94 marks and Alka scored 78 marks.
In the second test, Alisha scored 86 marks Alka scored 89 marks.
(a) Represent this information in the form of a matrix.
(b) Write down the special name of the matrix obtained in part(a).

4 A matrix has 6 elements. Write down all the possible orders of this matrix.

Operations on matrices

Remember

1. You can add or subtract matrices of same order.
2. A number outside a matrix can be multiplied with each term inside the matrix.
3. If matrix A has order of m x n and matrix B has order of n x p, then the product AB has of order m x p.
4. If A and B are two matrices, then AB = BA is not always true.
5. For multiplication of matrices, the number of columns of first matrix must equals the number of rows of second matrix.
6. In two equal matrices, their corresponding elements are equal.

Working rule

1) To add matrices, just add their corresponding elements.
2) To subtract matrices, just subtract their corresponding elements.
3) To multiply matrices, multiply each row of 1st matrix by each column of 2nd matrix.

Example

$A = \begin{pmatrix} 2 & 7 \\ -1 & 3 \end{pmatrix}$ and $B = \begin{pmatrix} -3 & 4 \\ 6 & 1 \end{pmatrix}$

Write down as a single matrix, (i) 2A + B, (ii) A − 3B, (iii) AB.

Working: (a) $2A + B = 2\begin{pmatrix} 2 & 7 \\ -1 & 3 \end{pmatrix} + \begin{pmatrix} -3 & 4 \\ 6 & 1 \end{pmatrix}$

$= \begin{pmatrix} 4 & 14 \\ -2 & 6 \end{pmatrix} + \begin{pmatrix} -3 & 4 \\ 6 & 1 \end{pmatrix} = \begin{pmatrix} 1 & 18 \\ 4 & 7 \end{pmatrix}$

To find 2A, multiply all elements of A by 2.

Add the corresponding elements.

(b) $A - 3B = \begin{pmatrix} 2 & 7 \\ -1 & 3 \end{pmatrix} - 3\begin{pmatrix} -3 & 4 \\ 6 & 1 \end{pmatrix}$

$= \begin{pmatrix} 2 & 7 \\ -1 & 3 \end{pmatrix} - \begin{pmatrix} 9 & -12 \\ -18 & -3 \end{pmatrix} = \begin{pmatrix} -7 & 19 \\ 17 & 6 \end{pmatrix}$

To find 3A, multiply all elements of A by 3.

Subtract the corresponding elements.

(c) $AB = \begin{pmatrix} 2 & 7 \\ -1 & 3 \end{pmatrix} \begin{pmatrix} -3 & 4 \\ 6 & 1 \end{pmatrix}$

$= \begin{pmatrix} 2 \text{ x } -3 + 7 \text{ x } 6 & 2 \text{ x } 4 + 7 \text{ x } 1 \\ -1 \text{ x } -3 + 3 \text{ x } 6 & -1 \text{ x } 4 + 3 \text{ x } 1 \end{pmatrix}$

$= \begin{pmatrix} 36 & 15 \\ 15 & -1 \end{pmatrix}$

Row 1 of first matrix x column 1 of second matrix.

Row 1 of first matrix x column 2 of second matrix.

Row 2 of first matrix x column 1 of second matrix.

Row 2 of first matrix x column 2 of second matrix.

Prastice exercise 34

1 $A = \begin{pmatrix} 2 & -3 \\ 4 & 5 \end{pmatrix}$, $B = \begin{pmatrix} 1 & 3 \\ 6 & -7 \end{pmatrix}$ and $C = \begin{pmatrix} 1 & 0 \\ 0 & 1 \end{pmatrix}$.

Write down each of the following as a single matrix.

(a) A + B (b) 2A + B (c) A + 4B (d) A + B + C
(e) A − B (f) 3A − B (g) A + B − 2C (h) AB
(i) BC (j) −AC (k) A(B + C) (l) B(C + 2A)
(m) A(B − C) (n) (2A − C) B (o) $\frac{-4}{5}C$ (p) $B - \frac{1}{2}C$

2 Calculate each of the following.

(a) $\begin{pmatrix} -3 & 7 & 4 \\ 9 & 2 & 5 \end{pmatrix} + \begin{pmatrix} 1 & 2 & 3 \\ -1 & 3 & 4 \end{pmatrix}$ (b) $\begin{pmatrix} 2 \\ 3 \\ 5 \end{pmatrix} - 2\begin{pmatrix} 1 \\ 0 \\ 1 \end{pmatrix}$ (c) $\begin{pmatrix} 3 & 4 \end{pmatrix}\begin{pmatrix} 1 & 7 \\ 9 & 5 \end{pmatrix}$ (d) $4\begin{pmatrix} 1 & 2 & 3 \end{pmatrix}\begin{pmatrix} 4 \\ 5 \\ 6 \end{pmatrix}$

3 Calculate the value of each letter.

(a) $\begin{pmatrix} a & b \\ 1 & 5 \end{pmatrix} = \begin{pmatrix} -3 & -2 \\ c & 5 \end{pmatrix}$ (b) $\begin{pmatrix} 1 & 3 \\ 5 & 7 \end{pmatrix} - \begin{pmatrix} x & 2 \\ 4 & y \end{pmatrix} = \begin{pmatrix} 0 & 1 \\ 1 & 0 \end{pmatrix}$

(c) $\begin{pmatrix} a-b & 10 \\ 12 & 14 \end{pmatrix} = \begin{pmatrix} 2 & 10 \\ 12 & a+b \end{pmatrix}$ (d) $\begin{pmatrix} 4 & -2 \\ 5 & 8 \end{pmatrix}\begin{pmatrix} 1 & 3 \\ 2 & 5 \end{pmatrix} = \begin{pmatrix} a & b \\ c & 2d \end{pmatrix}$

4 (a) $A = \begin{pmatrix} 6 & 7 \end{pmatrix}$ and $B = \begin{pmatrix} 5 & 4 \\ -6 & 3 \end{pmatrix}$. (a) Calculate

(i) matrix C such that $AB - 2C = \begin{pmatrix} 0 & 0 \end{pmatrix}$, (ii) matrix D such that $8D + A = \begin{pmatrix} 3 & -2 \end{pmatrix}$.

(b) Can you calculate matrix BA. Give a reason to your answer.

5 $M = \begin{pmatrix} x & 8 \\ 2 & 5 \end{pmatrix}$ and $N = \begin{pmatrix} 1 & 9 \\ -4 & y \end{pmatrix}$

(a) Calculate M^2. (b) Calculate the values of x, y and z if $MN = \begin{pmatrix} -28 & 48 \\ z & 28 \end{pmatrix}$.

6 $P = \begin{pmatrix} 3 & 6 & 9 \end{pmatrix}$ and $Q = \begin{pmatrix} 1 & 3 \\ 2 & 4 \\ 3 & -1 \end{pmatrix}$. Calculate matrix R such that $PQ = \frac{1}{2}R$.

7 $R = \begin{pmatrix} -1 & 2 \\ -3 & 4 \end{pmatrix}$, $S = \begin{pmatrix} -2 & 5 \\ 6 & 7 \end{pmatrix}$ and $T = \begin{pmatrix} -3 & 6 \\ 2 & 3 \end{pmatrix}$.

Show that (i) $R(S + T) = RS + RT$, (ii) $R(S - T) = RS - RT$, (iii) $(R + S)^2 = R^2 + S^2 + 2RS$.

8 (a) If $A = \begin{pmatrix} 0 & 1 \\ 2 & 6 \end{pmatrix}$, evaluate $A^2 - 2A$. (b) If $B = \begin{pmatrix} \sin 90^0 & \sin 30^0 \\ \tan 45^0 & \cos 60^0 \end{pmatrix}$, evaluate $10B - B^2$.

9 Find the 2 x 2 matrix A such that $\begin{pmatrix} 6 & 3 \\ 2 & 1 \end{pmatrix}\begin{pmatrix} 2 & 4 \\ 7 & 8 \end{pmatrix} - 2A = \begin{pmatrix} -5 & 12 \\ 0 & 23 \end{pmatrix}$.

The inverse of a matrix

Working rule

Given matrix $A = \begin{pmatrix} a & b \\ c & d \end{pmatrix}$.

1) The determinant value of A is $|A| = ad - bc$.

2) The inverse of A is $A^{-1} = \frac{1}{ad - bc}\begin{pmatrix} d & -b \\ -c & a \end{pmatrix}$.

Remember

1. A matrix is singular, if its determinant value is zero.
2. A matrix is non-singular, if its determinant value is not zero
3. You can find A^{-1}, if and only if A is a non singular matrix.
4. $AA^{-1} = I$ and $A^{-1}A = I$, where the identity matrix $I = \begin{pmatrix} 1 & 0 \\ 0 & 1 \end{pmatrix}$.
5. If $AB^{-1} = I$, then matrix A = matrix B.
6. The inverse matrix when given in a fraction followed by a matrix form, does not need to be further simplified.

Example

$A = \begin{pmatrix} 1 & -3 \\ 5 & 7 \end{pmatrix}$ (a) Find the determinant value of matrix A. (b) Find A^{-1}.

Working: $|A| = 1 \times 7 - -3 \times 5 = 22$

Here a = 1, b = −3, c = 5 and d = 7.

Substitute the values of a, b, c and d in the formulae for determinant and the inverse.

(b) $A^{-1} = \frac{1}{ab - bc}\begin{pmatrix} d & -b \\ -c & a \end{pmatrix} = \frac{1}{22}\begin{pmatrix} 7 & 3 \\ -5 & 1 \end{pmatrix}$

Prastice exercise 35

1 Find the determinant value of each matrix. State the matrices which are non-singular.

(a) $\begin{pmatrix} 1 & 2 \\ 3 & 6 \end{pmatrix}$ (b) $\begin{pmatrix} -2 & 3 \\ 7 & 8 \end{pmatrix}$ (c) $A = \begin{pmatrix} -1 & 0 \\ 0 & 1 \end{pmatrix}$ (d) $B = \begin{pmatrix} 3 & 12 \\ 5 & 20 \end{pmatrix}$ (e) $X = \begin{pmatrix} -1 & 2 \\ -3 & -4 \end{pmatrix}$

2 (a) $A = \begin{pmatrix} x & 6 \\ 2 & x \end{pmatrix}$. Find the two values of x for which $|A| = 13$.

(b) $M = \begin{pmatrix} 1 & 2x \\ -3 & x^2 \end{pmatrix}$. Find the two values of x for which $|M| = 7$.

3 $M = \begin{pmatrix} 7 & -9 \\ 2 & 3 \end{pmatrix}$ (a) Show that matrix M is non-singular. (b) Find the inverse of matrix M.

4 $A = \begin{pmatrix} x & 6 \\ 2 & x \end{pmatrix}$ (a) Find the determinant value of matrix A. (b) Find A^{-1}.

5 Find the inverse of each matrix.

(a) $P = \begin{pmatrix} 1 & 3 \\ 9 & 8 \end{pmatrix}$ (b) $Q = \begin{pmatrix} 7 & -4 \\ 2 & 3 \end{pmatrix}$ (c) $R = \begin{pmatrix} 20 & 100 \\ -10 & -45 \end{pmatrix}$

6 $T = \begin{pmatrix} 9 & 2 \\ -1 & 3 \end{pmatrix}$ and $I = \begin{pmatrix} 1 & 0 \\ 0 & 1 \end{pmatrix}$. Show that (a) $TT^{-1} = I$, (b) $T^{-1}T = I$.

7 $A = \begin{pmatrix} 2 & 1 \\ 3 & 4 \end{pmatrix}$ and $B = \begin{pmatrix} 5 & 6 \\ 7 & 9 \end{pmatrix}$. Show that $(AB)^{-1} = B^{-1}A^{-1}$.

8 $P = \begin{pmatrix} -7 & 24 \\ 16 & -10 \end{pmatrix}$ and $Q = \begin{pmatrix} 0 & -30 \\ -24 & 1 \end{pmatrix}$. Fin (i) $(P+Q)^{-1}$, (ii) $(Q-P)^{-1}$.

9 $X = \begin{pmatrix} 10 & 20 \\ 30 & 40 \end{pmatrix}$. Show that $(X^{-1})^{-1} = X$.

10 $A = \begin{pmatrix} 2 & 4 \\ 6 & 8 \end{pmatrix}$ and $B = \begin{pmatrix} 1 & 3 \\ 5 & 7 \end{pmatrix}$. Find the 2 x 2 matrix C such that AC = B.

3.08 Vectors

Core curriculum

Vector notation and its operations

- A vector has both length (or size) and direction.
- A column vector is in the form $\begin{pmatrix} x \\ y \end{pmatrix}$. Where x is the horizontal movement (+sign for right and –sign for left) and y is the vertical movement (+sign for up and –sign for down).
- The movement from A to B is the column vector $\overrightarrow{AB}$ or $\mathbf{a} = \begin{pmatrix} 2 \\ 3 \end{pmatrix}$.

  The movement from B to C is the column vector $\overrightarrow{BC}$ or $\mathbf{b} = \begin{pmatrix} 2 \\ -4 \end{pmatrix}$.

  The movement from A to C is the column vector $\overrightarrow{AC} = \begin{pmatrix} 4 \\ -1 \end{pmatrix}$.

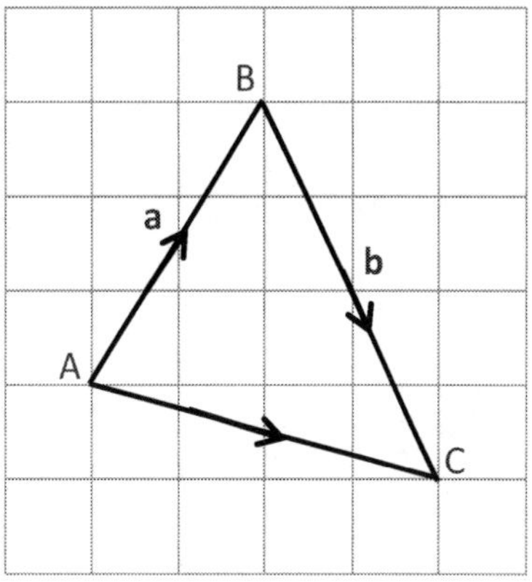

  Also $\mathbf{a} + \mathbf{b} = \begin{pmatrix} 2 \\ 3 \end{pmatrix} + \begin{pmatrix} 2 \\ -4 \end{pmatrix} = \begin{pmatrix} 4 \\ -1 \end{pmatrix}$. So $\overrightarrow{AC} = \mathbf{a} + \mathbf{b}$.
- Using a vector triangle, we can add or subtract vectors.

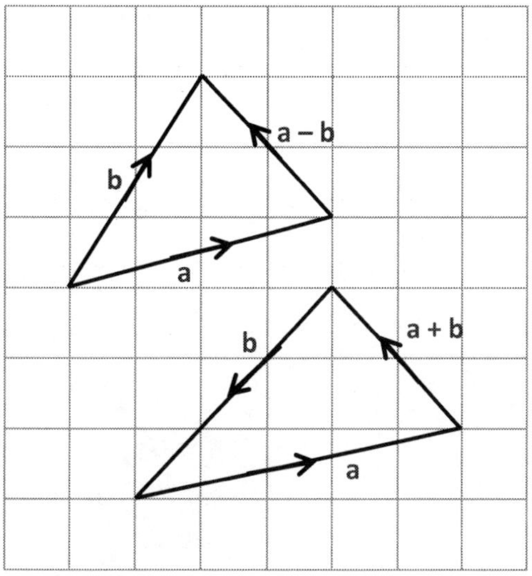

Working rule

1) To add column vectors, just add their corresponding vectors.
2) To subtract column vectors, just subtract their corresponding elements.
3) A number outside a vector can be multiplied by each element inside the vector.

Remember

1. If $\overrightarrow{AB} = \mathbf{a}$ then $\overrightarrow{BA} = -\mathbf{a}$.
2. If $\overrightarrow{AB} = n\overrightarrow{CD}$ (where n is a number), then $\overrightarrow{AB}$ is parallel to $\overrightarrow{CD}$.
3. If $\overrightarrow{AB} = n\,\overrightarrow{AC}$, then points A, B and C are on the same line.

Example

$\mathbf{a} = \begin{pmatrix} 1 \\ 2 \end{pmatrix}$, $\mathbf{b} = \begin{pmatrix} -3 \\ 4 \end{pmatrix}$ and $\mathbf{c} = \begin{pmatrix} -4 \\ -5 \end{pmatrix}$. Find (i) $\mathbf{a} + \mathbf{b}$, (ii) $\mathbf{a} - \mathbf{c}$, (iii) $5\mathbf{b}$, (iv) $3\mathbf{a} - 2\begin{pmatrix} 4 \\ -1 \end{pmatrix}$.

Working: (i) $\mathbf{a} + \mathbf{b} = \begin{pmatrix} 1 \\ 2 \end{pmatrix} + \begin{pmatrix} -3 \\ 4 \end{pmatrix} = \begin{pmatrix} -2 \\ 6 \end{pmatrix}$

Add the corresponding elements.

(ii) $\mathbf{a} - \mathbf{c} = \begin{pmatrix} 1 \\ 2 \end{pmatrix} - \begin{pmatrix} -4 \\ -5 \end{pmatrix} = \begin{pmatrix} 5 \\ 7 \end{pmatrix}$

Subtract the corresponding elements.

(iii) $5\mathbf{b} = 5\begin{pmatrix} -3 \\ 4 \end{pmatrix} = \begin{pmatrix} -15 \\ 20 \end{pmatrix}$

Each element is multiplied by 5.

(iv) $3\mathbf{a} - 2\mathbf{b} = 3\begin{pmatrix} 1 \\ 2 \end{pmatrix} - 2\begin{pmatrix} -3 \\ 4 \end{pmatrix}$

$= \begin{pmatrix} 3 \\ 6 \end{pmatrix} - \begin{pmatrix} -6 \\ 8 \end{pmatrix} = \begin{pmatrix} 9 \\ -2 \end{pmatrix}$

Each element of **a** is multiplied by 3 and each element of **b** is multiplied by 2.

Then subtract the corresponding elements.

Practise exercise 36

1 Using the points on the grid, write down each of the following column vectors.

(i) $\overrightarrow{AB}$ (ii) $\overrightarrow{BC}$ (iii) $\overrightarrow{CD}$ (iv) $\overrightarrow{DE}$

(v) $\overrightarrow{AE}$ (vi) $\overrightarrow{AD}$ (vii) $\overrightarrow{CB}$ (viii) $\overrightarrow{EB}$

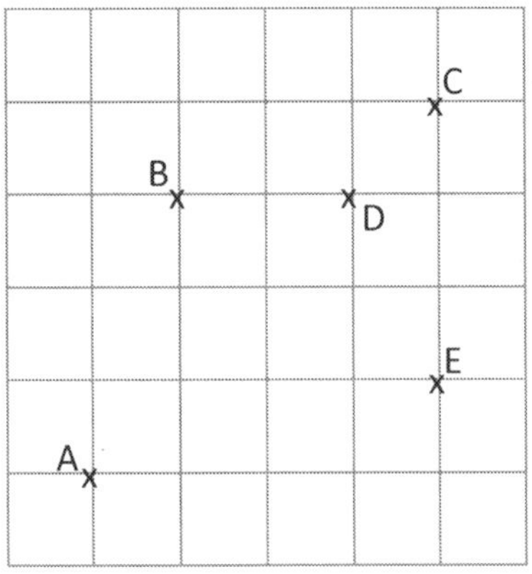

2 (a) On a grid, show the position of each vector.

(i) $\overrightarrow{OA} = \begin{pmatrix} 1 \\ 3 \end{pmatrix}$ (ii) $\overrightarrow{OB} = \begin{pmatrix} 6 \\ 3 \end{pmatrix}$ (iii) $\overrightarrow{OC} = \begin{pmatrix} 5 \\ 0 \end{pmatrix}$

(b) From your answer to part(a), write down
(i) the special name of the quadrilateral whose vertices are O, A, B and C,
(ii) $\overrightarrow{AC}$ as a column vector.

3 $\mathbf{a} = \begin{pmatrix} -3 \\ 2 \end{pmatrix}$, $\mathbf{b} = \begin{pmatrix} 0 \\ 5 \end{pmatrix}$ and $\mathbf{c} = \begin{pmatrix} -4 \\ -7 \end{pmatrix}$. Find

(i) $\mathbf{a} + \mathbf{b}$, (ii) $\mathbf{b} + \mathbf{c}$, (iii) $\mathbf{a} - \mathbf{b}$, (iv) $\mathbf{b} - \mathbf{c}$,
(v) $2\mathbf{a}$, (vi) $3\mathbf{b} + \mathbf{c}$, (vii) $7\mathbf{c} - 5\mathbf{b}$, (viii) $\mathbf{a} + \mathbf{b} + \mathbf{c}$,
(ix) $-6\mathbf{a} - 8\mathbf{b} + 3\mathbf{c}$, (x) $\frac{1}{4}\mathbf{c}$, (xi) $\frac{1}{2}\mathbf{a} + \mathbf{b}$, (xii) $8\mathbf{a} - \frac{3}{4}\mathbf{b}$.

4 $\mathbf{a} = \begin{pmatrix} 2 \\ 5 \end{pmatrix}$, $\mathbf{b} = \begin{pmatrix} -1 \\ 4 \end{pmatrix}$ and $\mathbf{c} = \begin{pmatrix} 6 \\ -3 \end{pmatrix}$.

Find and show the position of each of the following column vectors on a grid.
(i) $\mathbf{a} + \mathbf{b}$ (ii) $\mathbf{a} + \mathbf{b} + \mathbf{c}$ (iii) $2\mathbf{a} + 5\mathbf{b}$ (iv) $\mathbf{a} - \mathbf{b}$ (v) $\mathbf{c} - \mathbf{a} - \mathbf{b}$ (vi) $-4\mathbf{c} + 7\mathbf{b}$

5 (a) If $\mathbf{a} = \begin{pmatrix} 7 \\ -9 \end{pmatrix}$, $\mathbf{b} = \begin{pmatrix} 3 \\ 14 \end{pmatrix}$ and $\mathbf{a} + \mathbf{c} = \mathbf{b}$, find the column vector c.

(b) If $a = \begin{pmatrix} 0 \\ 6 \end{pmatrix}$, $\mathbf{b} = \begin{pmatrix} -2 \\ 3 \end{pmatrix}$ and $3\mathbf{a} - 4\mathbf{c} = 5\mathbf{b}$, find the column vector **c**.

6 (a) If $\overrightarrow{AB} = \begin{pmatrix} 4 \\ -10 \end{pmatrix}$, then show that $\overrightarrow{AB} + \overrightarrow{BA} = \begin{pmatrix} 0 \\ 0 \end{pmatrix}$.

(b) $\overrightarrow{MN} = \begin{pmatrix} -1 \\ 5 \end{pmatrix}$ and $\overrightarrow{PQ} = \begin{pmatrix} -4 \\ 20 \end{pmatrix}$. What is the special geometrical term that you can assign to the vectors $\overrightarrow{MN}$ and $\overrightarrow{PQ}$? State a reason for your answer.

(c) $\overrightarrow{AB} = \begin{pmatrix} 2 \\ 9 \end{pmatrix}$ and $\overrightarrow{AC} = \begin{pmatrix} 6 \\ 27 \end{pmatrix}$. Do the points A, B and C form a triangle? Show reason for your answer.

7 In the diagram, $\overrightarrow{AB} = \mathbf{a}$, $\overrightarrow{BC} = \mathbf{b}$ and $\overrightarrow{CD} = \frac{1}{2}\mathbf{a}$.

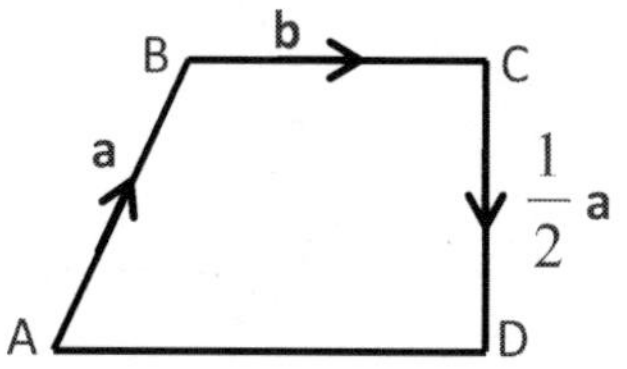

Write down the following vectors in terms of **a** and **b**.

(i) $\overrightarrow{DA}$ (ii) $\overrightarrow{AD}$ (iii) $\overrightarrow{AC}$ (iv) $\overrightarrow{BD}$

Extended curriculum

Magnitude of a vector

Working rule

If $\overrightarrow{AB} = \begin{pmatrix} x \\ y \end{pmatrix}$, then the magnitude of $\overrightarrow{AB}$ is $|\overrightarrow{AB}| = \sqrt{x^2 + y^2}$.

Example

(a) If $v = \begin{pmatrix} 3 \\ -4 \end{pmatrix}$, calculate $|v|$.

(b) The coordinates of A is (2, 5) and the coordinates of B is (–4, 7). Calculate $|\overrightarrow{AB}|$.

Working: (a) $\mathbf{v} = \sqrt{3^2 + (-4)^2} = 5$

Use Pythagoras theorem to calculate magnitude.

(b) $\overrightarrow{AB} = \begin{pmatrix} -4 \\ 7 \end{pmatrix} - \begin{pmatrix} 2 \\ 5 \end{pmatrix} = \begin{pmatrix} -6 \\ 2 \end{pmatrix}$

Change the coordinates to column vectors and then subtract.

$|\overrightarrow{AB}| = \sqrt{(-6)^2 + 2^2} = \sqrt{40}$

Use Pythagoras theorem to calculate magnitude.

Position vector

Working rule

If A has coordinates (x_1, y_1) and B has coordinates (x_2, y_2).

Then the position vector of B relative to A is $\overrightarrow{AB} = \begin{pmatrix} x_2 \\ y_2 \end{pmatrix} - \begin{pmatrix} x_1 \\ y_1 \end{pmatrix}$.

Remember

Two vectors are equal if they have same magnitude and same direction.

Example

Find the position vector of the points

(i) B relative to A, (ii) C relative to A,

(iii) D relative to B, (iv) E relative to C,

(v) F relative to A, (vi) A relative to O.

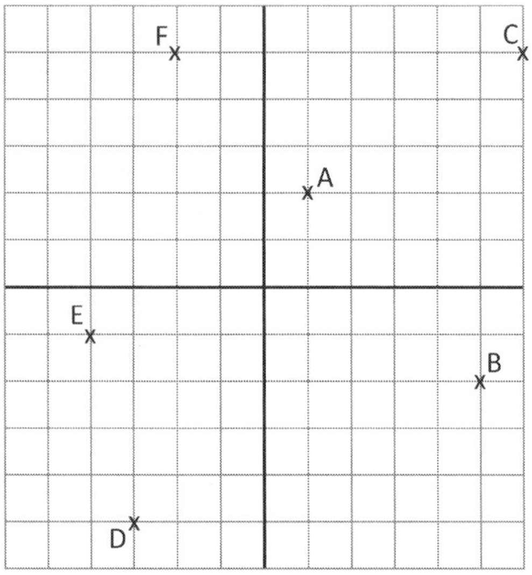

Working: (i) $\overrightarrow{AB} = \begin{pmatrix} 5 \\ -2 \end{pmatrix} - \begin{pmatrix} 1 \\ 2 \end{pmatrix} = \begin{pmatrix} 4 \\ -4 \end{pmatrix}$

(ii) $\overrightarrow{AC} = \begin{pmatrix} 6 \\ 5 \end{pmatrix} - \begin{pmatrix} 1 \\ 2 \end{pmatrix} = \begin{pmatrix} 5 \\ 3 \end{pmatrix}$

(iii) $\overrightarrow{BD} = \begin{pmatrix} -3 \\ -5 \end{pmatrix} - \begin{pmatrix} 5 \\ 2 \end{pmatrix} = \begin{pmatrix} -8 \\ -7 \end{pmatrix}$

(iv) $\overrightarrow{CE} = \begin{pmatrix} 6 \\ 5 \end{pmatrix} - \begin{pmatrix} -4 \\ -1 \end{pmatrix} = \begin{pmatrix} 10 \\ 6 \end{pmatrix}$

(v) $\overrightarrow{AF} = \begin{pmatrix} -2 \\ 5 \end{pmatrix} - \begin{pmatrix} 1 \\ 2 \end{pmatrix} = \begin{pmatrix} -3 \\ 3 \end{pmatrix}$

(vi) $\overrightarrow{OA} = \begin{pmatrix} 1 \\ 2 \end{pmatrix} - \begin{pmatrix} 0 \\ 0 \end{pmatrix} = \begin{pmatrix} 1 \\ 2 \end{pmatrix}$

Change each coordinates to a column vector and then use the formula

$\overrightarrow{AB} = \begin{pmatrix} x_2 \\ y_2 \end{pmatrix} - \begin{pmatrix} x_1 \\ y_1 \end{pmatrix}$.

Where $A = \begin{pmatrix} x_1 \\ y_1 \end{pmatrix}$ and $B = \begin{pmatrix} x_2 \\ y_2 \end{pmatrix}$.

Practise exercise 37

1 (a) If $\mathbf{a} = \begin{pmatrix} 9 \\ 12 \end{pmatrix}$, calculate $|\mathbf{a}|$. (b) If $\mathbf{b} = \begin{pmatrix} -20 \\ 15 \end{pmatrix}$, calculate $|\mathbf{b}|$.

(c) The coordinates of A is (–3, 4) and the coordinates of B is (7, 10).

Calculate (i) $|\overrightarrow{AB}|$, (ii) $|\overrightarrow{BA}|$.

2 A = (1, 2), B = (–6, 7) and C = (4, –3). Using these coordinates, calculate

(i) $|\overrightarrow{AB}|$, (ii) $|\overrightarrow{BC}|$, (iii) $|\overrightarrow{AC}|$, (iv) $|\overrightarrow{CB}|$.

3 The diagram shows the parallelogram OABC.

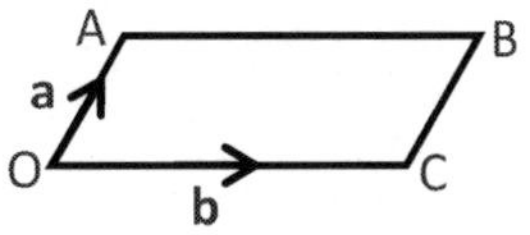

$\overrightarrow{OA} = \mathbf{a}$ and $\overrightarrow{OB} = \mathbf{b}$.

Express each vector in terms **a** and **b**.

(i) $\overrightarrow{CB}$ (ii) $\overrightarrow{BC}$ (iii) $\overrightarrow{AB}$ (iv) $\overrightarrow{BA}$
(v) $\overrightarrow{OB}$ (vi) $\overrightarrow{BO}$ (vii) $\overrightarrow{AC}$ (viii) $\overrightarrow{CA}$

4 ABCD is an isosceles trapezium.

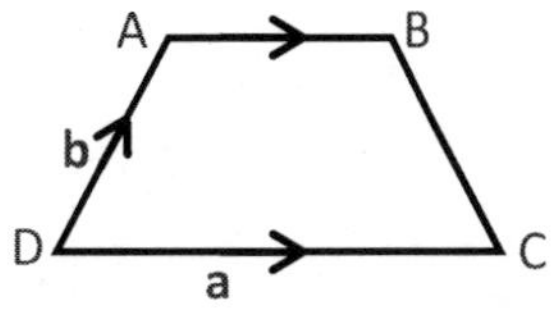

$\overrightarrow{DA} = \mathbf{b}$, $\overrightarrow{DC} = \mathbf{a}$ and $2\overrightarrow{AB} = \overrightarrow{DC}$.

Express each vector in terms of **a** and **b**.

(i) $\overrightarrow{AB}$ (ii) $\overrightarrow{AC}$ (iii) $\overrightarrow{BC}$ (iv) $\overrightarrow{BD}$

5 ABCD is a rectangle. E is the mid-point of AB, F is the mid-point of BC, G is the mid-point of CD and H is the mid-point of DA.

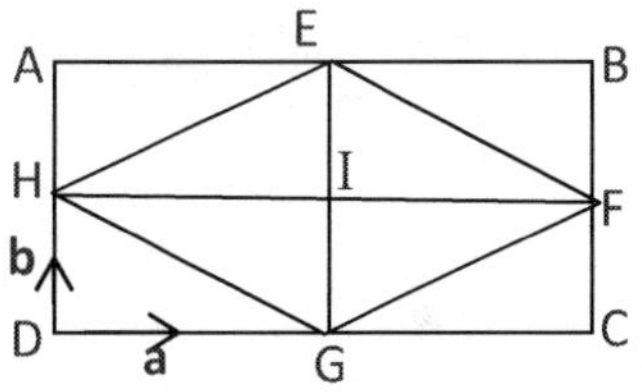

DG = **a** and DH = **b**.

(a) Write down the following vectors in terms of **a** and **b**.

(i) $\overrightarrow{DC}$ (ii) $\overrightarrow{CB}$ (iii) $\overrightarrow{BE}$ (iv) $\overrightarrow{HE}$

(v) $\overrightarrow{HF}$ (vi) $\overrightarrow{EG}$ (vii) $\overrightarrow{IH}$ (viii) $\frac{1}{4}(\overrightarrow{AC} + \overrightarrow{DB})$

(b) If $\mathbf{a} = \begin{pmatrix} 4 \\ 0 \end{pmatrix}$ and $\mathbf{b} = \begin{pmatrix} 0 \\ 2 \end{pmatrix}$, calculate (i) $|\overrightarrow{GH}|$ (ii) $|\overrightarrow{GA}|$ (iii) $\left|\frac{1}{2}\overrightarrow{EF}\right|$ (iv) $|\overrightarrow{AB} + \overrightarrow{BC}|$

3.09 Transformations

Core curriculum

Reflection

- The reflection of a shape is its image obtained in a mirror line and the image is congruent to the original shape.
- The perpendicular distance between any point on the object and the mirror line equals the distance between the corresponding point on the image and the mirror line.

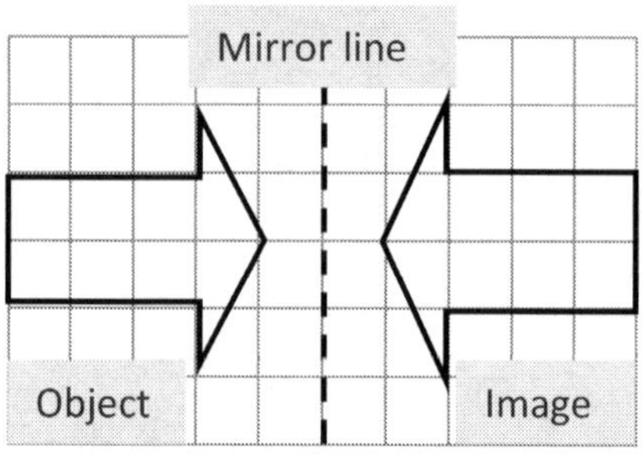

Remember

1. In reflection, the image must be drawn with the correct dimensions as the original shape.
2. To describe a reflection, you can state the line of reflection (e.g. x-axis or y-axis).

Rotation

- The rotation of a shape about a point (called centre of rotation) is obtained by changing its position through a given angle.

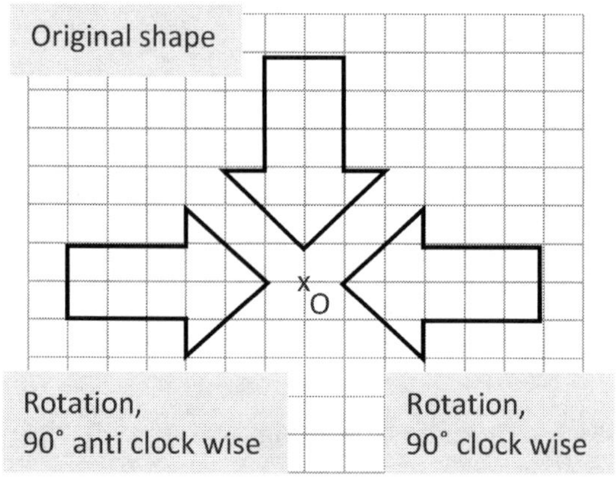

Remember

1. To describe a rotation use the elements centre of rotation, angle and direction(clock wise or anti-clock wise)
2. You can use tracing papers to find the centre of rotation accurately.
3. The centre of rotation is the point of intersection of the perpendicular bisectors of any two lines joining the corresponding vertices.

Translation

- The translation of a shape is obtained by changing its position using a given vector.

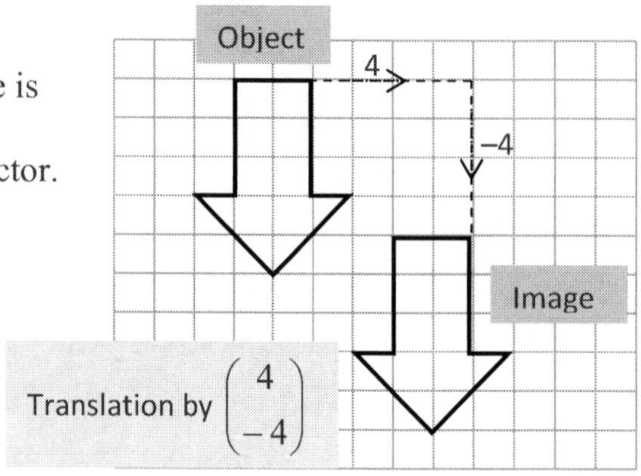

Remember

1. In translation all points of a shape move in the same direction for the same distance.
2. To describe a translation, write down the column vector.
3. To construct a translation, use the given column vector.

Enlargement

- An enlargement of a shape is obtained by increasing or decreasing its size using a given scale factor.

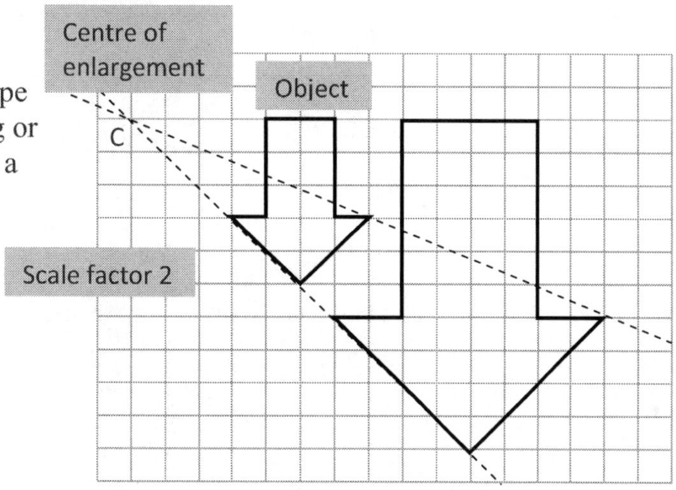

Remember

1. To describe an enlargement, use the elements scale factor and centre of enlargement.
2. In reflection, rotation and translation, the object and the image are congruent. But in enlargement, the object and the image are similar.
3. In an enlargement, the scale factor is the ratio between an image distance and the corresponding object distance.
4. The centre of enlargement is the point of intersection of any two pairs of corresponding vertices of the object and the image.

Example

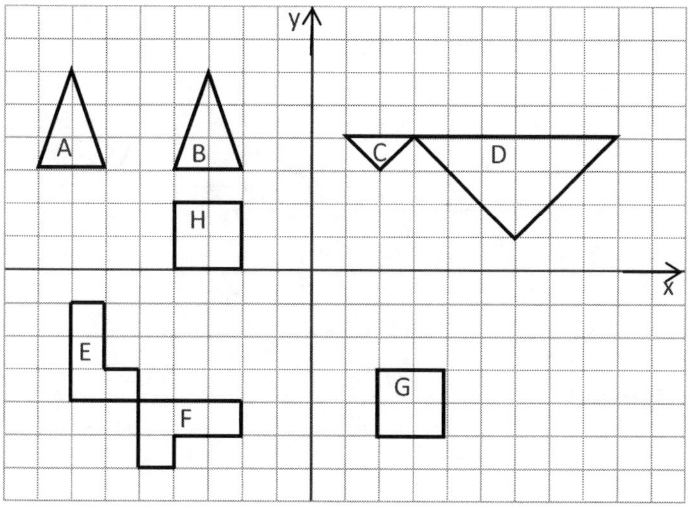

Describe fully the single transformation that maps
(i) shape A onto shape B,
(ii) shape C onto shape D,
(iii) shape E onto shape F,
(iv) shape G onto shape H.

Working: (i) Reflection in the line y = —4.
(ii) Enlargement, scale factor 3,
centre of enlargement (0, 4).
(iii) Rotation, 90^0 clock wise,
Centre of rotation (—6, —5).
(iv) Translation by the vector $\begin{pmatrix} -6 \\ 5 \end{pmatrix}$.

The mirror line is y = —4.
(0, 4) is the point of intersection of the lines joining the corresponding vertices.
(—6, —5) is the point of intersection of the perpendicular bisectors of any two lines joining the corresponding vertices.
$\begin{pmatrix} -6 \\ 5 \end{pmatrix}$ shows 6 units horizontal towards left and 5 units vertical toward up.

Practise exercise 38

1 Draw the reflection of each shape in the given mirror line.

(a)
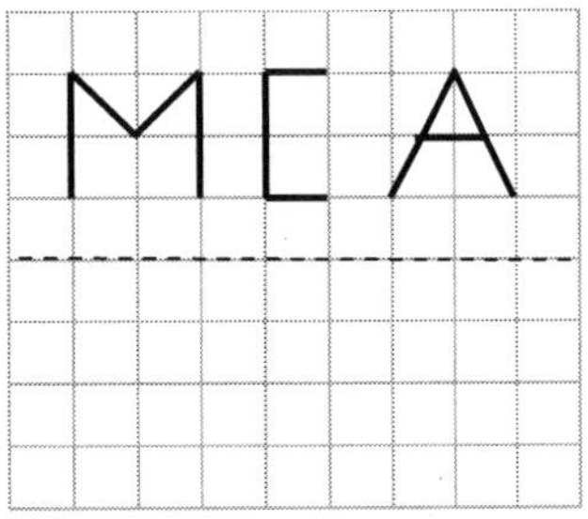
(b)
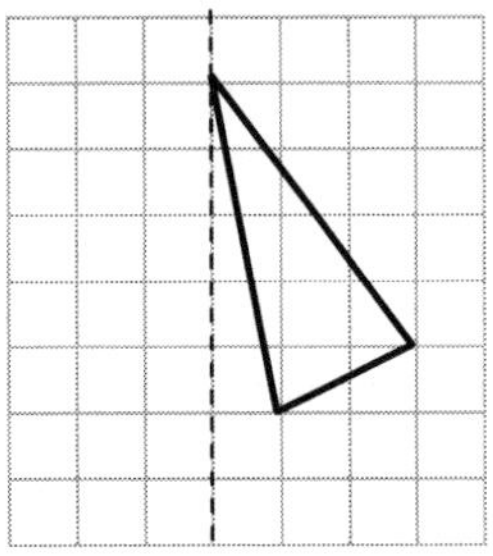
(c)
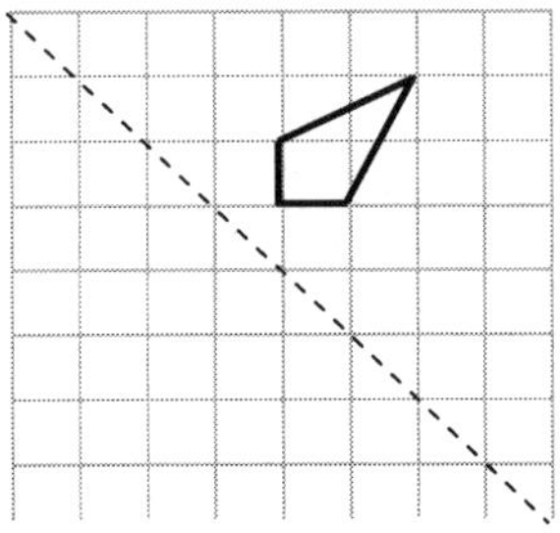

2 Draw the mirror line for each of the following reflections.

(a)
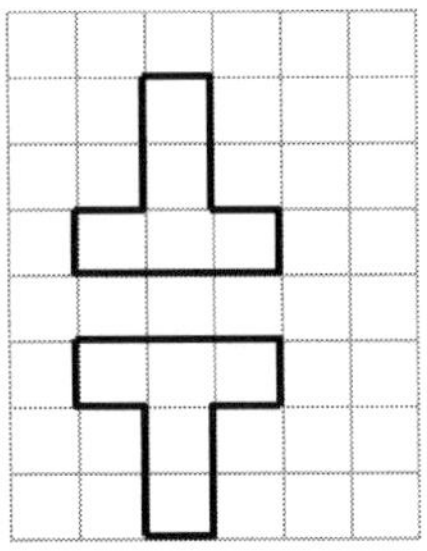
(b)
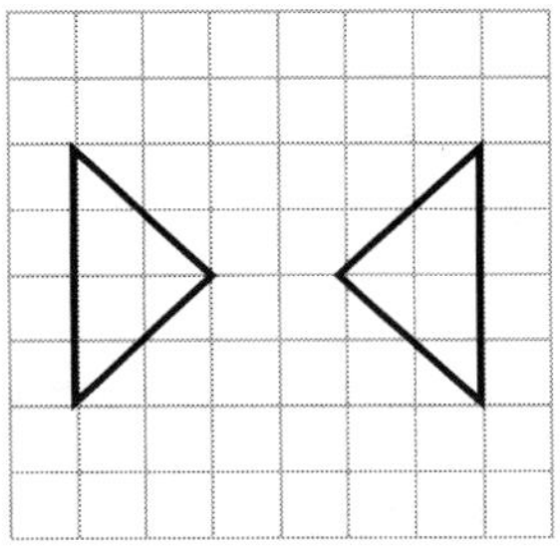
(c)
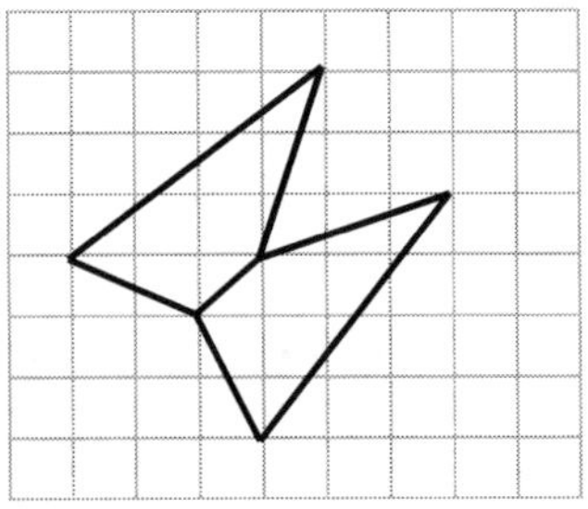

3 The diagram shows a shape and the point of rotation P.

Copy and rotate the shape about P through

(a) 90^0 clockwise,

(b) 90^0 anti-clock wise,

(c) 180^0 clockwise.

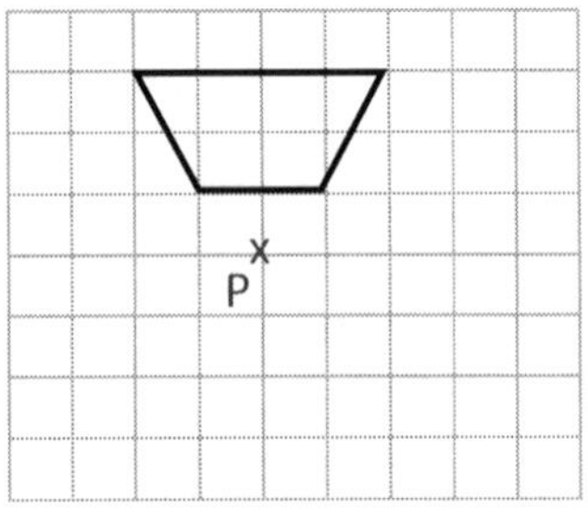

4 In the diagram, shape B is a rotation of shape A and shape D is rotation of shape C.

(a) Find the centre of rotation when

(i) shape B is an image of shape A,

(ii) shape D is an image of shape C.

(b) Describe fully the single transformation which maps

(i) shape A onto shape B, (ii) shape C onto shape D.

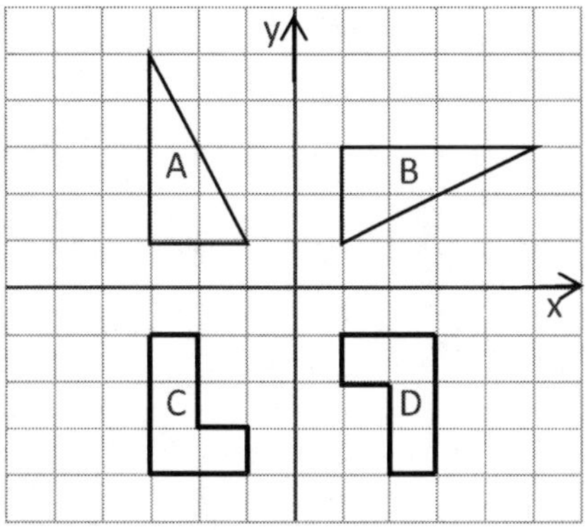

5 Using the diagram, describe the single transformation which maps

(a) shape A onto shape B,

(b) shape A onto shape C,

(c) shape A onto shape D.

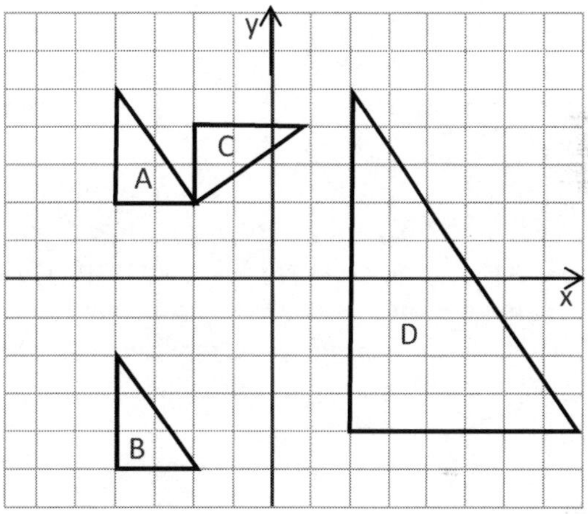

6 (a) Using a scale of 1 cm = 1 unit on both axes draw a pair of coordinate axes.

(b)(i) Draw quadrilateral ABCD with vertices A(—1, 0), B(0, 2), C(2, 2,) and D(1, 0).

(ii) What is the special name of the quadrilateral ABCD?

(c) Draw and label the images of quadrilaterals ABCD, using the following vectors.

(i) $T_1 = \begin{pmatrix} 3 \\ 1 \end{pmatrix}$ (ii) $T_2 = \begin{pmatrix} -3 \\ 2 \end{pmatrix}$ (iii) $T_3 = \begin{pmatrix} 1 \\ -2 \end{pmatrix}$ (iv) $T_4 = \begin{pmatrix} -2 \\ -3 \end{pmatrix}$

7 (a) Enlarge triangle T using a scale factor 3 and the centre of enlargement (—5, 3),

(b) Enlarge rectangle PQRS using a scale factor —2 and the centre of enlargement (—5, 1),

(c) Shape CDEF is a transformation of shape OABC. Describe fully the single transformation which maps shape OABC onto shape CDEF.

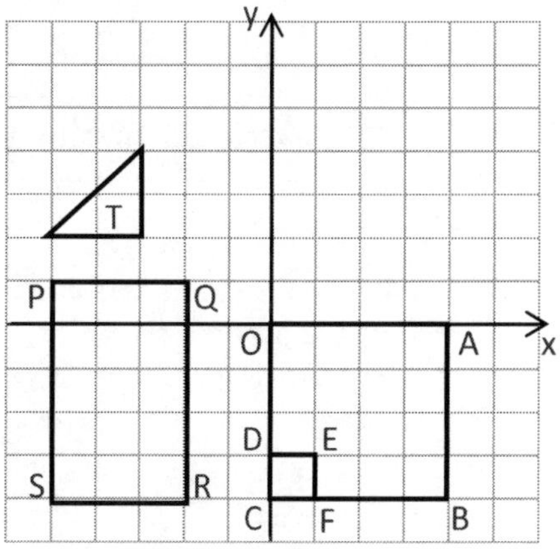

8

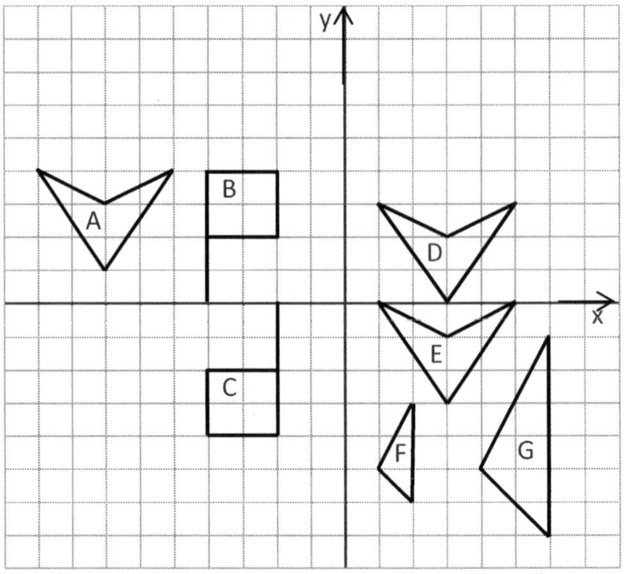

(a) Draw the reflection of shape A in the line y = 0. Label the image A^1.
(b) Describe fully the single transformation which maps
(i) shape B onto shape C, (ii) shape C onto shape B,
(ii) shape F onto shape G, (iv) shape G onto shape F.
(c) Find the translation vectors when
(i) shape A is translated to shape D, (ii) shape A is translated to shape E,
(iii) shape D is translated to shape E, (iv) shape D is translated to shape A,
(v) shape E is translated to shape A, (vi) shape E is translated to shape D.

9 A rectangle PQRS with vertices P(−8, 1), Q(−5, 1), R(−5, −1) and S(−8, −1) is enlarged to obtain its image $P^1Q^1R^1S^1$ with the coordinates $P^1(-3, 2)$, $Q^1(3, 2)$, $R^1(3, -2)$ and $S^1(-3, -2)$.
(a) On a pair of coordinate axes, draw the rectangle PQRS and $P^1Q^1R^1S^1$.
(b) Find the scale factor of the enlargement.
(c) Write down the coordinate of the centre of enlargement.

Extended curriculum

Shear

- A shear of a shape is obtained by moving the shape along a fixed line (called invariant line) using a shear factor.

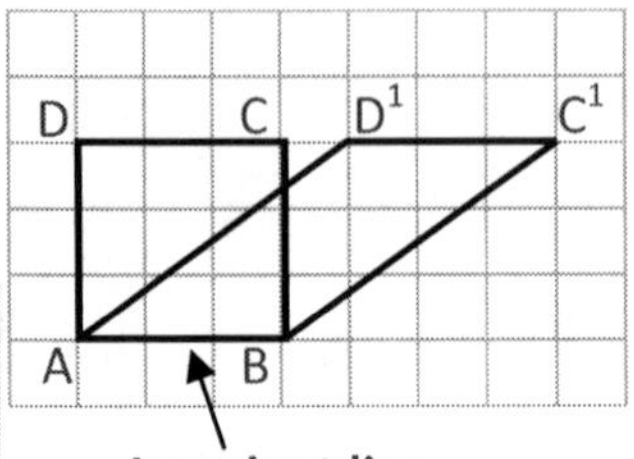

Shear factor = $\frac{CC^1}{BC} = \frac{4}{3}$

Remember

1. Shear factor = (distance a point moves due to the shear)÷(perpendicular distance of the point from the invariant line).
2. To describe a shear, use the elements shear factor, the invariant line and the direction of shear.

Stretch

- A stretch of a shape is obtained by lengthening the shape along a fixed line (called invariant line) using a scale factor.

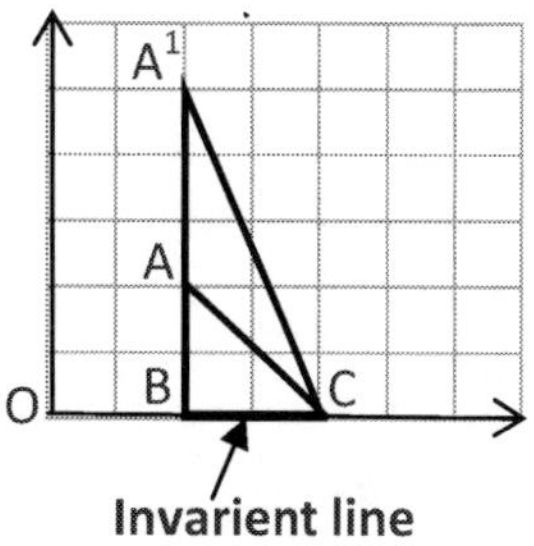

Scale factor $= \dfrac{A^1B}{AB} = \dfrac{5}{2} = 2.5$

Remember

1. For a Stretch,
   scale factor = (image distance of a point from the invariant line) ÷ (object distance of the corresponding point from the invariant line).
2. To describe a stretch, us e the elements scale factor, the invariant line and the direction of stretch.

Practise exercise 39

1 Shear each shape to the right by the given shear factor. AB is the invariant line.

(a) Shear factor 2

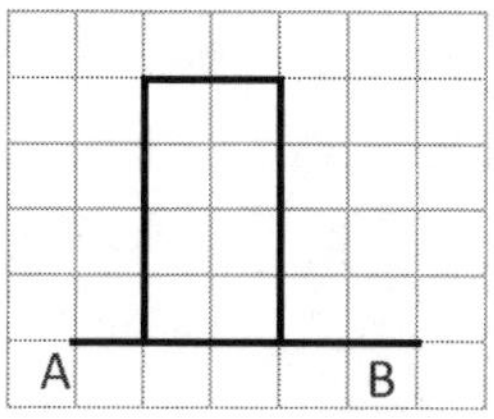

(b) Shear factor 0.5

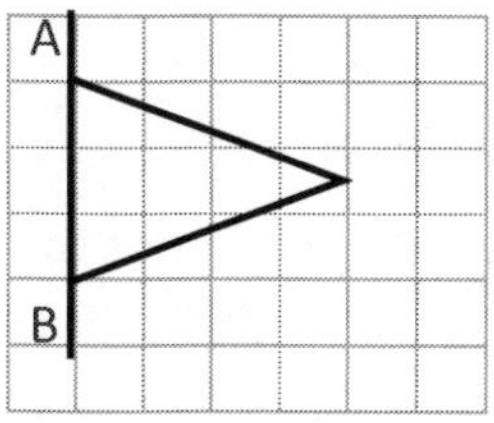

(c) Shear factor 3

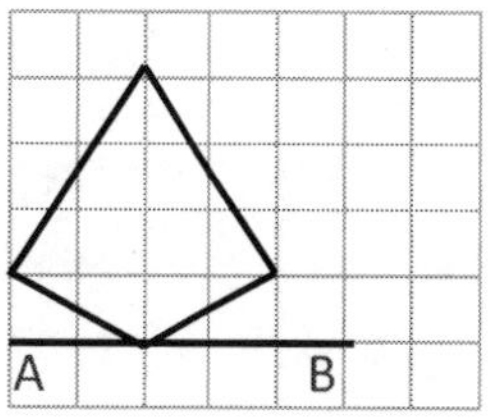

2 Describe fully the following transformations in which the object is shaded.

(a)

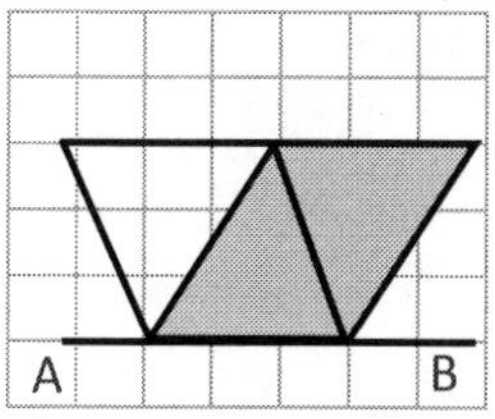

(b)

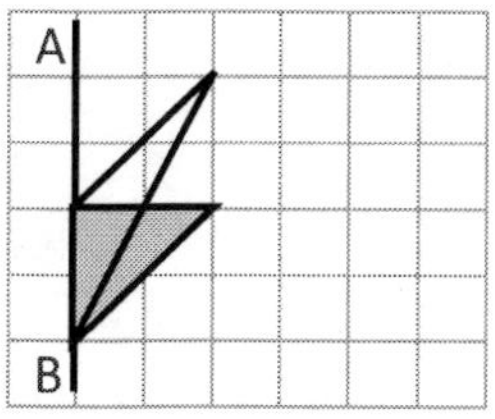

(c)

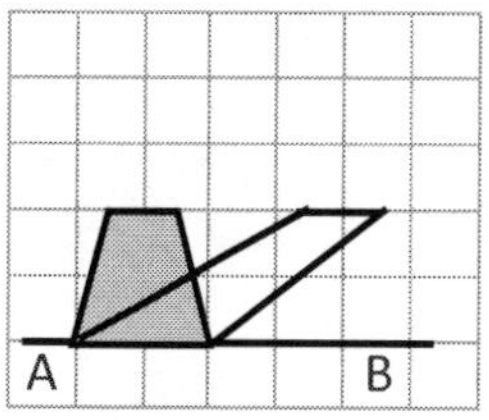

3 Stretch each shape to the specified direction by the given stretch factor.

(a) Parallel to y-axis, scale factor 2

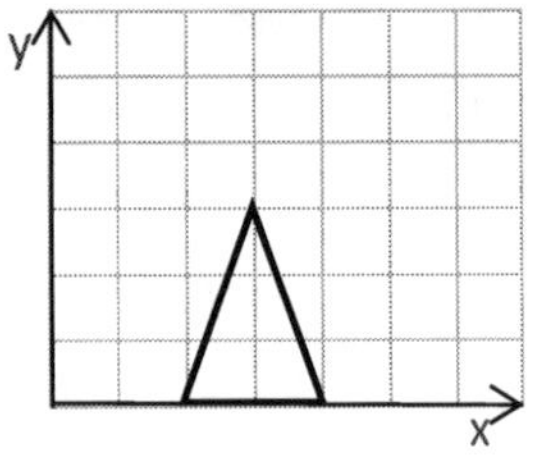

(b) Parallel to x-axis, scale factor 4

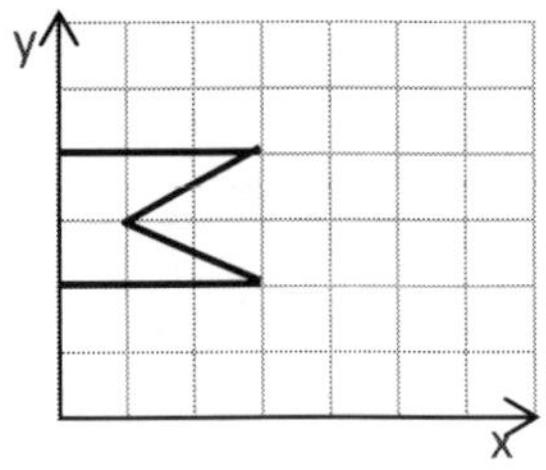

4 Describe fully each of the following transformations in which the object is shaded.

(a)

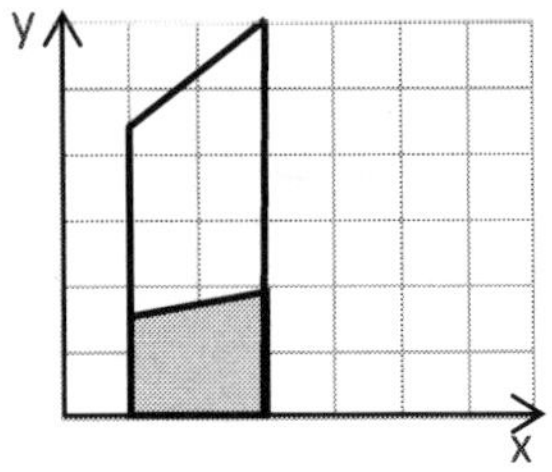

(b)

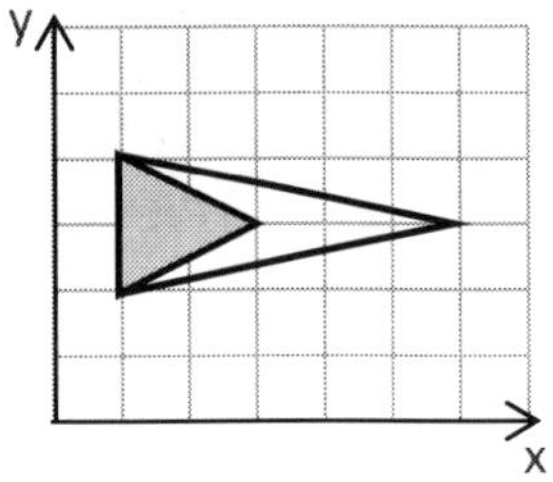

Combined transformations

- We may use more than one transformation to describe how a shape is moved from one place to another.

Example

(a) Draw the image of the letter E,

when it reflected in the line y = 3

and then translated by the vector $\begin{pmatrix} -8 \\ -5 \end{pmatrix}$.

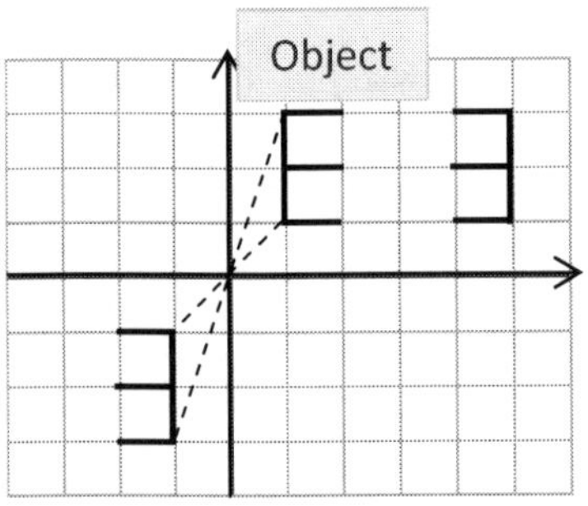

(b) Describe fully the single transformation that maps the object onto the final object.

Working: (b) Rotation, 180^0 clockwise (or anti-clockwise), centre of rotation (0, 0).

Transformation matrices

Remember

1. Some of the common transformations are described by the following matrices.

(i) $\begin{pmatrix} 1 & 0 \\ 0 & 1 \end{pmatrix}$ describes reflection in the x-axis.

(ii) $\begin{pmatrix} -1 & 0 \\ 0 & 1 \end{pmatrix}$ describes reflection in the y-axis.

(iii) $\begin{pmatrix} 0 & 1 \\ 1 & 0 \end{pmatrix}$ describes reflection in the line y = x.

(iv) $\begin{pmatrix} 0 & -1 \\ -1 & 0 \end{pmatrix}$ describes reflection in the line y = —x.

(v) $\begin{pmatrix} 0 & -1 \\ 1 & 0 \end{pmatrix}$ describes rotation, 90^0 anti-clockwise, centre(0, 0).

(vi) $\begin{pmatrix} 1 & 0 \\ 0 & 1 \end{pmatrix}$ describes rotation, 90^0 clockwise, centre(0, 0).

(vii) $\begin{pmatrix} -1 & 0 \\ 0 & -1 \end{pmatrix}$ describes rotation, 180^0, centre(0, 0).

(viii) $\begin{pmatrix} f & 0 \\ 0 & f \end{pmatrix}$ describes enlargement, scale factor f, centre(0, 0).

(ix) $\begin{pmatrix} 1 & f \\ 0 & 1 \end{pmatrix}$ describes shear, x-axis invariant, shear factor f.

(x) $\begin{pmatrix} 1 & 0 \\ f & 1 \end{pmatrix}$ describes shear, y-axis invariant, shear factor f.

(xi) $\begin{pmatrix} f & 0 \\ 0 & 1 \end{pmatrix}$ describes stretch, y-axis invariant, scale factor f.

(xii) $\begin{pmatrix} 1 & 0 \\ 0 & f \end{pmatrix}$ describes stretch, x-axis invariant, scale factor f.

2. If matrix A describes the transformation A onto B, then matrix A^{-1} describes the transformation B onto A.

Remember

1. The coordinates of the vertices of a shape can be represented by a matrix in column form.
2. A transformation can be represented by a matrix.
3. To calculate the coordinates of the vertices of an image, the best method is "pre-multiply the given matrix by the matrix formed from the vertices of the object(shape).
   i. e. (Transformation matrix) x (object matrix) = (image matrix).

Example

(a) The vertices of a triangle are A(1, 2), B(5, 2) and C(3, 4). Write down a matrix that represent triangle ABC.

(b) The matrix $\begin{pmatrix} 1 & 0 \\ 0 & -1 \end{pmatrix}$ represents a transformation.

(i) Draw the image of triangle ABC under this transformation and label $A_1B_1C_1$.

(ii) Describe fully this single transformation.

Working: (a) $\begin{pmatrix} 1 & 5 & 3 \\ 2 & 2 & 4 \end{pmatrix}$

(b) (i) $\begin{pmatrix} 1 & 0 \\ 0 & -1 \end{pmatrix}\begin{pmatrix} 1 & 5 & 3 \\ 2 & 2 & 4 \end{pmatrix} = \begin{pmatrix} 1 & 5 & 3 \\ -2 & -2 & -4 \end{pmatrix}$

(ii) Reflection in the x-axis.

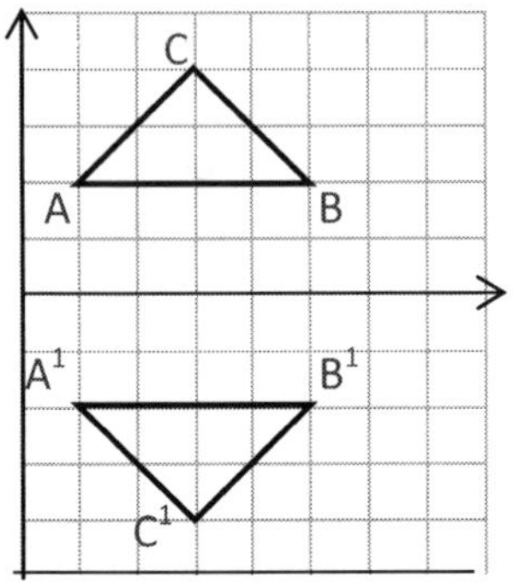

Practise exercise 40

1 The vertices of triangle ABC are A(1, 3), B(4, 3) and C(2, 6).

(a) On a grid, draw triangle ABC.

(b)(i) Draw the image of triangle ABC, when it is reflected in the line x = —1 and then translated by the vector $\begin{pmatrix} -3 \\ -5 \end{pmatrix}$.

(ii) Describe fully the single transformation that maps triangle ABC onto triangle $A_1B_1C_1$.

(c)(i) Represent triangle ABC in the form of a matrix.

(ii) The matrix $\begin{pmatrix} 1 & 0 \\ 0 & 2 \end{pmatrix}$ represents a transformation. Draw the image of triangle ABC under this transformation and label it $A_2B_2C_2$.

(iii) Describe fully this transformation.

2 (a)(i) Draw a pair of coordinate axes from –8 to 8 on both axes.

(ii) Draw quadrilateral ABCD with A(1, 0), B(3, 0), C(3, 2) and D(1, 2).

(iii) What is the special name of quadrilateral ABCD?

(b)(i) Rotate quadrilateral ABCD at 90° anti-clockwise about (–1, 0) and label it $A_1B_1C_1D_1$.

(ii) Reflect quadrilateral $A_1B_1C_1D_1$ in the line y = –1 and label it $A_2B_2C_2D_2$.

(iii) Describe fully the single transformation that maps quadrilateral ABCD onto quadrilateral $A_2B_2C_2D_2$.

(c)(i) The matrix $\begin{pmatrix} 1 & 0 \\ 3 & 1 \end{pmatrix}$ describes a transformation. Draw the image of quadrilateral ABCD under this transformation and label $A_3B_3C_3D_3$.

(ii) Describe fully this transformation.

3 (a)(i) Draw a set of coordinate axes from –7 to 8 on both axes.

(ii) Draw the shape PQRS with P(1, 1), Q(3, 1), R(4, 3) and S(2, 3).

(iii) Enlarge shape PQRS with a scale factor 2 and centre of enlargement (0, –1). Label it $P_1Q_1R_1S_1$.

(b)(i) Translate image $P_1Q_1R_1S_1$ by the vector $\begin{pmatrix} -5 \\ 0 \end{pmatrix}$ and label it $P_2Q_2R_2S_2$. Then translate $P_2Q_2R_2S_2$ by the vector $\begin{pmatrix} 3 \\ 6 \end{pmatrix}$ and label it $P_3Q_3R_3S_3$.

(ii) Describe fully the single transformation that maps shape PQRS onto the image $P_3Q_3R_3S_3$.

4 (a) On a pair of coordinate axes, draw triangle PQR with P(–3, 2), Q(–1, 3) and R(–2, 1).

(b)(i) Reflect triangle PQR in the line x = y and label $P_1Q_1R_1$.

(ii) Write down the matrix that represents this transformation.

(c) Reflect image $P_1Q_1R_1$ in the line y = 2 – x and label it $P_2Q_1R_2$.

(d)(i) The matrix $\begin{pmatrix} 0 & -1 \\ 1 & 0 \end{pmatrix}$ represents a transformation. Draw the image of triangle PQR under this transformation and label it $P_3Q_3R_3$.

(ii) Describe fully this transformation.

5 (a) On a suitable square grid, draw rectangle MNPQ with M(1, 1), N(4, 1), P(4, 3) and Q(1, 3).

(b)(i) The matrix $\begin{pmatrix} 1 & -2 \\ 2 & 0 \end{pmatrix}$ represents a transformation. Draw the image of MNPQ under this transformation and label it $M_1N_1P_1Q_1$.

(ii) Calculate the determinant of $\begin{pmatrix} 1 & -2 \\ 2 & 0 \end{pmatrix}$.

(iii) Calculate the area of $M_1N_1P_1Q_1$ using your answer to part(b)(ii).

(iv) Find the matrix that maps $M_1N_1P_1Q_1$ onto MNPQ.

CHAPTER 4 Statistics and Probability

4.01 Statistics

Core curriculum

Data

Remember

1. The frequency of a data is the number of times it repeats.
2. The range of some data is the difference between the highest value and the lowest value.
3. The mid-range of some data is the half of its range.

Example

The marks scored by 20 students in a Mathematics test are 41, 41, 48, 39, 27, 20, 15, 11, 9, 41, 9, 48, 39, 11, 27, 20, 27, 11, 20 and 15.
(a) Draw a simple frequency distribution table.
(b) Draw a grouped frequency distribution table with intervals $0 \leq m < 10$, $10 \leq m < 20$ etc.
(c) What is the range of these data? (d) What is the mid-range of these data?

Working: (a) **Simple frequency distribution table**

Mark	9	11	15	20	27	39	41	48
Frequency	2	3	2	3	3	2	3	2

(b) **Grouped frequency distribution**

Mark(m)	$0 \leq m < 10$	$10 \leq m < 20$	$20 \leq m < 30$	$30 \leq m < 40$	$40 \leq m < 50$
Frequency	2	5	6	2	5

(c) Range = 49 – 9 = 40

(d) Mid-range = 0.5 x 40 = 20

Range = Highest value – lowest value.
Mid-range = 0.5 x range.

Practise exercise 1

1 The ages of 15 pupils are 7, 11, 9, 7, 12, 12, 14, 16, 14, 18, 16, 12, 14, 12, 16 years.
(a) Draw a simple frequency distribution table.
(b) Draw a grouped frequency distribution table with intervals $0 \leq m < 5$, $5 \leq m < 10$ $10 \leq m < 15$ etc.
(c) What is the range of these data?
(d) What is the mid-range of these data?

2 The following are the number of hours that 25 students study in a day.
1, 4, 0, 4, 11, 14, 11, 11, 17, 4, 0, 1, 11, 12, 12, 12, 11, 14, 14, 9, 9, 17, 4, 6, 6.
(a) For these data find (i) range, (ii) mid-range.
(b) Represent these data in a simple frequency distribution table.

(c) Construct a grouped frequency distribution table using the class intervals $0 \leq h < 4$, $4 \leq h < 8$ etc.

3 The time(in minutes) taken by 20 pupils to solve a Mathematical puzzle is given below.
20, 12, 27, 42, 8, 20, 45, 42, 27, 16, 16, 27, 42, 30, 37, 30, 20, 20, 42, 12.
(a) Find (i) range, (ii) mid-range.
(b) Construct a simple frequency distribution table.
(c) Represent these data in a grouped frequency distribution table.

Mean, median and mode of individual data

Working rule

1) To calculate the mean, use the formula mean $= \dfrac{\text{Sum of all the values}}{\text{Total number of values}}$.
2) To find the median, (i) first arrange the data in order,
   (ii) then median = the middle value of the data..
   In case of two middle values, median = 0.5 (sum of both middle values).
3) To find mode, write down the data that occurs highest number of times.

Remember

1. The mean is the average of some data.
2. The median is the middle value of some data.
3. The mode is the highest occurrence value of some data.
4. Do not forget to arrange the data either in ascending or descending order to find median.
5. When all the data occurs same number of times, there is no mode.
6. The relationship among mean, median and mode is **mode = 3 median – 2 mean**.

Example

John records the number of words he writes each minute for 10 minutes.

15 21 17 10 18 25 25 12 25 27

(a) Calculate the mean. (b) Find the median. (c) Write down the mode.

Working: (a) Mean $= \dfrac{15+21+17+10+18+25+25+12+25+27}{10}$

$= 19.5$

(b) 10 12 15 17 18 21 25 25 25 27

Median = 0.5 (18 + 21) = 19.5

(c) Mode = 25

Use the formula for mean.
First arrange the data in order. Since there are two mid-values, find their average.
25 occurs highest number of times.

Practise exercise 2

1 For each set of the data,
(i) calculate the mean, (ii) find the median, (iii) write down the mode.
(a) 4, 2, 7, 1, 6 (b) 10, 12, 9, 6, 14 (c) 7, 0, 3, 1, 3, 4
(d) 8, 3, 5, 0, 7, 8 (f) 6, 4, 2, 10, 8, 9, 1, 6 (h) $\frac{1}{4}, \frac{3}{4}, \frac{1}{8}, \frac{3}{8}, \frac{7}{8}, \frac{5}{8}$

2 The marks scored by 12 students in a Mathematics test are given below.

97 94 71 77 64 69 54 49 17 97 81 80

(a) Calculate the mean mark. (b) Find the median mark. (c) Write down the mode mark.

3 The sum of 7 numbers is 91. Calculate the mean.

4 The mean of 10 numbers is 6.5. Find the sum of all the 10 numbers.

5 A set of numbers add up to 135 and the mean number is 7.5. How many numbers are there in the set?

6 (a) The mean of 17, 12, 14, 9, 6, 14, 12 and x is 12.125. Find the value of x.

(b) The mean of 2x, x, 3x and 5x is 22.

(i) Write down an equation in x. (ii) Solve the equation for x.

(iii) Write down the numerical values of this data.

(c) The mean of 9 numbers is 14.

(i) If each number is increased by 3, calculate the new mean.

(ii) If each number is reduced by 2, calculate the new mean.

7 (a) The median of 7, 3, 12, 8, 14 and x is 9. Find the value of x.

(b) Write down the modal value of each set of data.

(i) 2, 3, 2, 3, 4, 4 (ii) 7, 9, 8, 9, 7, 10 (iii) 0, 1, 2, 3, 4 (iv) 0.2, 0.22, 0.19, 2.3, 0.2

8 (a) The mean and median of a set of data are 25.5 and 20 respectively. Calculate the mode of this set of data.

(b) The mean and mode of a set of data are 54 and 42 respectively. Calculate the median.

Mean, median and mode of discrete data

- The mean, the median and the mode can be worked out from a simple frequency distribution table.

Working rule

1) To calculate the mean, use the formula Mean = $\dfrac{\sum xf}{\sum f}$.

2) To find the median, (i) first arrange the data in order,

(ii) then prepare the cumulative frequency table of the data,

(iii) median = the value having the cumulative frequency just more than $\dfrac{\sum f}{2}$.

3) To find the mode, write down the value with highest frequency.

Remember

When $\dfrac{\sum f}{2}$ equals the cumulative frequency of a value, then

median = 0.5(sum of the value and the next value).

Example

Mark	3	5	6	9	10
Frequency	4	3	7	8	3

The table above shows the marks obtained by 25 students in a test.

Work out (a) the mean mark, (b) the median mark, (c) the modal mark.

Working: (a) Mean = $\frac{\sum xf}{\sum f} = \frac{3 \times 4 + 5 \times 3 + 6 \times 7 + 9 \times 8 + 10 \times 3}{4+3+7+8+3} = \frac{171}{25} = 6.84$

(b) $\frac{\sum f}{2} = \frac{25}{2} = 12.5$ Median = 6

(c) Mode = 9

Use the formula for mean.
6 has the cumulative fi greater than 12.5.
The value 9 has highest frequency of 8.

Practise exercise 3

1 Work out (i) the mean, (ii) the median, (iii) the mode for the data given in each table.

(a)

Mark	17	19	25	29	30
Frequency	4	3	6	5	2

(b)

Age	7	9	10	12	14
Frequency	6	4	5	9	4

2

Time (hr)	2	4	6	8	10	12
Frequency	7	3	10	x	12	8

The mean time of the data shown in the above table is 7.6.
(a) Calculate the value of x. (b) Find the median time. (c) Write down the mode.

3

Height (cm)	140	148	150	152	156	160
Cumulative frequency	3	7	16	21	32	40

(a) Calculate the mean height. (b) Find the median height. (c) What is the modal height?

4 The marks of 20 students in a test were as follows.

97 94 84 84 81 76 76 76 74 74
71 60 60 60 60 58 58 54 50 47

(a) Calculate the mean mark. (b) Find the median mark. (c) What is the modal mark?

5 The frequency of the numbers 1, 3, 5, 7 and 9 are 4, 2, 3, x and x + 3 respectively. If the mean is 26.4, find the frequencies of the fourth and the fifth numbers.

6 The maximum daily temperature of two weeks are recorded in degree Celsius ($^{\circ}$C) as

31 31 37 37 37 34 34 34 34 32 32
33 38 38.

(a) Draw a frequency distribution table. (b) Calculate the mean temperature.
(c) What is the modal temperature? (d) Work out the median temperature.

Bar chart and pictograph

Remember

1. In a bar chart, the data are represented by parallel rectangular boxes of same length and the length of the bars are proportional to the given values or frequencies.
2. In a bar chart the bars may be separated from each other.
3. In a pictogram or pictograph, the data are represented by pictures and each picture shows a fixed number of units.

Example

The table shows the number of ball point pens produced by a factory in a period of 6 days.

Day	1st	2nd	3rd	4th	5th	6th
Number of pens	10 000	30 000	5 000	15 000	35 000	25 000

(a)(i) Draw a bar chart to show these data.
(ii) If the height of the bar for 5th day is 14 cm, find the height of the bar for 7500 pens.
(b)(i) Draw a pictograph to show these data using 1 picture = 5000 pens.
(ii) If one picture shows 10 000 pens, how many pictures are needed to draw the pictograph.

Working: (a) (i)

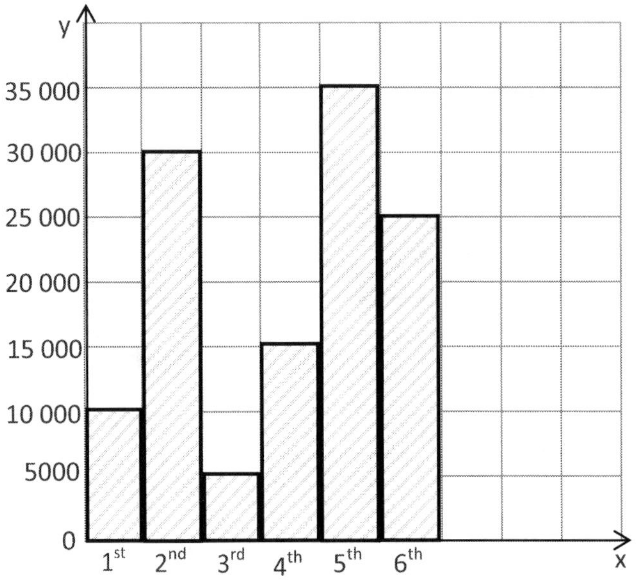

(ii) $35\,000\,x = 14 \times 7500$

Use direct proportion.

$x = 3$ cm.

(b)(i)

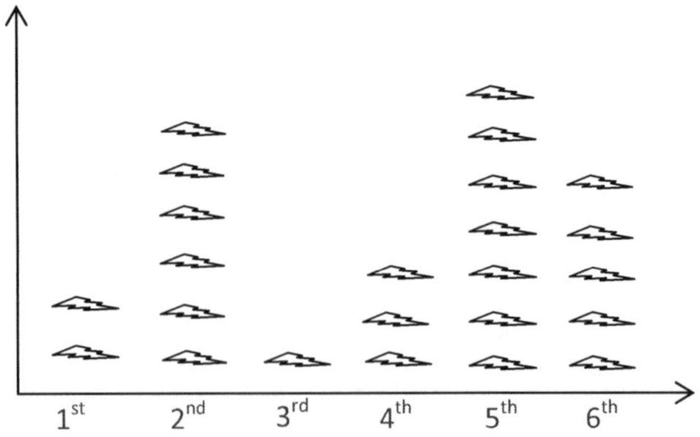

1 picture represents 5000 pens.

(ii) Total number of pens = 120 000
1 picture shows 10 000 pens.

Total number of pictures = 120 000 ÷ 10 000 = 12

Practise exercise 4

1 Reejana recorded his marks in 5 subjects obtained from a test.

Subject	English	Maths	Physics	Chemistry	Biology
Marks	50	90	60	70	80

(a)(i) Draw a bar chart to show these data.
(ii) If the height of the bar for Biology marks is 8 cm, find the height of the bar for 100 marks.
(b)(i) Draw a pictograph to show these data.
(ii) If one picture shows 5 marks, how many pictures are needed to draw the pictograph?

2 The table below shows the number of cars sold by a dealer in a period of five months.

Month	January	February	March	April	May
Number of cars	25	20	30	20	15

(a) Draw a bar chart to show this information. Show your scale on the vertical axis and label the bars.
(b) In a pictograph, one picture represents 5 cars.
(i) How many pictures are needed to show the number of cars sold in March?
(ii) How many pictures are needed to show the pictograph?

3

Category	A	B	C	D	E	F
Wages ($)	2700	2400	1800	1500	900	600

The above table shows the category and amount of wages fixed by a company for its employees.
(a)(i) Represent this information in a bar chart.
(ii) If height of a bar for the category B is 6 cm, find the height of the bar for category F.
(b)(i) Using $300 for one picture, draw a pictograph.
(ii) What are the total number of pictures that represent categories B, D and F?

Pie chart

Working rule

To draw a pie chart,
1) calculate the angle on the pie chart for each item by using the formula,

$$\text{angle on the pie chart} = \frac{360^0}{\text{Total value of all items}} \text{ x value of the item,}$$

2) joining any radius and draw the angles for all the items at the centre of the circle.

Remember

1. In a pie chart, the data are represented by sectors in a circle.
2. When the angles on a pie chart are not exact, give your answer up to 1 decimal place.

Example

120 students are participated in a survey of 'favourite sports'. The table below shows the results of the survey.

Games	Number of students	Angle on a pie chart
Basket ball	40	$\underline{120^{\circ}}$
Volley ball	50	$\underline{150^{\circ}}$
Badminton	20	$\underline{60^{\circ}}$
Chess	10	$\underline{30^{\circ}}$
Total	120	$\underline{360^{\circ}}$

$\frac{360^0}{120^0}$ x 40 = 120°,

$\frac{360^0}{120^0}$ x 50 = 150° etc.

(a) Complete the above table to draw a pie chart.
(b) Draw an accurate pie chart to show the data on the table and label the sectors.
(c) What is the percentage of students with Badminton as favourite game?

Working: (b)

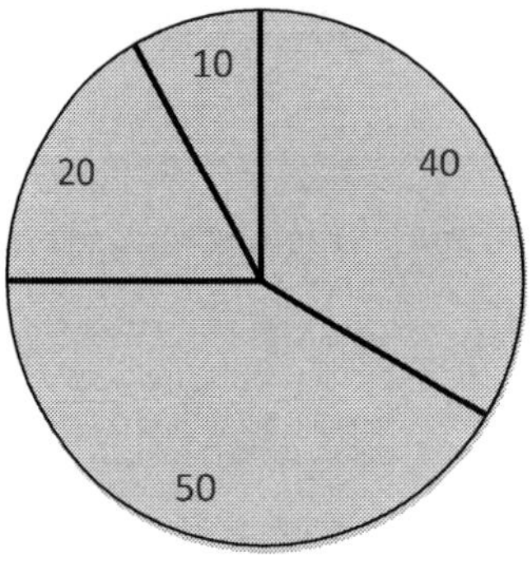

Draw the angles on a pie chart at the centre of the circle

(c) The number of students with Badminton as favourite game = $\frac{20}{120}$ x 100 = 16.7%

Practise exercise 5

1 The table below shows the marks scored by a student in different subjects.

Subject	Marks	Angle on a pie chart
Math	27	
English	20	
Commerce	18	
Accounting	25	
Total		

(a) Complete the table above to draw a pie chart.
(b) Draw an accurate pie chart to show the data on the table and label the sectors.
(c) What is the percentage of marks he scored in the subject English?

2 In a class of 45 students, the grades obtained in an examination are given on the table.

Grades	A^*	A	B	C	D
Number of students	7	12	15	6	5

(a) Draw an exact pie chart to show this information.

(b) If there are 10 students who obtained grade A, what will be the angle on the pie chart for grade A?

3

Year	2005	2006	2007	2008
Population(millions)	2	3.5	5	7.5

The table above shows the population of a City during four years.
(a) Represent this information on a pie chart accurately.
(b) What is the percentage of population increased during 2005 and 2008?

4 The pie chart shows how the students of Grades 8, 9 and 10 in a school are spread over.

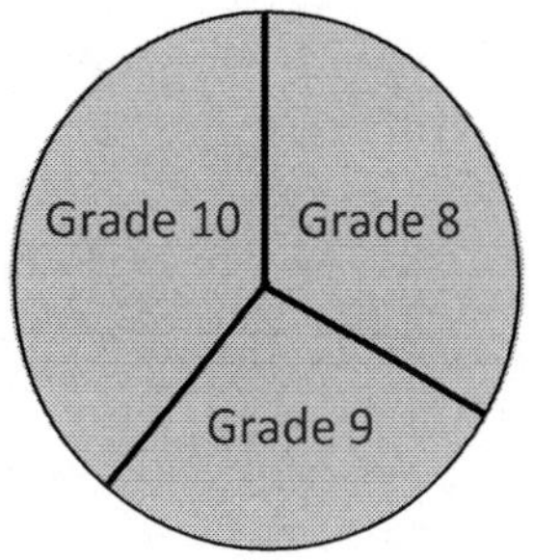

(a)

Class	Number of students	Angle on the pie chart
Grade 8		120
Grade 9	50	
Grade 10		140
Total	180	

(b) Write down in its simplest form the ratio of the students of Grades 8, 9 and 10.

5 The number of people that used an ATM during a week is given in the table.

Day	Mon	Tue	Wed	Thu	Fri	Sat	Sun
No. of people	64	47	85	90	77	40	36

Draw an accurate pie chart to show this information.

Scatter diagrams

- The relationship between two sets of data can be shown in a scatter diagram.
- A 'line of best fit' can be drawn to estimate the required value of a data item.
- A 'line of best feet' is a line that divides the points approximately two equal parts.

1) **Positive correlation**
When an increase in one data item results an increase in the corresponding data item, it is called positive correlation. That is the graph slopes upward from left to right.

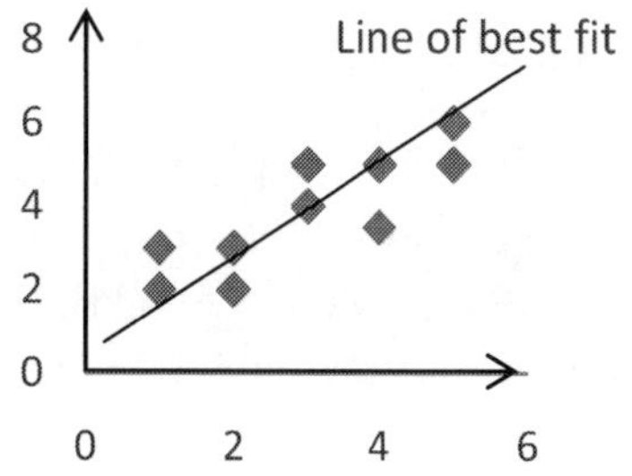

2) **Negative correlation**
When an increase in one data item results a decrease in the corresponding data item, it is called a negative correlation.
That is the graph slopes downward from left to right.

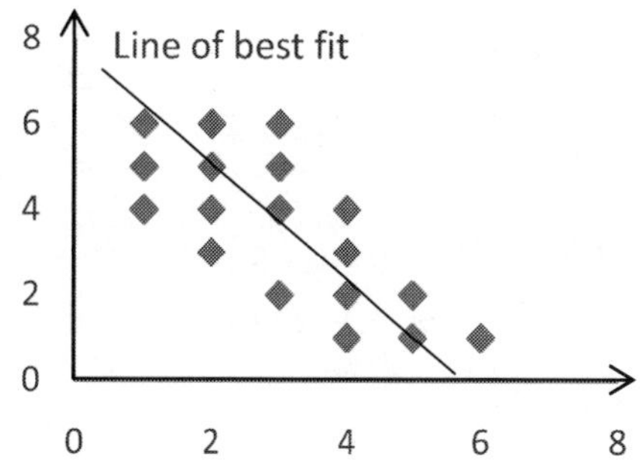

3) **Zero correlation**
When there is no direct or inverse relationship between the two sets of data, it called a zero correlation. There is no line of best fit for a zero correlation.

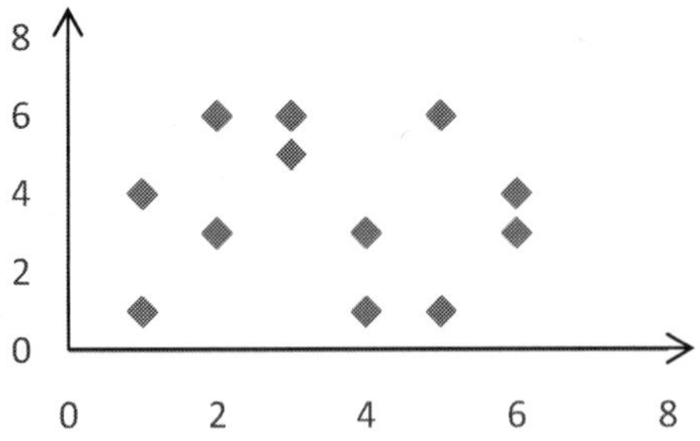

Example

The heights and weights of 12 students are measured and recorded as below.

Height (cm)	141	144	149	150	154	157	161	164	167	152	147	160
Weight (kg)	37	40	42	40	45	50	51	51	54	45	44	47

(a) Draw a scatter diagram to show this data.
(b) Which type of correlation does it describe?
(c) Estimate the weight of a student whose height is 170 cm.
(d) What does this indicate the relationship between the height and the weight?

Working: (a)

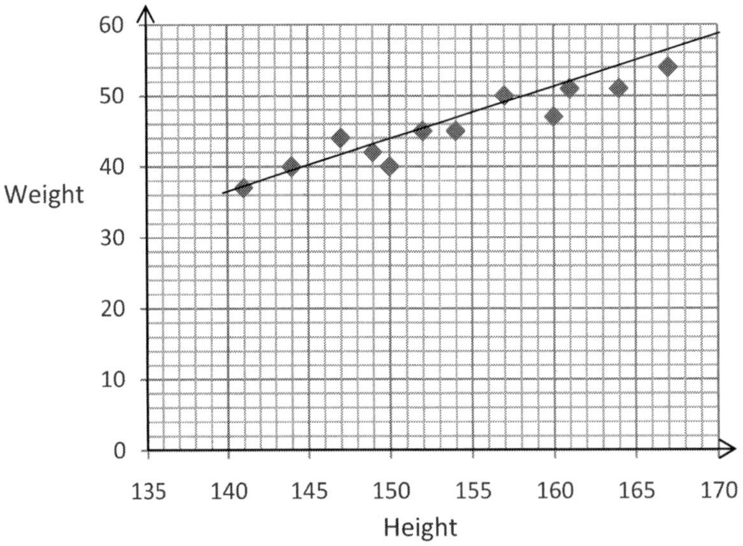

(b) It indicates positive correlation.
(d) The estimated weight of a student of height 170 cm is 58 kilograms.
(e) This indicates that when the height of a student increases then the weight also increases.

Practise exercise 6

1 The heights and shoe sizes of 10 children are measured and recorded as below.

Heights(cm)	125	130	133	137	140	142	146	150	152	160
Shoe sizes	1	1	2	3	4	4	5	6	6	7

(a) Draw a scatter diagram to show these data.
(b) Which type of correlation does it describe? (c) Draw the line of best fit on the grid.
(d) Estimate the shoe size of a child whose height is 165 cm.
(e) What does this indicate between the height and the shoe size?

2 The marks obtained by 8 students in a Mathematics examination and an English examination are recorded below.

Mathematics	50	47	20	11	8	25	34	20
English	12	17	32	40	35	22	32	30

(a) Draw a scatter diagram to show this information.
(b) Which type of correlation does it describe?
(c) A student scored 25 marks in English and did not write the Mathematics examination. Estimate his mark in Mathematics.
(d) What does this indicate between the Mathematics and English marks?

3 The number of penalties awarded and the number of goals scored from these penalties by a Football team during 12 tournaments is shown in the table below.

Goals	1	1	2	3	3	4	5	2	6	7	8	4
Penalties	4	10	2	8	12	6	8	12	12	10	12	14

(a) Represent this information on a scatter diagram.
(b) Which type of correlation does it describe?
(c) Can you draw a line of best fit on the scatter diagram? Give a reason for your answer.

4 (a) The regular polygons of sides 3, 4, 5, 6, 7, 8, 9, 10 have each interior angles of 60°, 90°, 108°, 120°, 128.6°, 90°, 135°, 90°, 140°, and 144° respectively.
Which type of correlation do the number of sides and each interior angle of regular polygons describe?
(b)The number of people and the time taken to complete a task can be described by a correlation. What is the name of this correlation?

Histogram

Working rule

To draw a histogram with equal classes, plot the frequency of each class on the vertical axis against the respective continuous class along the horizontal axis.

Remember

1. A histogram consists of adjacent rectangles.
2. For the continuous and equal classes,
   frequency of each class = vertical height of the rectangle.
3. To change a non-continuous class to a continuous class,
   find its true upper limit and true lower limit.
4. Do not leave gaps between the bars of a histogram.

Example

A group of children were asked how many hours they study in a day. The table below shows the results.

Hours	$0 \leq h < 4$	$4 \leq h < 8$	$8 \leq h < 12$	$12 \leq h < 16$	$16 \leq h < 20$
No. of students	8	20	15	5	1

(a) Draw a histogram to show this information.
(b) Using your histogram, find the number of students who study more than 12 hours a day.

(c) How many students study at least 8 hours a day? (d) Name the modal class.

Working: (a)

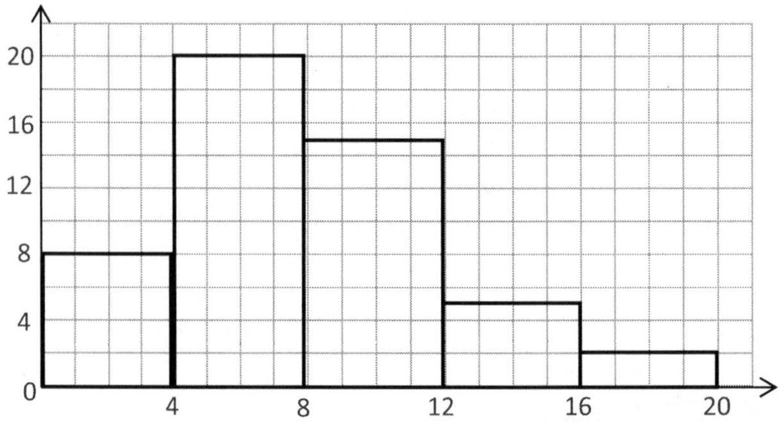

(b) 7 students study more than 12 hours a day.

2 squares rerpresent 1 student.

(c) 21

At least 8 hours a day means 8 or more than 8 hours.

(d) $4 \leq h < 8$

$4 \leq h < 8$ has highest frequency.

Practise exercise 7

1 A group of 40 students were asked how many problems they had solved in a week.

Problems	$0 \leq x < 10$	$10 \leq x < 20$	$20 \leq x < 30$	$30 \leq x < 40$	$40 \leq x < 50$
Number of students	4	7	13	10	6

(a) Draw a histogram to show this information. (b) Name the modal class.
(c) Using your histogram, find the number of students who solved more than 30 problems in a week.
(d) Find the number of students who had solved at least 20 problems in a week.

2 The heights of 60 pupils are measured and recorded in the table.

Heights	$100 \leq h < 110$	$110 \leq h < 120$	$120 \leq h < 130$	$130 \leq h < 140$	$140 \leq h < 150$
Number of pupils	16	20	9	10	5

(a) Draw a histogram using the above information.
(b) Using your histogram, find the number of pupils whose heights are
(i) at least 110 centimetres, (ii) less than 125 cnetimetres.
(c) Write down the modal class. State a reason for your answer.

3 The monthly expenditure on transportation of 100 employees of a city are given below.

Expenditure($x)	$50 \leq x < 100$	$100 \leq x < 150$	$150 \leq x < 200$	$200 \leq x < 250$
Number of employees	20	34	14	32

(a) Draw a histogram to show the above information.
(b) What percentage of the employees monthly expenditure on transportation is
(i) more than $150, (ii) less than $150, (iii) between $100 and $200 (both exclusive)?
(c) What is the range of the modal class?

4

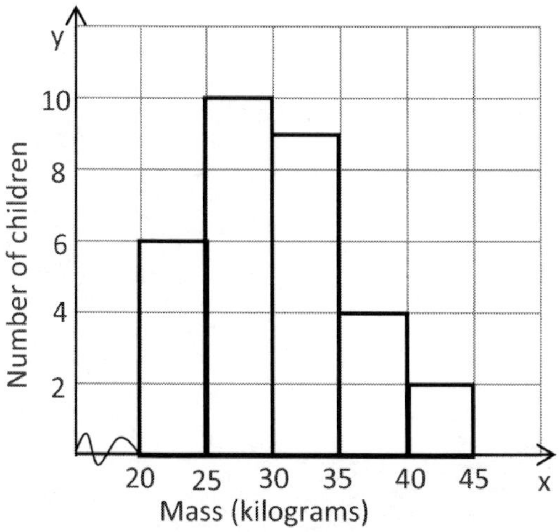

The above histogram shows masses on kilograms of a group of children.

(a) Find the total number of children.
(b) How many children are above the mass of 30 kilograms?
(c) How many children are below the mass of 35 kilograms?
(d) Find the class that has the highest frequency. What is the special name of this class?
(e) What percentage of the children has masses of 25 to 35 kilograms?
(f) Construct a frequency table from this histogram.

Extended curriculum

Histogram

Working rule

To draw a histogram with unequal classes, plot the frequency density of each class on the vertical axis against the respective continuous class along the horizontal axis.

Remember

1. Frequency density = $\dfrac{\text{frequency}}{\text{class width}}$.
2. Frequency of a class = area of the rectangle.
3. Area of a rectangle = length x breadth.

Example

A class of 25 students solved a mathematical puzzle in different intervals of time. The results are shown in the table.

Time (t minutes)	$0 < t \leq 2$	$2 < t \leq 5$	$5 < t \leq 10$	$10 < t \leq 22$
Frequency	4	5	10	6

(a) Using a scale of 1 cm to represent 1 minute on the horizontal axis and an area scale of 1 cm^2 to represent one student, draw a histogram for this data.
(b) The students who had solved the puzzle within the first 5 minutes are selected for a Mathematical competition.
  (i) How many students are selected for the Mathematical competition?

(ii) What percentage of the students is not selected for the Mathematical competition?

Working: (a)

Time (t minutes)	Frequency	Frequency density
$0 < t \leq 2$	4	2
$2 < t \leq 5$	5	1.7
$5 < t \leq 10$	10	2
$10 < t \leq 12$	12	3

FD = F/CW

4/(2 – 0) = 2,

5/(5 – 2) = 1.7 etc.

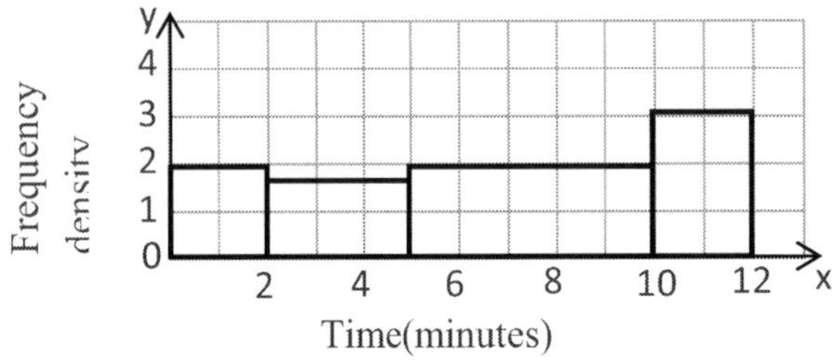

You must label the frequency density starting from zero.

(b) (i) 4 + 5 = 9 students

(ii) $\frac{16}{25} \times 100 = 64\%$

Practise exercise 8

1 The table below shows the number of days a class of 40 pupils were absent in the school during an academic year.

Days absent(A)	$0 < A \leq 5$	$5 < A \leq 7$	$7 < A \leq 15$	$15 < A \leq 20$	$20 < A \leq 30$
Number of pupils	12	17	6	2	3

(a) Using a scale of 1 cm to represent 1 day on the horizontal axis and an area scale of 1 cm^2 to represent one pupil, draw a histogram for this data.
(b) Calculate the fraction of pupils who were absent
(i) for longer than 15 days, (ii) for not more than 5 days.

2 The heights of 74 members of a Youth club are given on the table below.

Height (h cm)	$130 < h \leq 140$	$140 < h \leq 145$	$145 < h \leq 160$	$160 < h \leq 170$
Members	15	18	21	20

(a) Using a scale of 1 cm to represent 1 centimetres on the horizontal axis and an area scale of 1 cm^2 to represent one member, draw a histogram for this data.
(b)(i) What fraction of the members have heights more than 145 cm?
(ii) Calculate the percentage of the members with heights not less than 140 cm.

3 The masses of 1000 apples are shown on the following table.

Mass(m grams)	$0 \le m < 80$	$80 \le m < 120$	$120 \le m < 160$	$160 \le m < 200$	$200 \le m < 240$
Number of oranges	120	200	240	300	140

Using a scale of 2 cm to represent 40 grams, draw a horizontal axis for $0 \le m < 240$.
Using an area scale of 1 cm^2 to represent 20 oranges, draw a histogram.

4 The temperature (T °C) during a month of 30 days were recorded below.

4	7	4	8	8	3	3	3	5	5	7	10
10	7	7	12	12	8	8	10	10	7	2	2
2	1	1	1	2	1						

(a) Construct a frequency table using the class intervals as $0 \le T < 2$, $2 \le T < 5$, $5 \le T < 10$ and $10 \le T < 13$.

(b) Draw a histogram to show this information.

Cumulative frequency diagrams

Working rule

To construct a cumulative frequency diagram,

1) plot the cumulative frequency of each class against the upper class boundaries,
2) then for cumulative curve, join the successive points by free hand and for cumulative polygon, join the successive points by straight edge.

Remember

1. The cumulative frequency of a class
 = frequency of that class + all the frequencies of the previous classes.
2. A cumulative curve can be approximated by a cumulative diagram.

Example

The table below shows a Mathematics test results of 80 students.

Marks(m)	$0 < m \le 20$	$20 < m \le 40$	$40 < m \le 60$	$60 < m \le 80$	$80 < m \le 100$
Frequency	6	9	41	10	14

(a) Construct a cumulative frequency table for this data.

(b) Draw a cumulative frequency curve. (c) Draw a cumulative frequency polygon.

Working: (a)

Marks(m)	$0 < m \le 20$	$20 < m \le 40$	$40 < m \le 60$	$60 < m \le 80$	$80 < m \le 100$
Frequency	6	9	41	10	14
C. F.	6	15	56	66	80

(b)

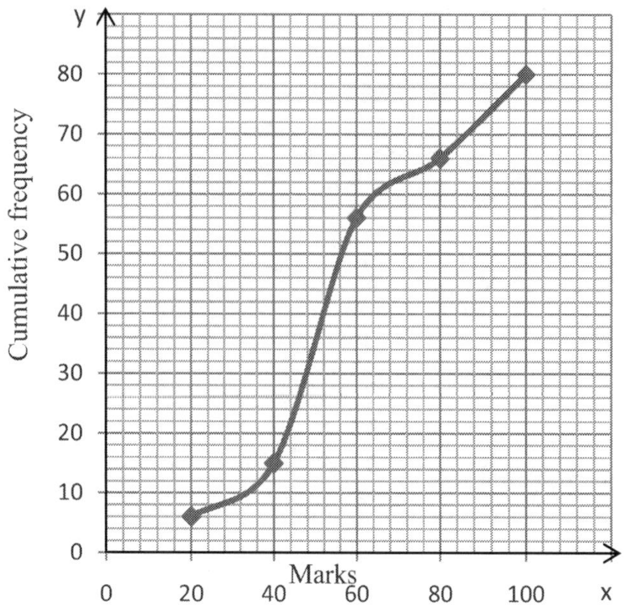

Join the successive points by free hand.

(c)

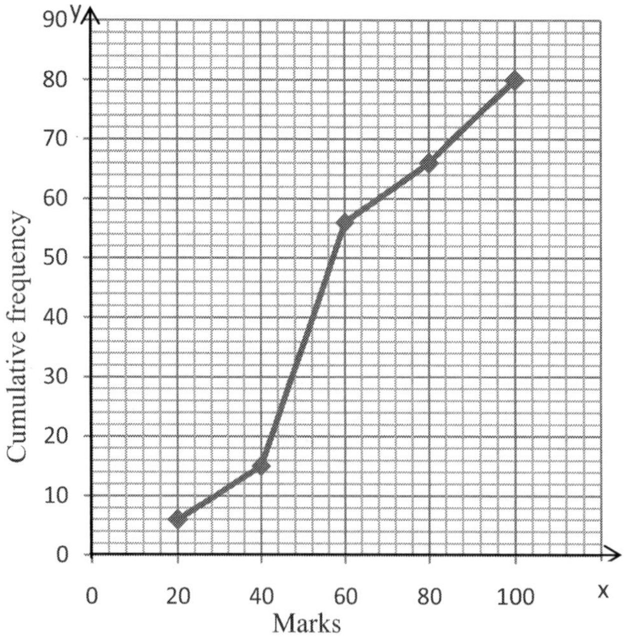

Join the successive points by a straight edge.

Practise exercise 9

1 The table below shows the weight of the luggage of 60 passengers in a plane.

Weight(w kg)	$0 < t \leq 10$	$10 < t \leq 20$	$20 < t \leq 30$	$30 < t \leq 40$	$40 < t \leq 50$
Frequency	10	25	20	17	8

(a) Draw a cumulative frequency curve to show this data.
(b) Draw a cumulative frequency polygon to show this data.

2 For each set of the following data, draw
(i) a cumulative frequency curve, (ii) a cumulative frequency diagram.

(a)

Classes	$20 < x \le 25$	$25 < x \le 30$	$30 < x \le 35$	$35 < x \le 40$	$40 < x \le 45$
Frequency	12	18	28	14	10

(b)

Age(A years)	Under 20	Under 40	Under 60	Under 80
Frequency	24	32	40	21

Estimation of mean

Working rule

To calculate an estimate of grouped and continuous data,
1) first find the mid-value(m) of each class,
2) then calculate fm for each class and 3) use the formula estimated mean = $\frac{\sum fm}{\sum f}$.

Remember

1. Mid-value =0.5(lower limit + upper limit).
2. Since the mid-value of each class is assigned with all the data in that class, the resulting mean is an estimate.

Example
The table below shows the marks of 150 pupils in a Physics test.

Marks	$0 < m \le 20$	$20 < m \le 40$	$40 < m \le 60$	$60 < m \le 80$	$80 < m \le 100$
Frequency	15	28	45	32	30

(a) Calculate an estimate of the mean mark.
(b) Explain why your answer to part(a) is an estimate.

Working: (a)

Marks	Frequency (f)	Mid-value(m)	fm
$0 < t \le 20$	15	10	150
$20 < t \le 40$	28	30	840
$40 < t \le 60$	45	50	2250
$60 < t \le 80$	32	70	2240
$80 < t \le 100$	30	90	2700

$\sum f = 150$ $\sum fm = 8180$

Estimated mean = $\frac{8180}{150} = 54.5$

Use the formula estimated mean = $\frac{\sum fm}{\sum f}$.

(b) Since the mid-value of each class is assigned with all the data in that class, the resulting mean is an estimate.

Practise exercise 10

1 Gresha writes 1000 words in different intervals of time. The table below shows the data.

Time(t min)	$0 < t \leq 5$	$5 < t \leq 10$	$10 < t \leq 15$	$15 < t \leq 20$
Frequency	220	260	280	240

(a) Calculate an estimate of the mean time.
(b) Explain why your answer to part(a) is an estimate.

2 Calculate an estimate of the mean for each set of the following distributions.

(a)

Marks(m)	$40 < m \leq 55$	$55 < m \leq 70$	$70 < m \leq 85$	$85 < m \leq 100$
Frequency	220	260	280	240

(b)

Mass(mg)	$10 < m \leq 20$	$20 < m \leq 30$	$30 < m \leq 40$	$40 < m \leq 50$	$50 < m \leq 60$
Cumulative frequency	15	28	45	62	70

(c)

Age(A years)	Under 10	Under 20	Under 30	Under 40
Number of pupils	16	25	38	50

Quartiles of individual data

- The three values that divide an order data into four equal parts are called quartiles. The first value is the lower quartile(Q_1), the second value is the median(Q_2) and the third value is the upper quartile(Q_3).

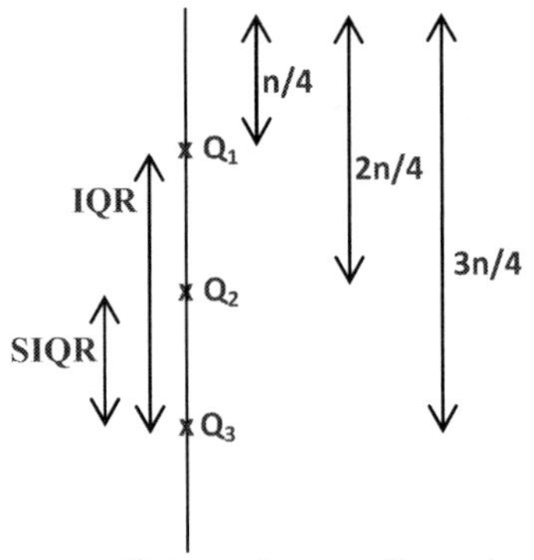

Data are in ascending order

Remember

1. Inter-quartile range(IQR) = $Q_3 - Q_1$
2. Semi inter-quartile range(SIQR) = $0.5(Q_3 - Q_1)$

Working rule

1) First arrange the data in ascending order and then find n (total number of values).

2) If $\frac{n}{4}$ = an integer, then (i) $Q_1 = 0.5\left[\left(\frac{n}{4}\right)^{th} \text{value} + \left(\frac{n}{4}+1\right)^{th} \text{value}\right]$ and

(ii) $Q_3 = 0.5\left[\left(\frac{3n}{4}\right)^{th} \text{value} + \left(\frac{3n}{4}+1\right)^{th} \text{value}\right]$.

3) If $\frac{n}{4} \neq$ an integer, then (i) $Q_1 = \left(\text{Integer next to } \frac{n}{4}\right)^{th}$ value and

(ii) $Q_3 = \left(\text{Integer next to } \frac{3n}{4}\right)^{th}$ value.

4) If $\frac{n}{2}$ = an integer, then $Q_2 = 0.5\left[\left(\frac{n}{2}\right)^{th} \text{value} + \left(\frac{n}{2}+1\right)^{th} \text{value}\right]$.

5) If $\frac{n}{2} \neq$ an integer, then $Q_2 = [0.5(n+1)]^{th}$ value.

Example

The first 11 odd natural numbers are given randomly as

5 1 9 13 11 3 7 19 17 15 21.

(a) Find the quartiles.

(b) Find (i) inter-quartile range, (ii) semi-inter quartile range.

Working: 1 3 5 7 9 11 13 15 17 19 21

(a) n = 11 and $\frac{n}{4} = \frac{11}{4} = 2.75$ ($\neq$ an integer)

Lower quartile, $Q_1 = 3^{rd}$ value = 5

Upper quartile, $Q_3 = 9^{th}$ value = 17

$\frac{n}{2} = \frac{11}{2} = 5.5$ ($\neq$ an integer)

Median, $Q_2 = 6^{th}$ value = 17

(b)(i) Inter quartile range = 17 – 5 = 12

(ii) Semi- inter quartile range = 0.5 x 12 = 6

> Arrange the data in ascending order.
> (Integer next to $2.75)^{th}$ value is 5.
> $\left(\text{Integer next to } \frac{3n}{4}\right)^{th}$ value is 17.
> Q_2 is $(0.5(n+1))^{th}$ value.
>
> $IQR = Q_3 - Q_1$.
> $SIQR = 0.5(Q_3 - Q_1)$.

Quartiles of simple frequency distribution

Working rule

1) Add up all the frequencies(let n).
2) Calculate the cumulative frequencies(c. f.).
3)Then Q_1 = The value with c.f. just greater than $\frac{n}{4}$,

Q_2 = The value with c.f. just greater than $\frac{2n}{4}$ and

Q_3 = The value with c.f. just greater than $\frac{3n}{4}$.

Remember

If $\frac{n}{4}$ or $\frac{2n}{4}$ or $\frac{3n}{4}$ = an exact c.f., then take the average of the corresponding value and the next value.

Example

The marks allotted for different questions in a test paper were shown in the table below.

Marks	1	2	4	6	9
No. of questions	16	10	4	5	2

(a) Find the median. (b) Find the inter quartile range.

Working:

Marks	1	2	4	6	9
No. of questions	16	10	4	5	2
C.F.	16	26	30	35	37

(a) n = 37

$\frac{2n}{4} = \frac{2 \times 37}{4} = 18.5$

Median, $Q_2 = 2$

(b) $\frac{n}{4} = \frac{37}{4} = 9.25$

Lower quartile, $Q_1 = 1$

$\frac{3n}{4} = \frac{3 \times 37}{4} = 27.5$

Upper quartile,, $Q_3 = 4$

Inter quartile range = 4 – 1 = 3

The c.f. just greater than 18.5 is 26.
The mark with c.f. 26 is 2, which is Q_2.
The c.f. just greater than 9.25 is 16.
The mark with c.f. 16 is 1, which is Q_1.

The c.f. just greater than 27.75 is 30.

The mark with c.f. 30 is 4, which is Q_3.
IQR = $Q_3 - Q_1$.

Practise exercise 11

1 For each of the following set of data, find
(i) the quartiles, (ii) the inter quartile range, (iii) semi- inter quartile range.
(a) 2, 9, 7, 8, 4, 6, 3 (b) 5, 3, 12, 6, 11, 11, 14, 10
(c) 14, 9, 9, 20, 38, 95, 64, 0 (d) 21, 22, 23, 24, 25, 26

2 The number of correct answers given by 9 participants in a quiz competition were 8, 7, 7, 12, 9, 9, 10, 9 and 10.
Find (i) the lower quartile, (ii) the upper quartile,
(iii) the inter quartile range, (iv) the range.

3 The time allotted for different speakers in a seminar were shown in the table below.

Time(minutes)	5	10	15	20	30
No. of speakers	8	13	10	6	2

(a) Find the median. (b) Find the inter quartile range.

4 For each set of the following distributions, find
(i) the upper quartile, (ii) the lower quartile,
(iii) the inter quartile range, (iv) median, (v) the range.

(a)

Marks	6	10	13	17	19
Frequency	4	8	11	6	4

(b)

Mass (grams)	100	150	200	220
No. of oranges	120	70	150	200

(c)

Variate	7	14	8	10	16	19
Frequency	6	11	15	6	11	2

(d)

Age(years)	9	10	11	12	13
Cumulative frequency	8	13	19	29	40

Quartiles of grouped frequency distribution

- The quartiles of a grouped frequency distribution can be estimated using cumulative frequency diagram.

Working rule

1) First find n (the c.f. of last value).
2) Then $Q_1 = \left(\frac{n}{4}\right)^{th}$ value, $Q_2 = \left(\frac{2n}{4}\right)^{th}$ value, $Q_3 = \left(\frac{3n}{4}\right)^{th}$ value.

Remember

1. The percentiles divide an ordered data in to 100 equal parts and named as 1st percentile, 2nd percentile, 3rd percentile etc.
2. $Q_1 = 25^{th}$ percentiles, $Q_2 = 50^{th}$ percentile, $Q_3 = 75^{th}$ percentile.

Example

The cumulative frequency curve shows the marks scored by 80 students in a Mathematics test. Use your graph to estimate

(a) the lower quartile,

(b) the upper quartile,

(c) the median,

(d) the inter quartile range,

(e) the 10th percentile,

(f) the number of students who scored more than 60 marks.

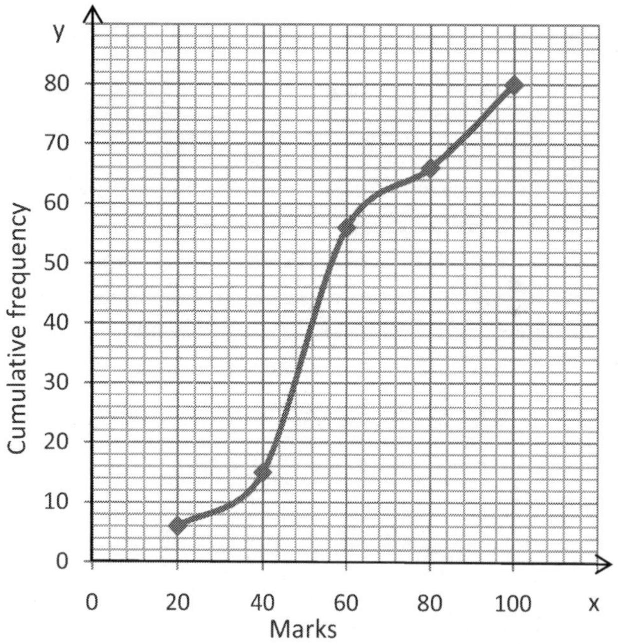

Working: n = 80

(a) Lower quartile, $Q_1 = 44$

(b) Upper quartile, $Q_3 = 66$

(c) Median, $Q_2 = 52$

(d) Inter quartile range = 66 – 44 = 22

(e) The 10th percentile = 28

(f) Scored more than 60 marks = 80 – 56 = 24

n is the c.f. of last value.

$\frac{n}{4} = 20^{th}$ value is 44, which is Q_1.

$\frac{3n}{4} = 60^{th}$ value is 66, which is Q_3.

$\frac{2n}{4} = 40^{th}$ value is 52, which is Q_2.

10th percentile $= \frac{80}{100} \times 10 = 8$ as c.f. = 28

Practise exercise 12

1 The age of 100 people are shown in the table below.

Age(A years)	$20 < A \le 30$	$30 < A \le 40$	$40 < A \le 50$	$50 < A \le 60$	$60 < A \le 70$
No. of people	22	16	33	17	12

(a) Construct a cumulative frequency table for these 100 people.
(b) Using a scale of 2 cm to represent 10 years on the horizontal axis and 1 cm to represent 10 people on the vertical axis, draw a cumulative frequency curve.
(c) Using your cumulative frequency curve, find
(i) the median age, (ii) the lower quartile, (iii) the upper quartile,
(iv) the inter quartile range (v) the 20th percentile, (vi) the 84th percentile.
(d) How many people are older than 45 years?

2

Speed(x km/h)	$30 < x \le 35$	$35 < x \le 40$	$40 < x \le 45$	$45 < x \le 50$	$50 < x \le 5$
No. of vehicles	20	15	30	25	20

(a) Draw a cumulative frequency curve for the above data.
(b) Use your cumulative frequency curve to find
(i) the median speed, (ii) the lower quartile, (iii) the upper quartile,
(iv) the inter quartile range, (v) the 15th percentile, (vi) the 64th percentile.
(c) How many vehicles are travelling at less than 42 km/h?

3

Class interval	0 - 15	15 - 30	30 - 45	45 - 60	60 - 75	75 - 90
Frequency	45	64	36	25	10	20

(a) Draw a cumulative frequency curve to show this data.
(b) Using your curve, find
(i) the median, (ii) the modal class, (iii) the 35th percentile,
(iv) the 80th percentile, (v) the 25th percentile, (vi) the 75th percentile,
(vii) the inter quartile range, (viii) the semi-inter quartile range.
(c) (i) What is the special name for 25th percentile?
(ii) What is the special name for 75th percentile?

4.02 Probability

Core curriculum

Probability scale

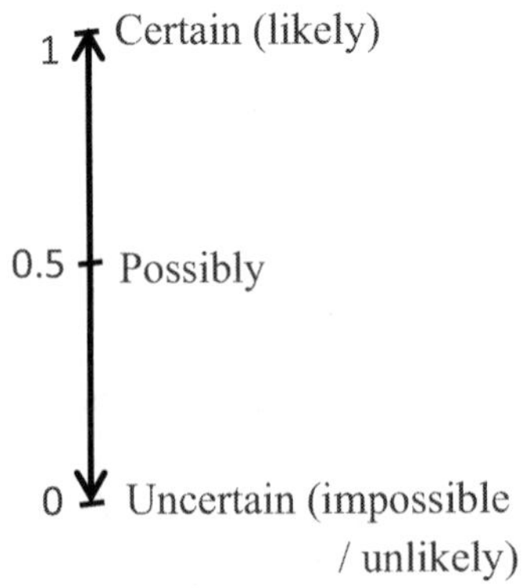

Probability of single event

Remember

1. The highest probability of an event is 1.

2. The lowest probability of an event is 0.

3. All probabilities must lie between 0 and 1.

4. The probability of an event can be written either as a fraction or decimal (not ratio).

8. The probability of an event never be negative or more than 1.

9. In a particular problem, the total probabilities add up to 1.

Working rule

1. The probability of an event $= \dfrac{\text{Number of possible outcomes}}{\text{Total number of possible out comes}}$.

2. The probability of an event occurring = 1 – probability of the event not occurring.

3. Relative frequency can be used as an estimate for the probability of an event.

Relative frequency $= \dfrac{\text{Frequency of the event}}{\text{Total frequency}}$.

Example 1

The probability that team A will win a watch is $\frac{2}{5}$. What is the probability of team A will not win a match?

P(not win) = 1 – P(win).

Working: The probability of team A will not win a match $= 1 - \frac{2}{5} = \frac{3}{5}$.

Example 2

A number is chosen from the list 10, 21, 23, 41, 49 and 59. Find the probability that the number is (a) even, (b) a multiple of 7, (c) less than 49, (d) a prime number.

Working: (a) P(even) $= \frac{1}{6}$

Since 1 even number out of 6 numbers.

(b) P(a multiple of 7) $= \frac{2}{6}$

Since 2 numbers are multiples of 7 out of 6 numbers.

(c) P(less than 49) $= \frac{4}{6}$

Since 4 numbers are less than 49 out of 6 numbers.

(d) P(a prime number) $= \frac{3}{6}$

Since 3 numbers are prime out of 6 numbers.

Practise exercise 13

1 (a) The probability of an event happening is 0.4. What is the probability of an event not happening?

(b) The probability of an event not occurring is $\frac{1}{3}$. What is the probability of this event not occurring?

(c) The probability that team A will not win a match is 0.25. What is the probability of team A will win a match?

(d) The probability that team A will win a match is 0.48 and that loss a match is half of it. What is the probability of team A will draw a match?

2 (a) The probability that Masa will solve a puzzle is 1.4. Explain why the value of this probability is not possible?

(b) The probability that William will win a champions title – 0.3. Explain why the value of this probability is impossible?

3 In a cricket match, the probability that team B will lose a match is 0.3 and that the probability it will draw the match is 0.1. What is the probability that team B will win the match?

4 A number is chosen from the list 4, 5, 6, 7, 8, 9, 10 and 11.

Find the probability that the number is

(a) an odd number, (b) an even number, (c) a prime number,

(d) not a prime number, (e) a multiple of 32, (f) not a multiple of 5.

5 From the letters of the word EXAMINATION, find the probability of choosing

(i) the letter M, (ii) the letter N, (iii) a vowel.

6 (a) A fair coin is tossed. What is the probability of getting (i) a head, (ii) not a tail?

(b) A fair die is thrown. What is the probability that the score is

(i) an odd number, (ii) a prime number, (iii) a square number, (iv) a cube number?

7 A box contains 14 roses and 10 jasmines. A flower is taken at random from the box.

What is the probability that (i) a rose will be taken, (ii) a jasmine will be taken?

8 Fifty tokens, numbered from 1 to 50 are placed in a box. A token is taken at random from the box. Write down the probability that the number on the token is

(a) a factor of 100, (b) a multiple of 10, (c) an even number,

(d) not a multiple of 5, (e) a square number, (f) not a cube number.

9 There are 20 girls and 25 boys in a class. Each student likes either Mathematics or English but not both. 15 girls like Mathematics and 12 boys like English.

(a) A student is chosen at random. Find the probability that the student is

(i) a girl, (ii) a boy.

(b) A girl is chosen at random. Find the probability that she likes English.

(c) A boy is chosen at random. Find the probability that he likes Mathematics.

10 A die is thrown for 200 times and the table below shows the results.

Out comes	1	2	3	4	5	6
Frequency	30	35	40	27	50	18

(a) Find the relative frequency of

(i) throwing a 3, (ii) throwing a prime number,

(iii) throwing a square number, (iv) throwing factors of 6,

(v) throwing at least 4, (vi) throwing at most 3.

(b) Is the die biased? Justify your answer.

Extended Curriculum

Probability of combined events

Remember

1. Addition rule is P(A or B) = P(A) + P(B). It is used for mutually exclusive events only.
2. Multiplication rule is P(A and B) = P(A) x P(B). It is used for independent events only.
3. Events are mutually exclusive, if they cannot occur at the same time.
4. Events are independent, if one event does not affect the other event.

Example

A box contains 7 red, 5 blue and 4 green pencils.

(a) Find the probability of picking out from the box either

(i) a red or a green, (ii) a green or a blue or a red.

(b) One pencil is picked out at random and replaced, and then a second pencil is picked out at random. What is the probability that

(i) both pencils are blue, (ii) one is red and the other is green.

Working: P(red) $=\frac{7}{16}$, P(blue) $=\frac{5}{16}$, P(green) $=\frac{4}{16}$.

(a) (i) P(red or green) $= \frac{7}{16} + \frac{4}{16} = \frac{11}{16}$

(ii) P(green or blue or red) $= \frac{4}{16} + \frac{5}{16} + \frac{7}{16} = 1$

Use addition rule for **or**.

Total probabilities add up to 1.

(b) (i) P(both are blue) $= \frac{5}{16} \text{ x } \frac{5}{16} = \frac{25}{216}$

(ii) P(one red and one green) $= \frac{7}{16} \text{ x } \frac{4}{16} = \frac{7}{64}$

Both are blue means blue and blue.

So use multiplication rule.

Practise exercise 12

1 A box contains 6 red and 3 yellow flowers.

Find the probability of picking out from the box

(a) a red or a yellow flower,

(b) both yellow flowers (if one is replaced before picking the other),

(c) a red and a yellow flower (if one is replaced before picking the other),

(d) both red flowers (if one is not replaced before picking the other).

2 A fair coin is tossed twice. Find the probability of obtaining

(a) a head or a tail, (b) a head and a tail.

3 A fair coin is tossed and a fair die is thrown. Find the probability of getting

(a) a head with the coin and an even number with the die,

(b) a tail with the coin and a prime number with the die,

(c) a head with the coin or a 4 with the die,

(d) a multiple of 3 with the die or a tail with the coin.

4 In a cricket tournament, the probability of team A winning is $\frac{1}{3}$, the probability of team B winning is $\frac{2}{5}$ and the probability of team C not winning is $\frac{11}{15}$.

Find the probability (a) A winning or B winning,

(b) C winning, (c) C winning or B winning,

(d) A winning or B winning or C winning.

5 A fair die is thrown two times. What is the probability of throwing

(a) a 4 on the first throw and a 3 on the second throw,

(b) both are odd numbers, (c) both are not prime numbers,

(d) exactly two fours in the both throws, (e) at least two sixes in both throws.

6 Using the Venn diagram, find the below probabilities.

(a) P(A) (b) P(B) (c) P(A or B)

(d) P(A and B) (e) P(not A) (f) P(universal set)

(g) P(B^1) (h) P(A^1 or B)

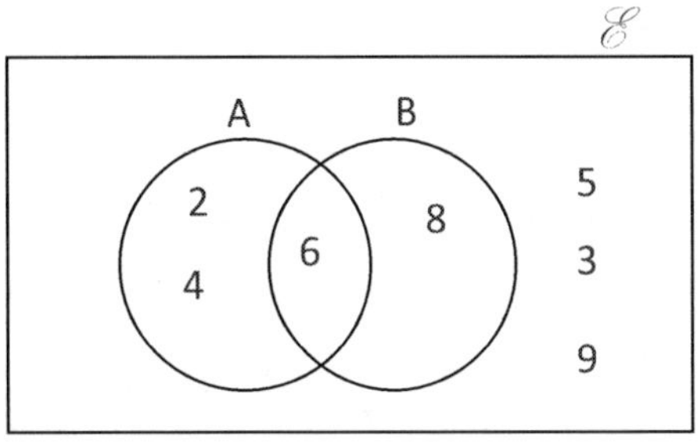

Probability diagrams

Remember

1. The probabilities of combined events can be calculated using possibility diagrams and tree diagrams.
2. In a possibility diagram, the outcomes are represented by points on a grid.
3. In a tree diagram, the outcomes are written at the end of the branches and the probabilities by the side of the branches.
   (i) Using tree diagrams, the probabilities of combined events will be calculated by multiplying along the branches.
   (ii) In a tree diagram, the sum of the probabilities on any set of branches is 1.

Example 1

A coin is tossed and a die is thrown.

(a) Represent all the outcomes of these events in a possibility diagram.

(b) Calculate the probability of obtaining a head with the coin and a prime number with the die.

Working: (a)

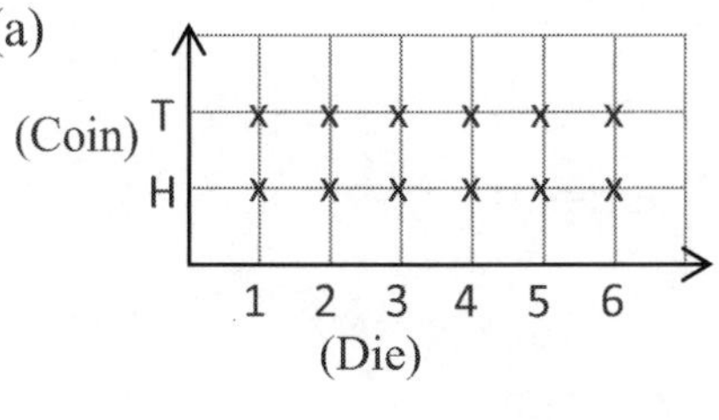

Represent the outcomes by points on a grid.

(b) P(a head and a prime number) $= \frac{2}{12}$

Example 2

There are 12 boys and 9 girls in a class. Two students are chosen at random from the class.

(a) Show this information in a tree diagram.

(b) Calculate the probability of choosing (i) two boys, (b) two girls, (c) at least one boy.

Working: (a)

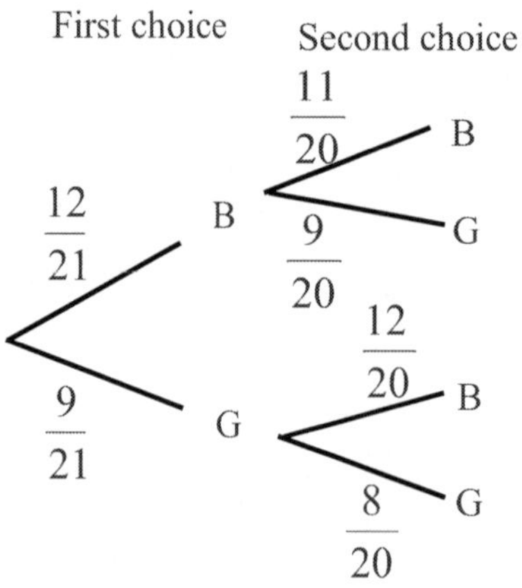

$P(BB) = \frac{12}{21} \times \frac{11}{20}$

$P(BG) = \frac{12}{21} \times \frac{9}{20}$

$P(GB) = \frac{9}{21} \times \frac{12}{20}$

$P(GG) = \frac{9}{21} \times \frac{8}{20}$

(b)(i) $P(\text{two boys}) = \frac{12}{21} \times \frac{11}{21} = \frac{11}{35}$ (ii) $P(\text{two girls}) = \frac{9}{21} \times \frac{8}{20} = \frac{6}{35}$

(iii) $P(\text{at least one boy}) = \frac{12}{21} \times \frac{11}{20} + \frac{12}{21} \times \frac{9}{20} + \frac{9}{21} \times \frac{12}{20} = \frac{29}{35}$

Multiply along the branches.

Or $P(\text{at least one boy}) = 1 - \left(\frac{9}{21} \times \frac{8}{20}\right) = \frac{29}{35}$

P(at least one boy) = 1 – P(all girls).

Practise exercise 15

1 A fair coin is tossed twice.

(a) Represent all the outcomes in a possibility diagram.

(b) Calculate the probability of obtaining (i) two tails, (ii) a head and a tail.

2 Two fair dice are thrown.

(a) Represent all the outcomes in a possibility diagram.

(b) Calculate the probability of obtaining

(i) a double 6, (ii) a total score of 10,

(iii) at least an odd number on one die, (iv) the same number on both throws,

(v) the different numbers on both throws, (vi) the score which is less than 8.

3 There are two spinners, one marked into equal sections numbered 1, 3, 5, 7, 9 and the other spinner marked into equal sections numbered 0, 2, 4, 6, 8.

The two spinners are spun at a time and the outcomes are recorded in a possibility diagram.

(a) Draw the possibility diagram on a grid.

(b) Calculate the probability of getting

(i) a 3 or a 6, (ii) both prime numbers, (iii) a prime number and an even number,

(iv) a total of 7, (iv) a total of more than 6, (v) a total of less than 12.

4 There are 10 green balls and 13 white balls in a bag. Two balls are chosen at random without replacement.

(a) Show this information in a tree diagram.

(b) Calculate the probability of choosing (i) both green balls, (ii) both white balls,

(iii) at least one green ball, (iv) a green ball followed by a white ball.

5 On the basis of past record, team A has the probability of winning a match is 0.7.

The team A played three consecutive matches.

(a) Draw a tree diagram to show this information and label the probabilities in the appropriate branches.

(b) Calculate the probability that team A will

(i) win all the three matches, (ii) lose all the three matches,

(iii) win at least two matches, (iv) win at most two matches.

6 In a mathematics mental test, the probability that Abraham will not score full mark is $\frac{2}{9}$.

Abraham participated in two mathematics mental tests.

(a) Draw a tree diagram showing all the probabilities in the correct branches.

(b) Using your diagram, calculate the probability that Abraham will

(i) score full marks in both tests, (ii) score full marks exactly in one test,

7 A competitive examination consists of a written test and a viva.

The probability that Alok will pass the written test is 0.8.

If Alok passes the written test, the probability that he will pass the viva is $\frac{7}{10}$.

Otherwise the probability reduces to $\frac{4}{10}$.

(a) Draw a tree diagram for Alok and label the probabilities in appropriate branches.

(b) Use your diagram to calculate the probability that Alok will

(i) pass both the tests,

(ii) pass only in the written test,

(iii) not pass the viva,

(iv) not pass any of the tests.

8 Xmala has two mathematics problems to solve.

The probability that she solves the first problem is $\frac{5}{7}$. If she solves the first problem, the probability that she will not solve the second problem is $\frac{1}{5}$.

If she does not solve the first problem, the probability that she will solve the second problem is $\frac{1}{3}$.

(a) Show this information in a tree diagram.

(b) Calculate the probability that Xmala solves both the problems.

(c) Calculate the probability that Xmala does not solve both problems,

(d) Calculate the probability that Xmala solves the first problem but does not solve the second problem.

(e) Calculate the probability that Xmala solves at least one problem.

Answers

Chapter – 1 **Number**

Practise exercise 1(A)

1. (a) 4 or $\sqrt{4}$ or 2 (b) 4 or $\sqrt{4}$ or 2 or –4 or ¼ (c) 4 or $\sqrt{4}$ or 2 (d) π (e) 4 or $\sqrt{4}$ or 2 or –4 (f) $\sqrt{4}$ or 2 2. (a) 29 or 31 (b) 28 (c) 25 (d) 30 (e) 27 (f) 32 3.(a) e (b) 4 or 64 (c) 64 (d) 4 (e) 72 (f) 4 or 72 4.(a) –2 (b) 2/3 (c) $\sqrt{2}$ (d) 2 (e) 4 (f) 8
5.(a) 1, 2, 3, 4, 6, 9, 12, 18 or 36 (b) 9 (c) 2 and 3 (d) 7 and 13 (e) 2 x 3 x 3 x 11 (f) 4
6.(a) 1, 3, 5, 15 (b) 15 (c) 150 (d) 525 (e) 525 000 7.(a) 2 x 2 x 3 x 5 (b) 3599
8.(a) 7 (b) 168 9.(a) 97 (b) 91 (c) 93

Practise exercise 1(B)

1.(a) 5 (b) 42 (c) 2, 3, 7 (d) 49 (e) 45 2.(i) $\sqrt{441}$, $\sqrt[3]{8}$ (ii) π , 0.020020002… (iii) 3.14, 22/7, $\sqrt{441}$, $\sqrt[3]{8}$, 0.020020002… (iv) 7 (v) $\sqrt[3]{8}$ (vi) $\sqrt{441}$ 3.(a) 1, 3, 5, 7, 9, 11, 13, 15, 17, 19 (b) odd numbers (c)(i) 2 prime number (ii) 16 (iii) square number (iv) 64 (v) square or cube number
4.(a)(i) 41, 43 (ii) 11, 13 (iii) 149 (b) No, 91 is not prime 5.(a) 103, 107, 119 (b) 2

Practise exercise 2(A)

1.(a) 42 (b) 15.913 (c) –35 (d) –2.27 (e) 91 (f) 0 (g) –1/6 (h) 9/22 (i) –57/40 (j) –6 (k) 15.45 (l) 5.15 2.(i) 16 (ii) –21 (iii) –65 (iv) 54 (v) 0 (vi) –21.15 (vii) 5 (viii) –5 (ix) 8 (x) 0 (xi) –3/8 (xii) 169/12 (xiii) 6/5 (xiv) –11/25 (xv) 8 3. 43.5 4. 7.2 5. 27716 6.(a) 1.6 (b) 5.8 7. –9.4 8. –4.2

Practise exercise 2(B)

1. 37 ^{0}C 2.(a) 32.7 ^{0}C (b)(i) 35.4 ^{0}C (ii) 31.7 ^{0}C 3.(a) 485 m (b)(i) 234 m (iii) –251 m
4. (a) 0.1 ^{0}C (b) 7 ^{0}C (c) Wednesday and Saturday (d) 27/70 or 0.386

Practise exercise 3(A)

1.(i) 9 , 11 (ii) –48, –60 (ii) 10, 12 (iv) 8, 4 (v) 17, 23 (vi) 45, 54 (vii) 100 000, 1000 000 (viii) 32, 44 (ix) 32, 50 (x) 16, 19 (xi) 2, 0 (xii) 13, 21 (xiii) –16. –20 (xiv) 10^1, 10^{-1} (xv) 5, 1 (xvi) 25, 36 (xvii) 125, 216 (xviii) 126, 217 (xix) 24, 35 (xx) 1/25, 1/36 (xxi) $6x^5$, $5x^4$ (xxii) $12a^{10}$, $14a^{12}$ 2.(a) 3, 5, 7, 9, 11 (b) 1, 4, 9, 16, 25 (c) 1, 8, 27, 64, 125 (d) 2, 5, 10, 17, 26 (e) –1, 6, 25, 62, 123 (f) 1, 3, 6, 10, 15 (g) 16, 49, 100, 169, 256 (h) 2, 14, 62, 254, 1022 (i) 38, 36, 34, 32, 30 (j) 1, 2, 4, 8, 16 (k) 1, 25, 625, 15625, 390625 (l) 0, 3, –9, 27, –81 (m) 2.25, 16, 56.25, 144, 306.25 (n) –2, 4, –8, 16, –32 (o) 1, 5, 14, 30, 55 3.(a) 46 (b) 94 (c) 8n–2 (d) 17 5.(a)(i) 300 (ii) 900 (iii) 15n (b) 330 6.(b) 64, 125, 216, 343 (c)(i) 1331 (ii) 1000 000 (iii) n^3 (d) 20

Practise exercise 3(B)

1.(a) 10, 15, 21, 28 (b)(i) 4 (ii) $(n+1)^2$ 2.(a) 1.8, 0.3, –1.5, –3.6 (b)(i) 39, 45, 6n + 9 (ii) 1, –6, 36 – 7n (iii) 8/17, 9/19, (n+3)/(2n+7) (iv) 6/19, 7/22, (n+1)/(3n+4) 3.(a) 32, 40, 48, 56 (b) 8n (c) 8(n+1) (d) 8 (e) 16 000 (f) 133 4.(a) 55 (b) $16x^{17}$ (c) 2048 5.(a) 22 – 7n/2 (b) –328
6.(a)(i) –5n –1 (ii) n^3 +2 (iii) n/(n+4) (iv) $(n+1)^3$ (v) $n^2 - 2$ (b) n^3 (c) 777

Practise exercise 4(A)

1.(a) 4.7 x 10^3 (b) 9.0 x 10^6 (c) 1.06 x 10^3 (d) 7.37 x 10^2 (e) 5.67 x 10^7 (f) 2.0 x 10^9 (g) 3.0 x 10^9 (h) 6.0 x 10^6 (i) 1.0 x 10^{-6} (j) 6.1 x 10^{-8} (k) 4.3 x 10^8 (l) 2.71 x 10^{21} (m) 4.52 x 10^4 (n) 9.3 x 10^{-11} (o) 4.0 x 10^{18} 2. (i) 2.2 x 10^3 (ii) 9.0 x 10^{10} (iii) 2.37 x 10^4 (iv) 8.0 x 10^3 (v)1.03 x 10^{10} (vi)5.15 x 10^{-10} (vii)3.0 x 10^0 (viii)1.64 x 10^{-3} 3.(a) 1.30 x 10^2 (b) 4.00 x 10^3 (c) 5.67 x 10^{-3} (d) 8.00 x 10^3 (e) 4.30 x 10^3 (f) 1.20 x 10^2 (g) 2.02 x 10^3 (h) 2.00 x 10^4 4. 1.47 x 10^7 5. 3.50 x 10^9 6. 2.00 x 10^{-7}
7. 1.44 x 10^{-15} 8.(a) 1.91 x 10^{-1} (b) 1.20 x 10^{-4} 9. 7.3 x 10^8

Practise exercise 4(B)

1.(a) 4.03 x 10^8 (b) –2.25 x 10^{-2} (c) –9.07 x 10^6 2.(a) 3.00 x 10^5 (b)(i) 1.50 x 10^5 (ii) 2.59 x 10^{10} (iii) 2.59 x 10^{12} 3.(a) 39 000 000 (b) 3.9 x 10^7 4.(a) 1.8 x 10^{24} (b) 6350 5. 3.4 x 10^8
6. 8. 25 x 10^{12} 7.(a) 0.045 (b) 4.5 x 10^{-2} 8.(a) 7.89 x 10^4 (b) 4.9 x 10^{-10}

Practise exercise 5(A)

1.(a) 0.3 (b) 0.047 (c) 0.28 (d) 1.008 (e) 1.625 (f) 5.02 (g) 0.667 (h) 0.714 2.(a) 2/5 (b) 9/100

(c) 17/100000 (d) 3/2 (e) 151/25 (f) 90617/5000 (g) 7519/625 (h) –186/25 3.(a) 6, 7 (b) 18, 300 (c) 16, 16, 1 (d) 25, 20, 2250 4.(a) 9/11 (b) 17/2 (c) 22/15 (d) 41/18 (e) 47/12 (f) 0.8 (g) 8.5 (h) 2/17 (i) 3.5 (j) –13/28 (k) 23/4 (l) 6/35 (m) 8/3 (n) 3 (o) 2 (p) 13/2 (q)1/4 (r) 62/9 (s) 10 (t) 12/5 6. 15 7. 20

Practise exercise 5(B)

1.(a) 18/5 (b) 12/5 (c) Text books and note books are of same mass 2.(a) 1301.25 (b) 1041 3.(a) 8/5 (b) 1369/9 4. 7074/5 5. 49/8, 77/8 6.(a) 81/64 (b) 1.265625

Practise exercise 6

1.(a) 75% (b) 820% (c) 1250% (d) 11304% (e) 25% (f) 37.5% g) 68% (h) 35% 2.(a) 6/25 (b) 29/20 (c) 69/200 (d) 457/400 (e) 1/500 (f) 603/800 (g) 2/1 (h) 0.00 0004 3.(a) 25 (b) 218.7 (c) 46.208 (d) 181 (e) 80 (f) 27.51 4.(a)(i) 600 (ii) 2/5 (iii) 60% (b) 40% (c) 324 5. 587.50 6. 7211.54 7. 19.05% 8. 25% 9. 406.4125, 1656.9125 10. 12% 11.(a) 1683.64 (b) 11183.64 12. 94.7% 13. 5328.60 14. 4488.75

Practise exercise 7

1. 202550 2. 410.40 3.(a) 70 (b) 73.50 4.(a)(i) 504 (ii) 525 (b) 8.51% 6.(a) 1080 (b) 1320

Practise exercise 8(A)

1.(a) 1 (b) 11 (c) 9/4 (d) 47/9 (e) 19/20 (f) 0 (g) 16.7 (h) 22 (i) –29.9 (j) 34 (k) 10 (l) –61 (m) 31 (n) –1 (o) 11

Practise exercise 8(B)

1. 54 2. 22 3. 26 4. 5 5. 20 6.(a) 0 (b) 27.1 (c) 136 (d) 0.40625 (e) 9/2 (f) 1/5

Practise exercise 9(A)

1.(a) < (b) < (c) < (d) > (e) < (f) > (g) = (h) < (i) > (j) = (k) = (l) < 2.(a) 0.157, 0.203, 0.23, 0.231, 2.00023 (b) 4.06, 6.04, 6.053, 6.053, 6.45 (c) 11/10, 5/4, 9/8, 7/6 (d) 1.02, 12.2%, 11/9, 4/3 (e) 0.21%, 0.021, 2.1%, 21%, 1/2 (f) 3999 ml, 4 l, 4002 cu. cm (g) 4200 g, 2400 kg, 24 t (h) 0.79, 4/5, 81%

Practise exercise 9(B)

1.(a) 5.1×10^3, 1.5×10^4, 1.2×10^5 (b) 6.4×10^{-2}, 6.27×10^{-2}, 4.6×10^6 (c) 1.7×10^{-2}, 17%, 17 x 17% (d) 2.4×10^{-4}, 2.4%, 24/1000, 0.24 (e) tan 75^0, cos 75^0, sin 75^0 (f) 1/125000, 8×10^{-5}, 0.0008, 0.8% 2.(a) $(0.4)^2$ (b) $(0.4)^{-1}$ (c) $(0.4)^{1/2}$, $\sqrt{0.4}$

Practise exercise 10(A)

1.(a) $60 2. 303 3.(a) 580 (b) 145 4.(a) 4.9 (b) 7.0 (c) 0.00 (d) 10.00 (e) 25.941 (f) 0.0001 5. (a) 60 (b) 500 (c) 4100 (d) 12200 (e) 6000 (f) 3000 (g) 300 000 (h) 5000 000 (i) 40 6.(a) 70 (b) 0.008 (c) 50 (d) 6.1 (e) 13.7 (f) 60.1 (g) 0.00408 (h) 7.051 (i) 0.01011 7.(a) 109.5 (b) 100 (c) 110 (d) 109 8.(a) 75100 (b) 75095 9.(a) (500 – 20 x 20)/(40 ÷ 10) (b) 25 (c) 30 10.(a) $(8 \times 5)/(39 \div 5 + \sqrt{10})$ (b) 8 (c) 8.03

Practise exercise 10(B)

1.(a) 2.2 (b) 431.98 2.(a)(i) 5 (ii) 7 (b)(i) 0.057142857 (ii) 0.05741 3.(a)(i) 30 (ii) 28 (b) 21952 (c) 52.0 4. 0.000 000 455 5. 174 6.(a) $(0.003 \times 3000)/(10 + 20)^2$ (b) 0.3

Practise exercise 11(A)

1.(a) 411.5, 412.5 (b) 45, 55 (c) 1050, 1070 (d) 850, 950 (e) 7150, 7250 (f) 18250, 18750 (g) 68000, 70000 (h) 19000, 29000 (i) 44.5, 45.5 (j) 7.0, 8.0 (k) 63.995, 64.005 (l) 26.5, 27.5 (m) 5.4695, 5.4705 (n) 14.49, 14.51 (o) 4.55, 4.65 (p) 9.735, 9.745 (q) 0.0065, 0.0075 (r) 6.25, 6.35 2.(a) 530 (b) 549 3.(i) 7.5 (ii) 8.49 4.(a) 112.5, 117.5 (b) $112.5 \leq h < 117.5$ 5. $339500 \leq P < 340500$ 6. $4150 \leq l < 4250$ 7. $46.5 \leq w < 47.5$ 8. $7.39 \leq T < 7.41$ 9. $4.15 \leq w < 4.25$ 10. $5.295 \leq \text{Cost} < 5.305$ 11. $16.235 \leq V < 16.245$ 12. $70.75 \leq A < 70.85$ 13.(a)(i) 8.5 (ii) 3.5 (iii) 9.49 (iv) 4.49 (b)(i) 29.75 (ii) 42.6101 (iii) 24 (iv) 27.96

Practise exercise 11(B)

1.(a) 1435, 1425 (b) 20500, 19500 (c) 1789.25, 1789.15 (d) 401.5, 400.5 (e) 18.0675, 18.0575 (f) 0.309, 0.305

2.(a) 200 (b) 240 3.(a)(i) 238.565, 238.555 (ii) 3556.9346, 3556.9296 (b) 0.005 4.(a)(i) 38.5 (ii) 37.5 (b) 1.45 5.(a) 216000 (b) 216.05, 215999.95 6.(a)(i) 9.61 (ii) 9.59 (b) 240.01, 239.99 7.(a)(i) 4974.0075 (ii) 4988.5975 (b) 583.2 (c) 12222.06 8.(a) 0.03255 (b) 0.205 (c) 0.00333 9.(a) 9255 (b) 9165

Practise exercise 12(A)
1.(a) 270 (b) 16200 (c) 4 hr 30 min 2.(a) 1.33 (b) 0.221 (c) 1 min 19.4 s 3. 2 hr 15 min 4. 4 hr 25 min 5. 01 25 6. 09 10 7.(a) 04 20 (b) 12 48 (c) 14.27 (d) 04 15 8.(a) 0.24 a.m. (b) 7.7.45 a.m. (c) 11.29 a.m. (d) 0.10 p.m. (c) 11.24 p.m. 9. 11 25 10. 7 hr 35 min 11. 2.37 p.m.

Practise exercise 12(B)
1.(a)(i) 5 min 10 s (ii) 4 min 58 s (b)(i) 10 04.183 (ii) 10.09.8 2.(a) 1 hr 40 min (b)(i) 10 15 (ii) 11 10 (c)(i) 2 hr 5 min (ii) 2.08 hr 3. 7 hr 43 min 4. 12 hr 25 min

Practise exercise 13(A)
1.(a) 15 (b) 2 (c) 198 (d) 0.25 (e) 240 (f) 43.4 2.(a) 244 (b) 14.7 (c) 0.244 3.(a) 4500 (b) 270 (c) 4.5 4.(a) 40.7 (b) 0.679 5.(a) 65.5 (b) 18.2 6.(a) 2.5 (b) 56 (c) 15.6 7.(a) 1.75 (b) 1 hr 45 min 8.(a) 162000 (b) 162 9.(a) 16.40625 (b) 62.9 10.(a)(i) 1065 (ii) 24 (iii) 44.375 (c)(i) 0.740 (ii) 740

Practise exercise 13(B)
1.(a) 1200 (b) 20 (c) 72 2.(a) 6.44×10^5 (b) 6.4×10^2 3. 0.278 min 4.(a) 50 (b) 0.833 5.(a) 968 (b) 4.25 (c) 228 (d) 39.1 6.(a)(i) 16.40625 (ii) 16 hr 24 min 22.5 s (b)(i) 1044 (ii) 72.9

Practise exercise 14(A)
1.(i) 74 (ii) 4440 2. 24 3. 6348 4. 1800 5. 144 6.(a) 1/4 (b)(i) 5/12 (ii) 1/6

Practise exercise 14(B)
1.(a)(i) 600 (ii) 10 (b)(i) 75 (ii) 1.25 (iii) 1250 2.(a) 2560 (b)(i) 10240 (ii) 4 hr 40 min 3.(i) 1260 (ii) 2520 4. 9 hr 22 min 30 s

Practise exercise 15(A)
1.(a) 2:7 (b) 5:12 (c) 10: 3 (d) 2:5 (e) 1:20 (f) 24:1 (g) 1:160 (h) 2:5 (i) 45:2 (j) 20:3 (k) 5:6:11 (l) 2500:50:1 2.(a) 5:6 (b) 6:5 (c) 5:11 3.(a) 1:4 (b) 1:7 (c) 1:40 (d) 1:6 4.(a)(i) 8:5 (b) 5:4 (c) 7:4:5 (b)(i) 7:13 (ii) 1:1 5.(a) 75, 120 (b) 41, 102.50, 184.50 6.(a) 330 (b) 220 7.(a) 78 (b) 91 8. (a) 51 (b) 204 9.(a) 7 (b) 5.6

Practise exercise 15(B)
1.(a) 10:11 (b)(i) 177:175 (ii) 275:262 2.(i) 80 (ii) 40 3. $60^0, 100^0, 80^0, 120^0$ 4.(a)(i) 720 (ii) 900 (b)(ii) 504 (iii) 126 5.(a)24 (b) 44 6.(a) 3:6:2 (b)(i) 30 (ii) 90

Practise exercise 16(A)
1. 32.76 2.(a) 31 (b) 91 3.(a) 10 min (b) 40 4.(a) 3 (b) 24 5. 5 6.(a)(i) 336 (ii) 112 (b) 14

Practise exercise 16(B)
1. 768000 2. 32 3. 40 4.(a) 110 (b) 55/12

Practise exercise 17(A)
1.(i) 100 000 (ii) 1000 (iii) 1 (iv) 100 2. 4.8 3. 26.4 4. 1:4000 5.(a) 1:80 000 000 (b) 2480 km 6. 3.3 m 7. 7762.5

Practise exercise 17(B)
1.(a) 400 (b) 0.04 (c) 0.000 000 04 (d) 0.008 2.(a)(i) 3500 000 (ii) 35000 (iii) 35 (b) 18 3. 150 4.(a) 27000 (b) 0.027 5.(a) 800 (b) 0.0008

Practise exercise 18
1.(a) q = (1/3)r (b)(i) 72 (ii) 37 2.(a) 441 (b) 1.36 3.(a) m = 100/n (b)(i) 4 (ii) 50 4.(a) T = 0.00625P (b) 1.2 (c) 560 5.(a) $p = 5/\sqrt[3]{q}$ (b)(i) 2.5 (ii) 125 6.(a) 5 (b) 6 7. 30 8. 96 9. 160 10.(a) 60 (b) 22.5 (c) both areas are equal 11. 24, 20

Practise exercise 19(A)

1. 870255.75 2.(a) 51957 (b) 17322.02 3.(a) 472.72 (b) 0.3184 4.(a) 53064.45 (b) 53064
5.(a) 563.60 (b) 564 6.(a)(i) 84.11410 (ii) 0.01189 (b)(i) 185051.02 (ii) 832.30 7. 50426.66

Practise exercise 19(B)

1.(a) 89650.545 (b) 89988 2.(a) 1.15 (b) 1269.35 3.(a)(i) 14250 (ii) 14278.25 (b) First option, 28.25 (c)(i) 0.70175 (ii) 0.69686 4.(a)(i) 1 £ = $ 1.93446 (ii) 1 $ = 0.51694 £ (c) 13543.15
5.(a) 1 € = £ 0.6408 (b) £ 2520

Practise exercise 20

1. –91 2. –670.3 3. 73.8 4. –1265.6 5. –28635 6. 65125 7. –1264.825 8. 50 9. 29
10. 65 11. –1.6 12. 3 13. 36 14. 2.57 15. 1020100 16. 55.12 17. 27 18. 1 19. 2
20. 1.90 21. 6.1 22. 49 23. –16 24. 144 25. 125 26. –729 27. 14641 28. 28 29. 343
30. 64 31. 1 32. 425 33. 0.25 34. 2401 35. 2 36. 6.7 37. 99 38. 4 39. 26 40. 100
41. 90300 42. 1 43. 0.5 44. 2 45. 1 46. 0.17 47. 1.467 48. 5 49. 1.56 50. 5142.9 51. 1.1
52. 10 53. 4 54. 6.17 55.(a) 0.2 (b) 0.25 (c) 1/9 (d) 0.0625 (e) 64 56.(a)(i) 1 hr 12 min 0 s (ii) 6 hr 30 min 0 s (iii) 0 hr 25 min 55.2 s (iv) 4 hr 17 min 24 s (b)(i) 3.42 (ii) 7.5925 (iii) 0.758

Chapter – 2

Algebra

Practise exercise 1

1. (i) 19a (ii) $7x^3+1$ (iii) x (iv) $-7x^2$ (v) 5 – 4t (vi) 6p – 1 (vii) $-15a - 10b^2$ (viii) 6ab
(ix) – 30xy (x) $3x^2y$ (xi) $\frac{5}{8}x^3y^3$ (xii) $5.925p^3q^4$ (xiii) 4y (xiv) – 3x (xv) –4p
(xvi) 0.5m (xvii) 2a (xviii) $-5x^2y^2z^2$ (xix) y – 4.5x – 1 (xx) $\frac{a}{4} + 0.15b$ (xxi) 10.1x – 0.101y

2. (a) 6 (b)15 (c) –1 (d) 5 (e) 2 (f) –7 (g) –5 (h) 1 (i) – 16 (j) 5.75 (k) –0.5 (l) –1 (m) 4 (n) –27 (o) 4 (p) 6 (q) 5 (r) 7 (s) 4.59 (t) 3 3. (a) –4 (b) 750 (c) 1.7 (d) –392 (e) 12.5 (f) $\frac{-3}{16}$ 4. (a) (i) 16 (ii) 16 (b) $(a + b)^2 = a^2 + 2ab + b^2$

5.(a) 74 (b) 0.125 (c) –19.99 (d) ±6 6.(a) 630 (b) 10.1 7. 28/45

Practise exercise 2

1.(a) b = x–a (b) c = d + e (c) p = x/q (d) m = ny (e) x = (y–c)/m (f) a = (y+4)/x (g) y = (c–ax)/b (h) b = (a–c)/2 (i) d = (9 – p)/q (j) c = 2m + d (k) e = 2(g–f) (l) x = 10(y–10) (m) b = (p–2l)/2 (n) l = p/4 (o) r = C/2π (p) b = 2s–a–c (q) z = 18 – y – x (r) l = c/πr

2.(a) p = 5(n – m) (b) b = 2a – w/c (c) $x = \sqrt{\frac{y}{2}}$ (d) $y = \sqrt{p+1}$ 3. (i) $x = y^2$ (ii) $x = (y/k)^2$ (iii) $x = (k/y)^2$ (iv) $x = \sqrt{\frac{k}{y}}$ 4.(a) $l = \sqrt{A}$ (b) h = 2A/b (c) a = (A/0.5h) – b (d) $l = \sqrt[3]{V}$

5. r = 4A/11π

Practise exercise 3

1. (a) $b = \sqrt{a - c^2}$ (b) $c = \sqrt{a - b^2}$ 2.(a) $y = \sqrt{x^2 + z^2}$ (b) $z = \sqrt{y^2 - x}$

3. $r = \sqrt[3]{\frac{3V}{4\pi}}$ 4. $b = \frac{ax^2 - c}{x}$ 5. (a) $l = \left(\frac{p - 2\sqrt[3]{m}}{3}\right)^2$ (b) $m = \left(\frac{p - 3\sqrt{l}}{2}\right)^3$ 6. $y = \sqrt{6(a - x)}$

7. (i) x = y/(1 – yp) (ii) y = x/(1 – xp) 8. (i) b = ac/(a – c) (ii) c = ab/(b – a)

9. $y = \sqrt[3]{\frac{5x + 25}{4}}$ 10. (i) $p = (6m + rt^4)/6q$ (ii) $t = \sqrt[4]{\frac{6pq - 6m}{r}}$

11. (a) a = (b + 4bc) / (c – 4) (b) b = (ac – 4a) / (1 + 4c)

12. (a) $s = \left(\frac{3t + 1}{9}\right)^3$ (b) (i) $v = \left(u^3 - \sqrt[3]{w}\right)^2$ (c) $w = \left(u^3 - \sqrt{v}\right)^3$

Practise exercise 4

1. (a) $2x + 6$ (b) $5x - 35$ (c) $32 - 8a$ (d) $-x + 4$ (e) $12x + 6$ (f) $30x - 20$
(g) $-4x + 12$ (h) $-7y - 35$ (i) $-72 + 12y$ (j) $x^2 + x$(k) $2x^2 - x$ (l) $20x - 15x^2$
(m) $-6x^2 + 8xy$ (n) $21a - 14b + 7$ (o) $-54x^3 + 6x^2y - 6x^2$ (p) $2x - 3y$ (q) $-9x^2\ _3 - 3y$
(r) $3x^3 - 2x^2 + 12x$ 2.(a) $2x$ (b) $10 - 9a$ (c) $13 - x$ (d) $4 - y$ (e) $2x - 1$
(f) $x - 5$ (g) $x^2 + 3x$ (h) $-7b - 10b^2$ (i) $x + 2y - 5$ (j) $13p - 9q$ (k) $-3x - 7y - 6xy$
(l) $4p^3 - 8p^2$ (m) $5x + 8$ (n) $11x + 13$ (o) $2x - 21$ (p) $-12a^2 + 4$ (q) $3x^2 - 9x$
(r) $3p^3 - 4p^2 + 3p$ (s) $-20t^4 - 2t^2$ (t) $2x^3 + 15x^2 + 4x$ 3.(a) $7(a+2)$ (b) $5(2 - x)$
(c) $x(x + 2)$ (d) $x(x - 7)$ (e) $-2(3a + 5)$ (f) $3(x + 2y - z)$ (g) $4(x^2 + y^2)$ (h) $2(a - b + c)$
(i) $0.2(p - 2q - 4r)$ 4. (a) $4a(1 - 2a)$ (b) $5x(x - 3)$ (c) $4ab(16a + 3)$ (d) $1.5y(x - 3z)$
(e) $x(2x^2 + x - 1)$ (f) $5y(2 - y - y^2)$ (g) $6x(x - 2 + 3y)$(h) $7x^2(1 - 7x)$ (i) $3y(3y - x)$ (j) $5xy(x + y)$
(k) $2a(2ab - 1)$ (l) $x^2y^2(x + 2y)$ (m) $12pq(2q - 5pr)$ (n) $4x^2y^3z^2(4x^2 + 3z^2)$ (o) $\frac{1}{4}x(x - \frac{1}{4}y)$
(p) $xy(x - y + 2xy)$ (q) $(x + y)(2x + 1)$ (r) $(q - 1)(6p - q)$

Practise exercise 5

1. (a) $x^2 + 3x + 2$ (b) $a^2 + 6a + 8$ (c) $x^2 + 2x - 3$ (d) $a^2 - a - 6$ (e) $x^2 - 6x + 8$ (f) $-a^2 + 8x - 15$
(g) $8x^2 + 22x + 15$ (h) $7a^2 + 26a - 8$ (i) $-3a^2 + 33a - 54$ (j) $x^4 - 16$ (k) $p^2 - 100$
(l) $m^4 - 4$ (m) $a^3 - b^3$ (n) $x^4 - x^3 - 5x^2 + 7x - 2$ (o) $8p^4 - 10p^3 + p^{2+22p} - 24$
2.(a) $x^2 + 2xy + y^2$ (b) $x^2 + 2x + 1$ (c) $a^2 + 20a + 100$ (d) $x^2 - 4x + 4$ (e) $x^2 - 16x + 64$
(f) $p^2 - 8p + 4$ (g) $x^2 + x + \frac{1}{4}$ (h) $a^2 - \dfrac{3}{2}a + \dfrac{9}{16}$ (i) $x^2 + 2x + \dfrac{1}{x^2}$ (j) $36t^2 - 60t + 25$
(k) $100 - 200q + 100q^2$ (l) $a^4 + 2a^2b + b^4$ (m) $x^4 - 8x^2 + 16$ (n) $256x^4 + 800x^2 + 625$(o) $4x^2 + 12x + 9$
(p) $x^2 - 8x + 9$ (q) $x^4 - 1$ (r) $a^2 - 1.5a + 0.25$

Practise exercise 6

1.(a) $(x - 2)(x - 5)$ (b) $(x^2 + 3)(x - y)$ (c) $(x + y)(2 + 5a)$ (d) $(x - 2y)(3 + a)$
(e) $(a - b)(x - y)$ (f) $(a + y)(y - 5)$ (g) $(2a - 5)(2 + 3p)$ (h) $(2x + 3y)(6 - 3a)$ (i) $(x + 2)(5y + 2)$
(j) $(3 - a)(2 + 5b)$ (k) $(2p + q^2)(p - 2)$ (l) $(m - n)(m + 2)$ (m) $(2x - 1)(3x + y)$
(n) $(1 - y)(\frac{1}{2}x + \frac{1}{4}p)$ 2.(a) $(a + b)(a - b)$ (b) $(a + 2)(a - 2)$ (c) $(m + 3)(m - 3)$
(d) $(2 + x)(2 - x)$ (e) $(x + 1)(x - 1)$ (f) $(7p + 9q)(7p - 9q)$ (g) $(x + \frac{1}{2})(x - \frac{1}{2})$
(h) $(3/5 + y)(3/5 - y)$ (i) $(x + 2 + y)(x + 2 - y)$ (j) $(2x + 4)(2x - 4)$ (k) $(x + 17)(x - 17)$
(l) $3(m + 29)(m - 29)$ 3.(a) $2(x + y)(x - y)$ (b) $6(10 + p)(10 - p)$ (c) $9(x + 4)(x - 4)$
(d) $(2a + bc)(2a - bc)$ (e) $(m + 13n)(m - 13n)$ (f) $5(b + 7)(b - 7)$ (g) $x(x + 7)(x - 7)$
(h) $(x^2 + 1)(x + 1)(x - 1)$ (i) $(8 + x^2)(8 - x^2)$ (j) $10(x^2 + y^2)(x + y)(x - y)$ (k) $(t + 9)(t - 9)$
(l) $25n + 23m)(25n - 23m)$ (m) $2(x + y - 1)(x - y + 1)$ (n) $(x/4 + y/2)(x/4 - y/2)$
(o) $(p + q/31)(p - q/31)$ (p) $2(x^2 + 16)(x + 4)(x - 4)$ (q) $3(2x + y)(2x - y)$ (r) $5(n + 16)9n - 16)$

Practise exercise 7

1.(a) $(x + y)(x + y)$ (b) $(p - q)(p - q)$ (c) $(x + 2y)(x + 2y)$ (d) $(a - 5)(a - 5)$
(e) $2a + 3b)(2a + 3b)$ (f) $(3x - 5y)(3x - 5y)$ (g) $(m + 17)(m + 17)$ (h) $(n - 8)(n - 8)$
(i) $(10x + 2y)(10x + 2y)$(j) $13 - 12f)(13 - 12f)$ (k) $11a - 3t)(11a - 3t)$ (l) $2(5x - 7)(5x - 7)$
(m) $(a + 1/a)(a + 1/a)$ (n) $x - 1/x)(x - 1/x)$ (o) $(x + 1/2x)(x + 1/2x)$ 2.(a) $(x + 4)(x + 5)$
(b) $(x + 3)(x + 3)$ (c) $(a + 25(a + 25)$ (d) $(x - 3)(x - 4)$ (e) $(y - 9)(y - 11)$
(f) $(m - 30)(m - 20)$ (g) $(x - 1)(x + 4)$ (h) $(x - 3)(x + 13)$ (i) $(b + 5)(b - 16)$
(j) $(x - 5)(x + 1)$ (k) $p - 24)(p + 3)$ (l) $(m - 20)(m + 5)$ 3.(a) $(x + 1)(3x + 4)$
(b) $(x + 1)(5x + 1)$ (c) $(x + 8)(2x + 9)$ (d) $(x - 3)(2x - 5)$ (e) $(x - 2)(4x - 7)$
(f) $(3x - 10)(3x - 2)$ (g) $(x + 3)(3x - 2)$ (h) $(x + 3)(5x - 3)$ (i) $(a + 4)(7a - 10)$
(j) $(2x - 3)(2x + 1)$ (k) $(a - 3)(2a + 5)$ (l) $(4b - 7)(2b + 1)$

Practise exercise 8

1.(a) $\dfrac{12}{y}$ (b) $\dfrac{14}{3}y^2$ (c) $\dfrac{x}{5y}$ (d) $\dfrac{p^2 - p}{2p - 1}$ (e) $\dfrac{2x - 4}{x + 2}$ (f) $\dfrac{2ax + 4a}{x + 4}$

(g) (h) $\dfrac{x^2}{25}$ (i) $\dfrac{5x}{72}$ (j) $\dfrac{a + 2}{5a}$ (k) $\dfrac{x - 1}{x^3 - x}$ (l) $\dfrac{x - y}{(x + y)^3}$

2.(a) $\dfrac{12 + x^2}{6x}$ (b) $\dfrac{7p}{12}$ (c) $\dfrac{x^2 + 8x + 8}{2x + 8}$ (d) $\dfrac{x^2 - 20x + 198}{14x - 140}$ (e) $\dfrac{p^2 - 4 + p}{p - 2}$

(f) $\frac{6x-5}{x^2-1}$ (g) $\frac{2x+3}{(x+1)(x-2)}$ (h) $\frac{x^2+6x+3}{(x+3)(x+4)}$ (i) $\frac{9x+28}{(x+2)(x+4)}$ 3.(b) $\frac{m}{2}$

(c) $\frac{75-x^2}{10x-50}$ (d) $\frac{p^2+14p-245}{14p+98}$ (e) $\frac{4x-48}{x^2-36}$ (f) $\frac{2x-3}{x-2}$ (g) $\frac{2x^2-2x-7}{2}$

(h) $\frac{1+2x-x^2}{(x+1)(x+2)}$ (i) $\frac{x^2-6}{(x-1)(x-2)}$ (j) $\frac{x+5}{x(x+1)}$ (k) $\frac{4x+4}{x(x+2)}$

(l) $\frac{x^2-6x+25}{4x-12}$ 4.(a) $\frac{x}{(x-2)(x-1)}$ (b) $\frac{1}{x+3}$ (c) $\frac{1}{x+2}$ (d) $\frac{1}{x-7}$

(e) $\frac{x}{x+15}$ (f) $\frac{x}{x-5}$ (g) $\frac{2}{x-20}$ (h) $\frac{x-2}{x-6}$ (i) $\frac{x+5}{2x+5}$ (j) $\frac{2x^2}{2x-5}$

(k) $\frac{3x+4}{x+3}$ (l) $\frac{3}{x-3}$

Practise exercise 9

1.(a) a^5 (b) 1 (c) 2^{x+y} (d) 5 (e) p^{a+b+1} (f) a^2 (g) a^8 (h) 3^{x-y} (i) 1 (j) x^{-2} (k) $(x/y)^5$
(l) $(2/3)^3$ (m) $(yz)^a$ (n) 6^4 (o) x^{a^2-1} (p) $20x^5$ (q) $3x^2$ (r) x^{18} (s) a^{64} (t) $1/x^{20}$
(u) 1 (v) n^8 (w) $k^2/64$ (x) x^{-24} 2.(a) 16 (b) 1/27 (c) 1/11 (d) 1
(e) 1 (f) –1 (g) 16/81 (h) 49/484 (i) 4096 (j) 1 (k) 27/343 (l) 3125
(m) 1/243 (n) 1/16807 (o) 6 (p) 4096 3.(a) 3 (b) 4 (c) 4 (d) 8 (e) 5 (f) 2
(g) 2 (h) –3 (i) 3 (j) 0 (k) 2 (l) 2 (m) 1 (n) 6 (o) –1 (p) 1 (q) –1 (r) –4
(s) –2 (t) –4

Practise exercise 10

1.(a) 49 (b) 4 (c) 1/5 (d) 4 (e) 343 (f) 59049 (g) 1/27 (h) 1/64 (i) 1/4096
(j) 1/16 (k) 2/3 (l) 3/4 (m) 8/27 (n) 16/49 (o) 15625/729 (p) 12.8 (q) 2/3
(r) 0.168 (s) 2.40 (t) 1/343 2.(a) a/121 (b) a/13 (c) $x^6/6859$
(d) $1/32x^5$ (e) $0.354/x^4$ (f) $81/x^{36}$ (g) 1/p (h) x^6/y^{12} (i) $x^{1/4}$ (j) $x^{1/2}$ (k) $1/x^3$
(l) $1/x^{1/2}$ (m) b^4/a^5 (n) $1/x^{4/5}y^{19/15}$ (o) $x^{18}/9$ (p) x – 4 3.(a) 2/3
(b) 3/2 (c) 40/3
(d) 5 (e) 7 (f) 2 (g) 3/2 (h) 5/3 (i) –5/2
(j) 8/3 (k) –2/3 (l) 2 (m) 0 (n) 1/3 (o) –1/3 (p) 5/2

Practise exercise 11

1.(a) –4 (b) –1/4 (c) –20/3 (d) 49 (e) –5 (f) –5/6 (g) 11
(h) –60 (i) –1/27 (j) 36 (k) –128 (l) 121/12 2.(a) 3 (b) –4.2
(c) 1/9 (d) 1 (e) –1 (f) –2/49 (g) 13.6 (h) –98 (i) 20
(j) 3 (k) 0.5 (l) –61 (m) 0.5 (n) –6 (o) –0.0625 (p) 2
(q) –100/7 (r) –8/3 (s) 3/4 (t) –8 (u) –0.2 (v) –2.7 (w) –14
(x) 8.5 3.(a) –1.5 (b) 6.5 (c) 1/3 (d) 9.5 (e) 22/7 (f) 27/7
(g) 1 (h) –2.5 (i) 1/9 (j) –10/9 (k) 2/9 (l) 22 (m) 5/11
(n) –13/15 (o) –24/11 (p) 6.4 (q) 1.2 (r) 4.5 4. 30^0, 45^0, 105^0
5.(a) 21 (b) 42 6.(a) $28x - 20^0 = 540^0$ (b) 20^0 7. 15 8.(a) 30/x
(b) 1.50x (c) 1.50x – 30 9.(a) 6y/5 = y + 8 (b) 40 10.(a) x – 2 = 96 (b) 94 11. 112^0, 58^0

Practise exercise 12(A)

1.(a) x =3, y =1 (b) x +2, y = 4 (c) x = 2, y = 1 (d) x = 6, y = 3 (e) x = 7, y = 4 (f) x = 1, y = 1
(g) x = 7, y = 4 (h) x = 5, y = 9 (i) x = –1, y = 2 (j) x = 8, y = 5 (k) x = 2, y = 2/3 (l) x = 3, y = 1
(m) x = 3, y = –2 (n) x = –4, y = 4 (o) m = 12, n = 9 (p) p = –5.53, q = 8.16 (q) x = –3, y = 5
(r) a = 9, b = 2 (s) x = 1/6, y = –0.5 (t) c = 2.46, d = 1.14 (u) x = 40/11, y = 30/11 3. (a) l + w + 14, 2w – l = 1 (b) l = 9, w = 5 4.(a) 5x + 4y = 12, 3x + 2y = 7 (b) x = 2, y = 0.5
5.(a) 3x + 5y, x + 3 = 2y (b) x = 15, y = 9 6.(a) m – n = 3, m + 2n + 15 (b) m = 7, n = 4 (c) 47
7.(a) x – 2y = – 6, 2x – y = 9 (b) x = 8, y = 7 (c) 13, 11, 13, 11

Practise exercise 12(B)

1.(a) x = 6, y = 1 (b) x = 2, y = 5 (c) m = 60, n = 15 (d) p = 35, q = 133/6 (e) x = 5.2, y = –2.4 (f) x = 5/19, y = –35/19 (g) x = –2/75, y = 17/75 (h) x = –31, y = 1 (i) x = 10, y = 3 (j) x = 8, y = 6 (k) x = 8, y = 1 (l) x = 0.25, y = –0.2 3.(a) $\frac{1}{2}p + \frac{1}{3} = 15, 2p - \frac{1}{6}q = 42$ (b) p = 22, q = 12

4.(a) b + g = 60, $\frac{2}{3}b + \frac{1}{9}g = 25$ (b) g = 27

Practise exercise 13

1.(a) ±4 (b) ±7 (c) ±3 (d) 1.55 (e) x = 0 or 6 (f) x = 0 or –11 (g) 4 or –2 (h) 10.6 or 5.35
2.(a) 2 or 5 (b) 3 or –7 (c) –2 or –1.5 (d) 1 or 4/3 (e) 3 or –1 (f) 1 or 3/7 (g) 5 or 10/3 (h) 5 or –1 (i) –12 or –14 (j) 1 or 8 (k) 4 or –3 (l) 10 or –6
3.(a) 1.16 or or –5.16 (b) 6.16 or –0.163 (c) 3.80 or 0.789 (d) –3/2 (e) 0.225 or –2.22 (f) 0.39 or –5.13 (g) 1 or 0.6 (h) –0.5 or 4 (i) 1.60 or 0.627 (j) 2.21 or –0.377 (k) 10.5 or –9.46 (l) 0.795 or –1.93 (m) 4 or –1 (n) 4 or 1.5 (o) 3.1 or or –14.2 4.(b) 2.47 or –4.47 5. 28, 29
6. 22 or –11 7.(a)(i) x^2 (ii) $x^2 + 3x + 2$ (b) $2x^2 + 3x - 44 = 0$ (c) 4 (d) 4, 5, 6
8.(a) 640/x (b) 9.25 (c) 32, –40 (d) 32 9.(a) $x^2 + 2x - 101 = 0$ (b) 9.25 (c) (i) 75.5625 (ii) 21.5 10.(a) $34x^2 - 30x - 1672 = 0$ (b) 3.56 or –1.56 (c) 14.8, 10.68 (d) 79.032

Practise exercise 14

1.(a) x > 7 (b) x < –2 (c) x ≥ 5 (d) x ≤ 5 (e) x < 0.75 (f) t ≤ –7.25 (g) x > –7/3 (h) x ≤ 8/17 (i) x < 4 (j) x > 13 (k) 1/3 ≤ x < 3 (l) –1.5 ≤ x < 5 (m) y ≥ 0.5 (n) 26 > x > –10 (o) 1/3 > x > 5/3 2.(a) x ≤ 28 (b) x ≥ –1/24 (c) x > –7.5 (d) x < 0.25 (e) a ≤ 39/70 (f) x < 12/13 (g) x ≤ 7.5 (h) x > –1.5 (i) x < 0.62 (j) 12 > x ≥ 0 (k) 2/11 < x < 2/9 (l) x < 19 (m) x > –4/7 (n) x > –1 (o) 4 < x < 6 3.(a) –7 (b) 1

Practise exercise 15

2.(a) x + y ≥ 3, x + y < 6, x > 2, y ≥ 0
3.(a) g + b < 120, g ≥ 48, b ≤ 60
4. (b) x + y ≤ 20
5.(a) A + B ≤ 1000, A > 400, B < 350
6.(a) p + 2b > 60, 3p + 5b ≤ 300

Practise exercise 16

1.(a) –7 (b) 5 (c) –37 (d) –5 (e) 7.6 (f) 2
2.(a) $\frac{-x}{-x+1}$ (b) $\frac{2x}{2x+1}$ (c) $\frac{x}{x+3}$ (d) $\frac{x^2}{x^2+1}$ (e) $\frac{x+1}{x+2}$ (f) $\frac{2-x}{3-x}$
3.(a) –112.67 (b) 2/3 (c) 10 5.(a) ±3 (b) –2 (c) –2 or 3 (d) 25 (e) 2 or –6 (f) –4 or 5 6.(a) ±2.5 (b) ±2.625 7.(i) –1 or –2 (ii) –3 or 4 (iii) –2
8.(a)(i) 4.6875 (ii) 13.21875 (iii) 0.99375 (iv) –8.7 (b) The function will be undetermined. (c)(i) 1.865, –12.865 (ii) 4 or –6 9.(a)(i) 1 (ii) 2187 (iii) 46656 (iv) 0.25 (v) 1/243 (b)(i) 3 (ii) 0 (iii) 4 (iv) 6 (c)(i)The square root of negative number is not defined. (ii) 0^0 is undeterminant. 10.(a) 25 (b) –24/7

Practise exercise 17

1.(a) (x+15)/4 (b) 6 (c) 3 (d) (2a+15)/4 (e) (x + 60)/16 (f) 2 2.(a)(i) x – 4 (ii) 4 (b)(i) 2 – x (ii) –6 c.(i) (2x + 1)/6 (ii) 2.83 d.(i) (10 – x)/5 (ii) 0.4 (e)(i) 3x – 6 (ii) 18 (f)(i) 7/(1 – x) (ii) –1 (g)(i) (12 – y)/36 (ii) 1/9 (h)(i) $\sqrt{x+9}$ (ii) 4.12 (i) (i) $\sqrt[4]{x}$ (ii) 1.68
3.(a) 5x + 14 (b) 5x + 2 (c) 14 (d) 8 4.(a) $6 - x^2$ (b) 3.75 (c) $x^2 - 12x + 36$ (d) 30.25 5.(a) (1 – x)/2 (b) (–x – 4)/2 (c) 11 – 2x (d) (–x – 9)/2 (e) 2
6.(a)(i) (x + 25)/8 (ii) 64x – 225 (iii) – 225 (iv) 3.171875 (b) 3 7.(a)(i) 10 (ii) –6 (b) 6, –2 8.(a) (x + 7)/3 (b) 3 (C) – 277 9.(a) x (b) 81x – 50 10.(a) 5.5 (b) 2 (c) 64 (d) –1.17

Practise exercise 18

1.(a) $\in$ (b) $\in$ (c) $\notin$ (d) $\not\subset$ (e) $\subset$ (f) $\subset$ (g) $\in$ (h) $\subset$ 2.(a) A = {2, 4, 6, 8, 10, 12, 14}
(b) B = {1, 3, 5, 7, 9, 11, 13, 15} (c) P = {2, 3, 5, 7, 11, 13} (d) M = {5, 10, 15} (e) F = {1, 2, 3, 4, 6, 12}
3. (a)(i) {10, 12, 13, 14, 15} (ii) {12, 14} (iii) {11, 13, 15, 16, 17} (iv) {10, 11, 16, 17}
(v) {11, 13, 14, 15, 16, 17} (vi) {10} (vii) {11, 16, 17} (viii) {11, 16, 17} (b)(i) 3
(ii) 4 (iii) 8 (iv) 5 (v) 2 (vi) 5 (vii) 6 (viii) 1 (ix) 6 4.(a)(i) {a, b, c, d}
(ii) {c, d, e, f, g} (iii) {a, b, c, d, e, f, g, h, k} (iv) {a, b, c, d, e, f, g} (v) {c, d}
(vi) {e, f, g, h, k} (vii) {a, b, h, k} (viii) {c, d, e, f, g, h, k} (ix) {a, b, c, d, h, k} (x) {h, k}
(xi) {h, k} (xii) {a, b, e, f, g} (b)(i) 4(ii) 5 (iii) 9 (iv) 7 (v) 2 (vi) 5 (vii) 4 (viii) 7
(ix) 6 (x) 3 (xi) 2 (xii) 7 (xiii) 2 (xiv) 7 (xv) 2 (xvi) 0 5.(a) A (b) Ø (c) A
(d) A (e) A (f) ξ (g) A (h) $A \cup (B \cap C)$ (i) $A \cup B$ (j) A (k) B (l) Ø (m) Ø (n) ξ

Practise exercise 19

1.(b)(i) 13 (ii) 12 (iii) 3 (iv) 31 2.(b) 5 (c)(i) 8 (ii) 6 (iii) 5 3.(b) (i) 43
(ii) 13 (iii) 73 4.(a)(i) 10 (ii) 40 (b)(i) 70 (ii) 80 (iii) 70 (iv) 80 5.(b) 21

Practise exercise 20

2.(a) M(1, –3) N(–2, 0) P(1, 3) (c) (i)(1, 0) (ii) (–0.5, 1.5) 3. (–3, –2) 4. 4 cm^2

Practise exercise 21

1.(a) m = 2, c = 3 (b) m = 1, c = –1 (c) m = – 1, c = 6 (d) m = 2, c = 7 (e) m = ¼, c = 8
(f) m = –7, c = ½ (g) m = –6, c = 8 (h) m = –3, c = –5 (i) m = –4, c = 12 (j) m = 1, c = 2
(k) m = –1, c = 5 (l) m = –1/3, c = 1/9

Practise exercise 22

1. –1.25, y = –1.25x + 5 2. 3.75, y = 3.75x + 4.5 3. Not defined, x = 3 4. 0, y = 4
5. 1, y = x + 1 6. –2, y = –2x + 10

Practise exercise 23

1. a, d 2. y = 12x – 52 3. y = 2x – 3 4. y = –1.5x + 3.5 5. y = 6x + 5 6. –2x – 2/3
7. y = 2 8. y = –5x 9. y = 1.25x + 1.75 10. y = (11/18)x + 11.8

Practise exercise 30

1.(a) 1 (b) –1 (c) –0.1 (d) 5/3 (e) –1 (f) 1.25 (g) –91/265 (h) –99/35 2. 8 3. a = 10, b = 5
4. 115/3 5. 1 6.(b)(i) 1 (ii) Because the tangent is drawn using a single point. (c) y = x
7.(b) (i) –5 (ii) 2.87 (c)(i) y = –5x – 15 (ii) y = 2.87x – 10.4 8.(b) –1.625
(c) y = –1.625x + 6.5 9.(b) –5.6, –5.6 (c) –5.6 (d) Yes. Because gradients of both tangents are equal.

Practise exercise 31

1.(a) 10 (b) 1 (c) 7.21 (d) 8 (e) 9.97 (f) 3.13 2. 8.1
3.(a) A(2,5), B(–2,3), C(3, –2) (b) AB = 4.47, BC = 7.01, CA = 7.01 (c) (i) 5.39
(ii) 3 5.(a)(i) 3 (ii) 2 (iii) 3.61 (b) 3.61 (c) 4.90

Practise exercise 32

1.(a) (2, 6) (b) (4, 3) (c) (4, –0.5) (d) (–6, –2.5) (e) (–0.4, –0.4) (f) (7/12, 19/154)
2.(i) (3, 5.5) (ii) (1.5, 6.5) (iii) (–1.5, 7) (iv) (0, 6) 3. –4 4. (4, –3) 5. (2, 4)
6. (2, 4)

Practise exercise 33

1.(b)(i) 9.38 (ii) 6.67 (c) 180 2.(a) 48, 80 (c)(i) 56 (ii) 1.25 (d) 50 3.(b)(i) 25.2
(ii) 3.9 (c) 18 4.(b)(i) 15 (ii) 425.00 5.(b)(i) 75¢ (ii) 15 min 50 sec

Practise exercise 34

1.(a) 07 00 (b) 15 (c) 08 00 (d) 2 hrs (e) 10 km (f)(i) 6.67 (ii) 1.85
2.(a) 12 (b)(i) 0.75 km/min (ii) 45 km/h (iii) 12.5 m/s (c) 3.8 km (d)(i) 5 min
36 sec
(ii) 5.6 (iii) 0.093 (e) City bus

Chapter – 3 **Shape and Space**

Practise exercise 1

1.(a) angle (b) perpendicular lines (c) diameter (d) pentagon (e) hexagon (f) heptagon (g) equilateral triangle (h) isosceles triangle (i) right angles triangle (j) rhombus (k) trapezium (l) kite (m) parallelogram 2.(a) acute angle (b) acute angle (c) right angle (d) obtuse angle (e) straight angle (f) reflex angle (g) reflex angle (h) reflex angle (i) acute angle (j) reflex angle (k) acute angle (l) complete angle 3.(a) F (b) T (c) F (d) F (e) T (f) T (g) F (h) T

Practise exercise 2

1. a and c, d and f 2. X = 13 cm, y = 10 cm, z = 52^0 3.(a) Yes (b) AB = CD, BC = DA and CA = AC 4. x = 38^0, y = 6 cm 5. x = 13.3, y = 9.33 6.(I) x = 50^0 (ii) y = 12 (iii) 4 7.(b)(i) 4.45 (ii) 7.9 8.(ii) 3, 1

Practise exercise 3

1. 3. 90625 cm 2.(a) 1.125 m^2 (b) 16.875 m^2 3.(a) 10 cm (b) 1:9 (c) 8:1 5.(a) 1;16 (b) 320 cm^2 6.(a) 5 (b) 2y 7.(a) 1:5 (b) 4:5 (c) (i) 36 cm^2 (ii) 45 cm^2 (d) 1/9

Practise exercise 4

1. x = 20 cm, y = 12 cm, z = 2 cm 2. (a) 216 cm^3 (b) 1:4 3.(a) 5.00 (b) 1:4 4.(a) 16.8 (b) 64.1 (c) 12 5.(a) 6 cm (b) 1/3

Practise exercise 9(A)

1.(a) 1:15000 (b)(i) 315 m (ii) 960 m (iii) 555 m 2.(a) 1:900 (b) (ii) 12.7 (c)(ii) 18 m 3.(a) 10000 m (b) 4800m (c)(i) 11 00000 cm (ii) 11000 m (iiii) 11 km 4.(c) 49.5 m 6.(a) 48 km

Practise exercise 9(B)

1.(a)(i) 15.2 m (ii) 154 m^2 (iii) $346.50 3.((b)(ii) 1590.4 km^2 4.(a) AB = 360 m, BC = 240 m (c)(ii) 1256.6 m^2

Practise exercise 10

1.(a) 74° (b) 40° (c) 96° (d) 74° (e) 22.5° 22.(a) a = 48°, b = 132° (b) q = 109°, p = 71° (c) x = m = p = 54°, y = z = n = 126° (d) a = 120°, t = 30° 3. l = 45°, m = 90° 4.(i) x + 4y + 85° + 91° = 36° (ii) x + 4y = 184° (iii) 33°

Practise exercise 11

1.(a) 45° (b) 31° (c) a = b = 65° (d) y = 131° (e) m = 128.5° (f) a = 51°, b = 39.75° 2.(a) a = 70° (b) b = 97° (c) a = 65°, b = 130° 3.(i) 74° (ii) 70° 4.(a) Isosceles triangle (b)(i) 28° (ii) 62° (iii) 56° 5. e = 50°, f = 43.3°, g = 28.5° 6.(a) 30° (b) 90° 7.(a) 10t – 1 = 360° (b) t = 36.1° (c) 75.2°, 108.3°, 120.4°, 56.1° 8.(a) 2x + y = 180°, x + y = 124° (b) x = 56°, y = 68°

Practise exercise 12

1. (a) x = y = 120° (b) x = y = 54° (c) x = z = 49°, y = 131° (d) b = d = f = 110°, a = c = e = g = 70° (e) x = 75°, y = 80° (f) d = b = 77°, c = 49°, a = 54° 2.(a) x + 2y = 180°, 3x + y = 180° (b) x = 36°, y = 72° 3.(a) 40° (b) 55° (c) 85° 4.(a) 51° (b) 129° 5. a = b = 45° 6.(a) 69° (b) 62° 7.(a) 65° (b) 53° (c) 62°

Practise exercise 13

1. (a) a = b = 45° (b) x = 30° (c) a = 60 °, b = c = 45 ° 2.(a) 45 ° (b) a = b = 23 °, c = 46 ° (c) 140° (d) x = 60 °, y = 55 ° 3. x = 27 °, y = 63°, z = 31 °, t = 59 ° 4.(a) isosceles triangle (b) kite 5.(a) rectangle (b)(i) 20 ° (ii) 140 ° 6.(a)(i) 12 (ii) 12 (b) PQ = PR (c) tangents from an external point are equal. 7.(a) 2x + y = 180°, x + 2y = 180° (b) x = 22.5 °, y = 45°

Practise exercise 14

1.(a) 540° (b) 108° (c) 72 ° 2.(a)(i) 720 ° (ii) 120° (iii) 60° (b)(i) 900° (ii) 128.6° (iii) 51.4° (c)(i) 1440° (ii) 144° (iii) 36° (d)(i) 17640° (ii) 176.4° (iii) 3.6° 3.(a) (n – 2)180° = 1800° (b)(i) 12 (ii) 30° 4.(i) 9 (ii) 40° 5.(a) 60 (b) sexagon 6.(a) 90° (b) square 7.(a) (n – 2)180° /n – 360° /n = 132° (b) n = 15 8. b = 54°, a = 72°, c = 216°

Practise exercise 15
1.(a) $18x = 540^\circ$, $x = 30^\circ$ (b)(i) 27° (ii) 216° 2.(a) $41x + 4^\circ = 1260^\circ$ (b) 34.6°, 61.2°, 91.8°, 153°, 183.6°, 214.2°, 244.8°, 275.44° (c) 145.4° 3.(i) 51.4°, 205.6° 4. 143°

Practise exercise 16
1.(a) $x = 45^\circ$, $y = 90^\circ$ (b) $a = b = 55^\circ$ (c) $a = 42^\circ$, $b = 84^\circ$ (d) $x = 36^\circ$, $y = 108^\circ$, $z = 54^\circ$ (e) $x = y = z = 61^\circ$ (f) 110° 2.(a) $a = 74^\circ$, $y = 60^\circ$ (b) $a = 90^\circ$, $b = 84^\circ$ (c) $x = 60^\circ$, $y = 120^\circ$ (d) $x = 135^\circ$, $y = 45^\circ$ (e) $a = 50^\circ$, $b = 20^\circ$, $c = 66^\circ$ 3.(a) angle A = angle B, angle D = angle E (b) angle A and angle E, angle A and angle D, angle B = angle E, angle B = angle D 4.(a) trapezium (b) $x = 130^\circ$, $y = 70^\circ$, $z = 70^\circ$ 5.(a) rectangle (b) 90° 6.(b) 12.5 cm

Practise exercise 17
1.(a)(i) 1 (ii) 0 (b)(i) 3 (ii) 3 (c)(i) 1 (ii) 0 (d)(i) 2 (ii) 2 (e)(i) 2 (ii) 2 (f)(i) 0 (ii) 0 2.(a) Hint: sketch a parallelogram (b)(i) Hint: sketch an isosceles triangle (ii) isosceles triangle 4.(a)(ii) 0 (b)(ii) 3 (c)(ii) 0 (d)(ii) 5 (e) 0 (f)(ii) 4 5.(ii) perpendicular bisector 6.(a) C, E (b) I and S (c) I

Practise exercise 18
1.(a)(i) 4 (ii) 9 (b)(i) 2 (ii) 3 (c)(i) infinite (ii) infinite (d)(i) 3 (ii) 4 (e)(i) 3 (ii) 3 2.(a) 2 (b) 2 3. 2

Practise exercise 19
1.(a) 16 cm (b) 6 cm (c) 16 cm 2. 14.1 cm 3.(b) kite 4.(i) 143° (ii) 71.5° (iii) 108.5° (iv) 18.5° 5. $A = 40^\circ$, $b = 70^\circ$, $c = 40^\circ$ 6.(a) 30 cm (b) 6 cm (c) 18 cm

Practise exercise 20
1.(a) 16 (b) 9 (c) 30 (d) 16.5 (e) 30.8 (f) 13.0 2.(a) $x = 14.4$, $y = 4.2$ (b) $x = 13.2$, $y = 19.3$ (c) $x = 9.85$, $y = 11.5$ (d) $x = 8.31$, $y = 32.6$ (e) $x = 21.9$, $y = 12.4$ (f) $y = 5$, $x = 13$ 3.(a) 39.6 cm (b) 144.2 cm (c) 21.1 cm (d) 9 m 4.(a) 18.3 cm (b) 24.3 cm 5. Right angled triangle

Practise exercise 21
1.(a) AC = 21.7, BC = 12.5 (b) AC = 36.9, AB = 41.1 (c) AC = 33.6, AB = 39.6 (d) QR = 5.14, PQ = 10.4 (e) PN = 18.2, PM = 35.6 (f) MN = NL = 24.7 2.(a) 53.1° (b) 46.2° (c) 60.6° 3.(a) 0.755 (b) 0.886 (d) 0.792 (e) 2.75 (f) 32.9 (g) 50.7 (h) 57.7 (i) 1 (j) -1 (k) 1 (l) 3.49 4.(a) 28.0° (b) 37.6° (c) 76.8° (d) 42.2° (e) 35.4° (f) 27.8° 5.(a) 22.3 (b) 37.6° 6.(a) 12.1 cm (b) 4.66 cm (c)(i) 6.27 (ii) 48 7.(a) (i) 53.9 m (ii) 39.5 (b) 35°

Practise exercise 22
1. 035° 2. 056° 3.(b) 316 km (c) 30° 4.(b) 14 cm (c) 315° 5.(b) 100 km 6.(a) 21.2 km (d)(i) 045° (ii) 225° (iii) 315° (iv) 270°

Practise exercise 23
1. 41.6 m 2. 75.1° 3.(a) 37.8 m (b) 35.5° 4. 9.77 m 5. 97.8 m 6. 256.59 m 7.(a) 85.9 m (b) 63.3 m

Practise exercise 24
1.(a) 10.4 (b) 10.3 (c) 36.8° (d) 96.8° (e) 8.85 (f) 88.3° 2.(a) 42.2 cm^2 (b) 30.6 cm^2 3.(a)(i) 13.9 cm (ii) 8.5 cm (b) 35.9 cm^2 4.(a) 17.7 cm (b) 33.8° (c) 128 cm^2 6.(a)(i) 62.1° (ii) 106.1° (iii) 73.9° (b)(i) 15.2 cm (ii) 20.1 cm (c)(i) 74.0 cm^2 (ii) 148 cm^2 8.(a)(i) 19.5 cm (ii) 30.5 cm (iii) 34.4 cm (b) 502.9 cm^2

Practise exercise 25
1.(i) 23.6° or 156.4° (ii) -30° (iii) 90° (iv) 61.6° or 118.4° (v) 45° or -135° 2.(i) 36.9° or -143.1° (ii) 0° or 180° (iii) 11.5° or -168.5° (iv) 138.6° or -41.4° (v) 30° or -150°

Practise exercise 26
1.(a) 18.4 cm (b) 19.6 cm (c) 20.2° 2.(a)(i) 17.0 cm (ii) 20.8 cm (b) 35.2° (c) 144 cm^2 3.(a)(i) 11.3 cm (ii) 12.4 cm (b) 62.8° (c) 70.0° 4.(a) 9.90 cm (b) 45° (c) 54.7° (d) 84.9 cm^2 (e) 535.5 cm^2 5.(a)(i) 21.6 cm (ii) 10.8 cm (iii) 21.4 cm (b)(i) 63.3°, 26.7° (ii) 74.3°, 15.6° (c)(i) 22.4° (ii) 15.6° 6.(a) 8.66 cm (b)(i) 30.4 cm (ii) 31.6 cm (c) 986.6 cm 7.(a)(i) 72° (ii) 54° (b)(i) 44.1° (ii) 50.2°, 39.8°

Practise exercise 27

1.(a)(i) 24.8 cm (ii) 38.44 cm^2 (b)(i) 28 cm (ii) 40 cm^2 (c)(i) 28 cm (ii) 30 cm^2 (d)(i) 20 cm (ii) 14 cm^2 (e)(i) 44 cm (ii) 97.0 cm^2 (f)(i) 46 cm (ii) 92.3 cm^2 (g)(i) 48 cm (ii) 110.3 cm^2 (h)(i) 33.6 cm (ii) 70 cm^2 (i)(i) 58 cm (ii) 108 cm^2 2.(a) 78 cm^2 (b) 186 cm^2 (c) 48 cm^2 (d) 414.3 cm^2 3.(a) 1022 cm^2 (b) 279.9 cm^2 (c) 425 cm^2 4.(a)(i) 13.4 cm (ii) 53.6 cm (b)(i) 20 cm (ii) 600 cm^2 (c) 495 cm^2 (d) 64 cm (e) 18 cm 5.(a)(i) 112 cm (ii)735 cm^2 (b)(i) 9 cm (ii) 15 cm (c)(i) 315 cm^2 (ii) 382.5 cm^2 6.(a) $12x^2 = 432$ (b) $x = 6$ (c) 11664 m^2 7.(a) $8x - 3y = 3$, $2x - y = 5$ (b) $x = 4.5$, $y = 11$ (c) 37, 11 (d) 311.8 cm^2

Practise exercise 28

1.(a)(i) 804 cm^2 (ii) 101 cm (b)(i) 254 cm^2 (ii) 56.5 cm (c)(i) 226 cm^2 (ii) 61.7 cm 2.(a)(i) 346 cm^2 (ii) 66.0 cm (b)(i) 60.0 m (ii) 11309.7 m^2 (c)(i) 64.0 m (ii) 201 m 3. 179.1 cm^2 4.(a) 40 cm (b) 686.7 cm^2 5.(a) 215 cm^2 (b) 64.3 cm 6.(a)(i) 778.3 m (ii) 72.0 m (b) 62.6 m^2 7. 363

Practise exercise 29

1.(a)(i) 5040 cm^3 (ii) 1200 cm^2 (iii) 1572 cm^2 (b)(i) 1728 cm^3 (ii) 576 cm^2 (iii) 864 cm^2 (c)(i) 480 cm^3 (ii) 448 cm^2 (iii) 648 cm^2 (d)(i) 9500 cm^3 (ii) 1583 cm^2 (iii) 2488 cm^2 2.(a) 216 cm^3 (b)(i) 7.35 cm (ii) 397 cm^3 (iii) cube, 181 cm^3 3.(a)(i) 672 cm^2 (ii) 992 cm^2 (iii) 24.7 cm (b)(i) 6.68 cm (ii) 140 cm^2 (iii) 322 cm^2 4.(a) 72 cm^2 (b) 1224 cm^3 5.(a) 531 cm^2 (b) 33984 cm^3 6.(a) 7 cm (b)(i) 8 m (ii) 512 m^3 (c)(i) 6.00 cm (ii) 1169 cm^2 (d) 44544 cm^3

Practise exercise 30

2.(c)(ii) rectangle 3.(b)(i) 31.2 cm^2 (ii) 125 cm^2 6.(a) isosceles trapezoidal prism (b)(I0 208 cm^2

(ii) 168 cm^3

Practise exercise 31

1.(a)(i) 41.9 cm (ii) 503 cm^2 (iii) 89.9 cm (b)(i) 41.9 cm (ii) 314 cm^2 (iii) 71.9 cm 2.(a)(i) 17.8 cm (ii) 107 cm^2 (b)(i) 48.9 cm (ii) 489 cm^2 (iii) 57.8 cm (iv) 768 cm^2 3.(a)(i) 71.2 cm (ii) 142 cm (b)(i) 1211 cm^2 ii) 2421 cm^2 (iii) 501 cm^2 (iv) 1920 cm^2 4.(a) isosceles triangle (b)(i) 16.7 cm (ii) 50.1 cm (iii) 88.2 cm^2 (iv) 35.02 cm^2 5.(a) 535.6 cm^2 (b)(i) 118 cm (ii) 17.28 cm 6. 7803 cm^2 7.(a) 70.7 cm^2 (b) 7069 cm^3 (c) $ 4241.40

Practise exercise 32

1.(a) 96 cm^3 (b) 540 cm^3 2.(a) 8.66 cm (b) 423.6 cm^2 (c) 289 cm^3 3.(a) 1716 m^3 (b) 14 m (c)(i) 73.18 m^3 (ii) 292.72 m^2 4.(i) 16.2 cm (ii) 305.4 cm^2 (iii) 113.1 cm^2 (iv) 418.5 cm^2 (v) 565.5 cm^3 5.(a)(i) 23.0 cm (ii) 21.6 cm (iii) 779.11 cm^2 (iv) 1447.65 cm^3 (b)(i) 5 (ii) 204 cm^2 (iii) 283 cm^2 (iv) 314 cm^3 6.(a)(i) 5542 cm^2 (ii) 38792 cm^3 (b)(i) 8.00 cm (ii) 804 cm^2 (c)(i) 8.5 cm (ii) 454 cm^2 (iii) 681 cm^2 (iv) 1286 cm^3 (d)(i) 44602 cm^3 (ii) 13304 cm^3 (iii) 26608 g (iv) 117 7.(a) 147 cm^2

Practise exercise 33

1.(a) 2x2 (b) 1x3 (c) 3x1 (d) 3x3 2.(i) $\begin{pmatrix} 24 & 10 \\ 18 & 9 \\ 7 & 11 \end{pmatrix}$ (ii) $\begin{pmatrix} 24 & 18 & 7 \\ 10 & 9 & 11 \end{pmatrix}$ 3.(a) $\begin{pmatrix} 94 & 78 \\ 86 & 89 \end{pmatrix}$

(b) square matrix 4. 1x6, 6x1, 2x3, 3x2

Practise exercise 34

1.(a) $\begin{pmatrix} 3 & 0 \\ 10 & -2 \end{pmatrix}$ (b) $\begin{pmatrix} 5 & -3 \\ 14 & 3 \end{pmatrix}$ (c) $\begin{pmatrix} 6 & 9 \\ 28 & -23 \end{pmatrix}$ (d) $\begin{pmatrix} 4 & 0 \\ 10 & -1 \end{pmatrix}$ (e) $\begin{pmatrix} 1 & -6 \\ -2 & 12 \end{pmatrix}$ (f) $\begin{pmatrix} 5 & -12 \\ 6 & 22 \end{pmatrix}$

(g) $\begin{pmatrix} 1 & 0 \\ 10 & -4 \end{pmatrix}$ (h) $\begin{pmatrix} -16 & 27 \\ 34 & -23 \end{pmatrix}$ (i) $\begin{pmatrix} 1 & 3 \\ 6 & -7 \end{pmatrix}$ (j) $\begin{pmatrix} 2 & -3 \\ 4 & 5 \end{pmatrix}$ (k) $\begin{pmatrix} -14 & 24 \\ 38 & -18 \end{pmatrix}$ (l) $\begin{pmatrix} 17 & 15 \\ 7 & -85 \end{pmatrix}$

(m) $\begin{pmatrix} -18 & 30 \\ 30 & -28 \end{pmatrix}$ (n) $\begin{pmatrix} -33 & 51 \\ 62 & -39 \end{pmatrix}$ (o) $\begin{pmatrix} -4/5 & 0 \\ 0 & -4/5 \end{pmatrix}$ (p) $\begin{pmatrix} 0.5 & 3 \\ 6 & -7.5 \end{pmatrix}$

2.(a) $\begin{pmatrix} -2 & 9 & 7 \\ 8 & 5 & 9 \end{pmatrix}$ (b) $\begin{pmatrix} 0 \\ 3 \\ 3 \end{pmatrix}$ (c) $\begin{pmatrix} 39 & 41 \end{pmatrix}$ (d) $\begin{pmatrix} 128 \end{pmatrix}$ 3.(a) a = –3, b = –2, c = 5 (b) x = –1, y = –7

(c) a = 8, b = 6 (d) a = 0,, b = 2, c = 21,, d = 27.5 4.(a)(i) (–6 22.5) (ii) (–0.375 –1.125) (b) No. Since the number of columns of first matrix is not equal to the number of rows of second matrix.

5.(a) $\begin{pmatrix} x^2+16 & 8x+40 \\ 2x+4 & 65 \end{pmatrix}$ (b) x = 4, y = 2, z = –18 6. (66 48) 8.(a) $\begin{pmatrix} 2 & 4 \\ 8 & 26 \end{pmatrix}$ (b) $\begin{pmatrix} 8.5 & 4.25 \\ 8.5 & 4.25 \end{pmatrix}$

9. $\begin{pmatrix} 19 & -18 \\ 5.5 & -3.5 \end{pmatrix}$

Practise exercise 35

1.(a) 0 (b) –37, non-singular (c) –1, non-singular (d) 0 (e) 10, non-singular 2.(a) 5, –5 (b) 1, –7

3.(b) $\frac{1}{39}\begin{pmatrix} 3 & 9 \\ -2 & 7 \end{pmatrix}$ 4.(a) $x^2 - 12$ (b) $\frac{1}{x^2-12}\begin{pmatrix} x & -6 \\ -2 & x \end{pmatrix}$ 5.(a) $\frac{1}{-19}\begin{pmatrix} 8 & -3 \\ -9 & 1 \end{pmatrix}$

(b) $\frac{1}{29}\begin{pmatrix} 3 & 4 \\ -2 & 7 \end{pmatrix}$ (c) $\frac{1}{100}\begin{pmatrix} -45 & -100 \\ 10 & 20 \end{pmatrix}$ 8.(i) $\frac{1}{15}\begin{pmatrix} -9 & 6 \\ 8 & -7 \end{pmatrix}$ (ii) $\frac{1}{-2083}\begin{pmatrix} 11 & 54 \\ 40 & 7 \end{pmatrix}$

10. $\begin{pmatrix} 1.25 & -0.25 \\ 0.25 & 0.75 \end{pmatrix}$

Practise exercise 36

1.(i) $\begin{pmatrix} 1 \\ 3 \end{pmatrix}$ (ii) $\begin{pmatrix} 3 \\ 1 \end{pmatrix}$ (iii) $\begin{pmatrix} 1 \\ 1 \end{pmatrix}$ (iv) $\begin{pmatrix} 1 \\ -2 \end{pmatrix}$ (v) $\begin{pmatrix} 4 \\ 1 \end{pmatrix}$ (vi) $\begin{pmatrix} 3 \\ 3 \end{pmatrix}$ (vi) $\begin{pmatrix} -3 \\ -1 \end{pmatrix}$ (vii) $\begin{pmatrix} -3 \\ 2 \end{pmatrix}$

2.(b)(i) parallelogram (ii) $\begin{pmatrix} 4 \\ -3 \end{pmatrix}$ 3.(i) $\begin{pmatrix} -3 \\ 7 \end{pmatrix}$ (ii) $\begin{pmatrix} -4 \\ -2 \end{pmatrix}$ (iii) $\begin{pmatrix} -3 \\ -3 \end{pmatrix}$ (iv) $\begin{pmatrix} -4 \\ 12 \end{pmatrix}$ (v) $\begin{pmatrix} -6 \\ 4 \end{pmatrix}$

(vi) $\begin{pmatrix} -4 \\ 8 \end{pmatrix}$ (vii) $\begin{pmatrix} -28 \\ 74 \end{pmatrix}$ (viii) $\begin{pmatrix} -7 \\ 0 \end{pmatrix}$ (ix) $\begin{pmatrix} 6 \\ -73 \end{pmatrix}$ (x) $\begin{pmatrix} -1 \\ -7/4 \end{pmatrix}$ (xi) $\begin{pmatrix} -1.5 \\ 6 \end{pmatrix}$ (xii) $\begin{pmatrix} -24 \\ 12.25 \end{pmatrix}$

5.(a) $\begin{pmatrix} -4 \\ 23 \end{pmatrix}$ (b) $\begin{pmatrix} -2.5 \\ 8.25 \end{pmatrix}$ 6.(b) parallel lines, Since PQ is a scalar product of MN. (c) N0. Since AB and AC must be on a straight line. 7.(i) b – 1/2a (ii) 1/2a – b (iii) a + b (iv) 1/2a + b

Practise exercise 37

1.(a) 15 (b) 25 (c)(i) 14.1 (ii) 11.7 2.(i) 8.60 (ii) 14.1 (iii) 5.83 (iv) 14.1 3.(i) a (ii) –a (iii) b (iv) –b (v) b + a (vi) –a–b (vii) b – a (viii) a – b 4.(i) 1/2a (ii) a – b (iii) –b (iv) b + 1/2a 5.(a)(i) 2a (ii) 2b

(iii) –a (iv) b + a (v) 2a (vi) –2b (vii) –a (viii) 1/2a – 1/2b (b)(i) $\begin{pmatrix} 4 \\ 2 \end{pmatrix}$ (ii) $\begin{pmatrix} 4 \\ -4 \end{pmatrix}$ (iii) $\begin{pmatrix} -2 \\ 1 \end{pmatrix}$

(iv) $\begin{pmatrix} 8 \\ -2 \end{pmatrix}$ (b)(i) 4.47 (ii) 5.66 (iii) 2.24 (iv) 8.25

Chapter – 4 **Statistics and Probability**

Practise Exercise 1

1.(a)

Ages	7	9	11	12	14	16	18
Frequency	2	1	1	4	3	3	1

(b)

Ages	$0 \leq x < 5$	$5 \leq x < 10$	$10 \leq x < 15$	$15 \leq x < 20$
Frequency	0	3	8	4

(c) 11 (d) 5.5

2.(a)(i) 17 (ii) 8.5

(b)

No. of hours	0	1	4	6	9	11	12	14	17
Frequency	2	2	4	2	2	5	3	3	2

(c)

No. of hours	$0 \leq h < 4$	$4 \leq h < 8$	$8 \leq h < 12$	$12 \leq h < 16$	$16 \leq h < 20$
Frequency	4	6	7	6	2

3. (a)(i) 37 (ii) 18.5

(b)

Time(min)	8	12	16	20	27	30	37	42	45
Frequency	1	2	2	4	3	2	1	4	1

(c)

Time(min)	$0 \leq T < 10$	$10 \leq T < 20$	$20 \leq T < 30$	$30 \leq T < 40$	$40 \leq T < 50$
Frequency	1	4	7	3	5

Practise Exercise 2

1.(a)(i) 4 (ii) 4 (iii) No mode (b)(i) 10.2 (ii) 10 (iii) No mode (c)(i) 3 (ii) 3 (iii) 3 (d)(i) 5.17 (ii) 6 (iii) 8 (e)(i) 5.75 (ii) 6 (iii) 6 (f)(i) 1/2 (ii) 1/2 (iii) No mode 2.(a) 70.8 (b) 78.5 (c) 97 3. 13 4. 65 5. 18 6.(a) 13 (b)(i) $11x/4 = 22$ (ii) 8 (iii) 16, 8, 24, 40 (c)(i) 17 (ii) 12 7.(a) 9 (b)(i) 2, 4 (ii) 9 (iii) No mode (iv) 0.2 8.(a) 9 (b) 50

Practise Exercise 3

1.(a)(i) 24 (ii) 25 (iii) 25 (b)(i) 10.4 (ii) 15 (iii) 12 2.(a) 5 (b) 6 (c) 10 3.(a) 152.95 (b) 152 (c) 156 4.(a) 66.7 (b) 72.5 (c) 60 5. 5, 8

6.(a)

Temperature	31	32	33	34	37	38
Frequency	2	2	1	4	3	2

(b) 34.4 (c) 34 (d) 34

Practise Exercise 4

1.(a)(ii) 10 cm (b)(ii) 70 2.(b)(i) 6 (ii) 22 3.(a)(ii) 1.5 cm (b)(ii) 15

Practise Exercise 5

1.(a) 108^0, 80^0, 72^0, 100^0 in order (c) 22.2% 2.(b) 83.7^0 3.(b) 275% 4.(a) 60, 70 in order (b) 6:5:7

Practise Exercise 6

1.(b) Positive correlation (e) Height and shoe size are in direct proportion

2.(b) Negative correlation (d) Math and English marks are in inverse proportion

3.(b) Zero correlation (c) No. Because there is no direct or inverse relationship between goals and penalties

4.(a) Positive correlation (b) Negative correlation

Practise Exercise 7

1.(b) $20 \leq x < 30$ (d) 29 2.(c) $110 \leq h < 120$. It is the class with highest number of pupils.

3.(c) 50 4.(a) 31 (b) 15 (c) 25 (d) 25-30, modal class (e) 61.3%

Practise Exercise 8

1. (ii) 3/10 2.(b)(i) 41/74

4.

Temperature(^{0}C)	$0 \leq T < 2$	$2 \leq T < 5$	$5 \leq T < 10$	$10 \leq T < 13$
Frequency	4	9	11	6

Practise Exercise 10

1.(a) 10.2 (b) Since the mid-value of each class is assigned with all the data in that class 2.(a) 70.6 (b) 33.6 (c) 14.7

Practise Exercise 11

1.(a)(i) $Q_1 = 3$, $Q_2 = 6$, $Q_3 = 8$ (ii) 5 (iii) 2.5 (b)(i) $Q_1 = 5.5$, $Q_2 = 10.5$, $Q_3 = 11.5$ (ii) 6 (iii) 3 (c)(i) $Q_1 = 9$, $Q_2 = 17$, $Q_3 = 51$ (ii) 42 (iii) 21 (d)(i) $Q_1 = 22.5$, $Q_2 = 23.5$, $Q_3 = 25.5$ (ii) 3 (iii) 1.5 2.(i) 8 (ii) 10 (iii) 2 (iv) 5 3.(a) 10 (b) 5 4.(a)(i) 17 (ii) 10 (iii) 7 (iv) 13 (v) 13 (b)(i) 220 (ii) 150 (iii) 200 (iv) 70 (v) 120 (c)(i) 16 (ii) 8 (iii) 8 (iv)10 (v) 12 (d)(i) 13 (ii) 11 (iii) 2 (iv) 12 (v) 4

Practise Exercise 13

1.(a)0.6 (b) 2/3 (c) 0.75 (d) 0.28 2.(a) 1.4 > 1. The probability lies between 0 and 1. (b) –0.3 < 0. The probability lies between 0 and 1. 3. 0.6 4.(a) 4/8 (b) 4/8 (c) 3/8 (d) 5/8 (e) 2/8 (e) 6/8 5.(i) 1/11 (ii) 2/11 (iii) 6/11 6.(a)(i) 1/2 (ii) 1/2 (b)(i) 3/6 (ii) 3/6 (iii) 2/6 (iv) 1/6 7.(i) 14/24 (ii) 10/24 8.(a) 8/50 (b) 5/50 (c) 25/50 (d) 40/50 (e) 7/50 (f) 47/50 9.(a)(i) 20/45 (ii) 25/45 (b) 5/45 (c) 13/45 10.(a)(i) 40/200 (ii) 125/200 (iii) 57/200 (iv) 123/200 (v) 95/200 (vi) 105/200 (b) No. Because the outcomes do not have a large difference.

Practise Exercise 14

1.(a) 1 (b) 1/9 (c) 2/9 (d) 10/27 2.(a) 1 (b) 1/4 3.(a) 1/4 (b) 1/4 (c) 1/12 (d) 1/6 4.(a) 11/15 (b) 4/15 (c) 2/3 (d) 1 5.(a) 1/36 (b) 1/4 (c) 3/4 (d) 1/36 (e) 1/36 6.(a) 3/7 (b) 2/7

(c) 5/7 (d) 6/49 (e) 4/7 (f) 3/7 (g) 5/7 (h) 6/7

Practise Exercise 15

1.(b)(i) 1/4 (ii) 2/4 2.(b)(i) 1/36 (Ii) 3/36 (iii) 27/36 (iv) 6/36 (v) 30/36 (vi) 19/25

4.(a)

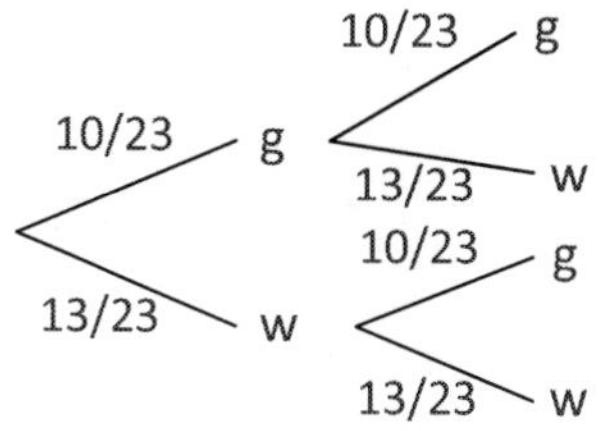

(b)(i) 100/529 (ii) 169/529 (iii) 360/529 (iv) 130/529

5.(a)

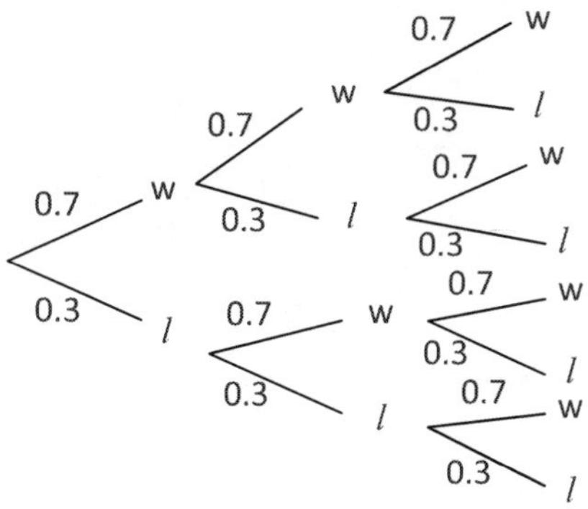

(b)(i) 0.343 (ii) 0.027 (iii) 0.784 (iv) 0.657

6.(a)

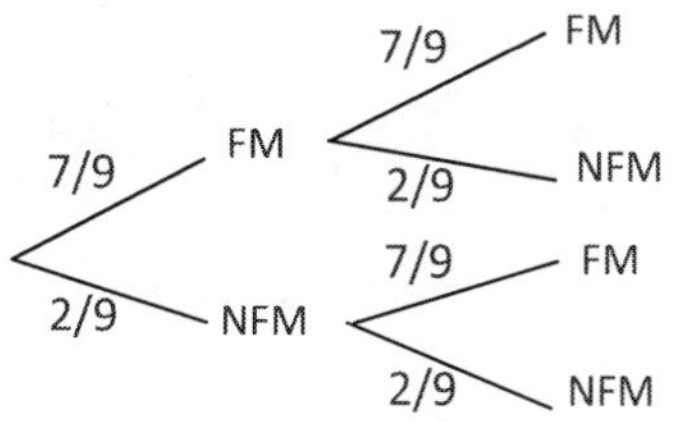

(b)(ii) 49/81 (iii) 28/81

7.(a)

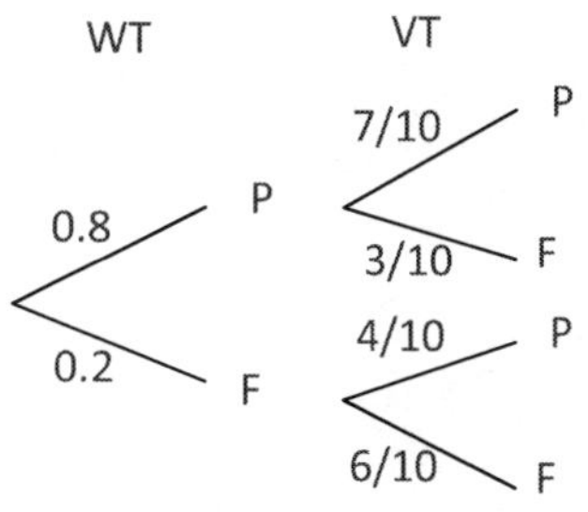

(b)(i) 0.56 (ii) 0.24 (iii) 0.36 (iv) 0.12

8.(a)

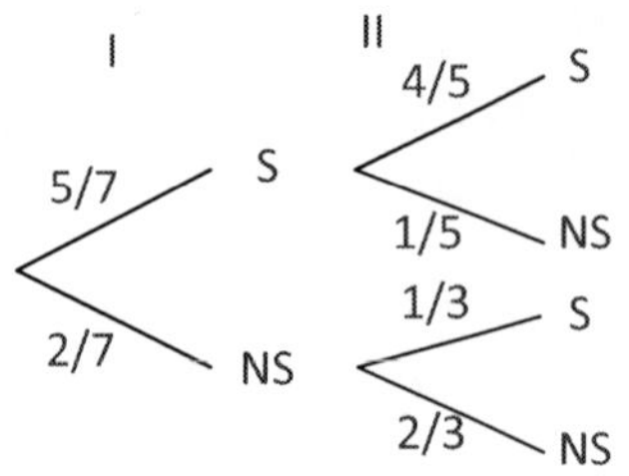

(b) 4/7 (c) 3/7 (d) 1/7 (e) 17/21

Made in the USA
Monee, IL
11 November 2026

67384904R00136